D0073063

Advance Praise for *Principles of Emergency Management*

A true professional, Mike Fagel arrived at FDNY WTC Incident Command Post on Duane Street, a short distance from Ground Zero, as chaos was still not contained. He organized, directed, and cajoled until order again appeared in our health and safety efforts for the thousands of personnel struggling at rescuing and recovering the victims of 9/11. Many of the Ground Zero workers have their health still intact because of Mike's courage and efforts. The Fire Department was well served by his knowledge and expertise.

—Charles R. Blaich (Ret.), ***Deputy Chief FDNY, Logistics Chief, WTC ICP***

Dr. Michael Fagel has assembled a group of experts in a variety of areas of emergency management and has edited a highly usable book that belongs on the desks of EM professionals. Most emergency operations plans have appendices relating to specific critical events. The organization of Fagel's book around hazard-specific issues makes it easy to find useful guidance when planning for a wide range of critical incidents from agroterrorism to pandemics to active shooters to large-scale public events. The coverage is very up to date, as evidenced by references into 2011 and coverage of such modern topics as the impact of social media on emergency management. Having taught with Dr. Fagel, I see in this book the effective classroom style that I associate with his work, but translated into a very practical and useful manual. All in all, it's a book that's easy to recommend.

—Frank K. Cartledge, ***Alumni Professor of Chemistry Emeritus, Louisiana State University***

I have worked beside Dr. Mike Fagel for more than three years. He is a professional in every sense, a committed emergency responder/manager, an agribusiness expert, an educator, and a good friend. The diversity of thinking, working, and experience Dr. Fagel offers on the subject matter of agroterrorism can only be matched by a very select group of experts. Dr. Fagel has meticulously detailed all the important aspects associated with preventing, responding, and recovering from an attack on agribusiness and the food supply. Mike introduces the subject by showing the immense scope and size of the number-one industry in the United States, agriculture and the allied industries of food production. He outlines the complexity of the farm-to-table continuum, making a special effort to point out where security should be improved. He goes on to point out the recognition of agriculture as critical infrastructure formally recognized by the federal government. Several presidential directives and the Department of Homeland Security place an updated emphasis on the importance of agricultural infrastructure. And finally, Dr. Fagel effectively emphasizes the devastating psychological and economic consequences of an agroterrorism event.

—Stan W. Casteel, DVM, PhD, ***Professor of Veterinary Pathobiology, Veterinary Medical Diagnostic Laboratory University of Missouri College of Veterinary Medicine***

Effective emergency and crisis management requires vigilance across a panoply of evolving threats, hazards, technologies, and operational capabilities. No matter how experienced one is as an emergency responder or emergency manager, there are new lessons to be learned every day. This book complements earlier treatments of EOC design and operations by Dr. Fagel and offers the practitioner new confidence-building measures for confronting a range of public health, agroterrorism, and active shooter incidents that can impact a community and shake the confidence of the populace to return to normalcy. His focus on best use of social media and other communications modalities is timely and important in shaping contemporary planning and community resilience. Maintaining the trust and confidence of the general population from incident onset through long-term recovery is an essential element of effective emergency management, and this book is a tool kit for best practices in citizen-centric preparedness.

—Robert J. Coullahan, CEM, CPP, CBCP
President, Readiness Resource Group

In this book, Mike Fagel provides emergency management guidance based upon his three decades of experience as an emergency responder and emergency manager, including operations on numerous natural and technological disaster sites, including the 9/11 World Trade Center incident site. These are critical lessons in emergency management from one who has responded to several of the most devastating incidents in recent U.S. history, who provided his expertise at those scenes and prevented further disaster through well-planned operations and safe practices, and who still provides his expertise in developing the emergency management procedures and protocols to be used by the next generation of emergency managers.

—J. Howard Murphy, CEM, ***Senior Homeland Security Analyst***

Whether new to the field of emergency management or a person of experience, this is a "must have" resource for individuals with crisis and emergency management responsibilities. Dr. Fagel easily translates his education and practical experience, as well as that of other chapter authors, into a well-written document of substance which emphasizes a pragmatic approach to emergency management and emergency operations center management. The reader will find many concepts, ideas, and "best practice" suggestions which can easily be incorporated within many organizations. As a result of his work, Dr. Fagel has contributed an excellent resource document for all emergency management personnel.

—Edward J. Krueger, ***Director***
Criminal Justice Center for Innovation, Fox Valley Technical College

Mike Fagel demonstrates in this new textbook his on-the-job expertise as an emergency manager. As someone who has known Mike for many years, I highly recommend his approach and his concepts. He continues to pursue the professional development of the field of Emergency Management and this is demonstrated in his most recent work. Dr. Fagel is committed to using his real-world "on the job" approach to making the rest of us safer.

—Edward Plaugher, Fire Chief (Ret.), ***Arlington County Fire Department***
Arlington, Virginia

...a must read for emergency managers, planners, first-line responders, plus faculty and students involved in the study of emergency response, homeland security, and public health. Mike Fagel has a rare combination of both superb academic and hands-on, first-responder credentials.

—Colonel Randall J. Larsen, USAF (Ret.), ***Director, Institute for Homeland Security***

As depicted by the authors of this text, we live in a complex world that is rife with risks and hazards. As such, it is imperative that officials at all three levels of government understand the risks, threats, vulnerabilities, and consequences that face their communities and to take the necessary actions to mitigate against, prepare for, respond to, and recover from those hazards and risks. This book provides helpful information to emergency management and homeland security professionals with taking those actions and achieving those goals.

—Edward G. Buikema, ***Director of Preparedness and Response for Armada Ltd.***
Former Regional Director, FEMA Region V

Published information relating to emergency management and emergency operations centers is extensive but scattered among a plethora of sources. Dr. Michael Fagel has brought the salient elements together in one comprehensive treatise. Further, the relevant information is presented in a direct, easy to read format that any individual associated with this evolving field can assimilate. This updated and expanded text should be a principal resource to all emergency management personnel.

—Edward H. Stephenson, ***National Center for Biomedical Research***
and Training, Louisiana State University

Principles of Emergency Management

Hazard Specific Issues and Mitigation Strategies

Principles of Emergency Management

Hazard Specific Issues and Mitigation Strategies

Michael J. Fagel, PhD, CEM

CRC Press is an imprint of the
Taylor & Francis Group, an **informa** business

CRC Press
Taylor & Francis Group
6000 Broken Sound Parkway NW, Suite 300
Boca Raton, FL 33487-2742

Printed in the United States of America on acid-free paper
Version Date: 20110830

International Standard Book Number: 978-1-4398-7120-1 (Hardback)

Library of Congress Cataloging-in-Publication Data

Principles of emergency management : hazard specific issues and mitigation strategies / [edited by] Michael J. Fagel.
p. cm.
Includes index.
ISBN 978-1-4398-7120-1 (hardcover : alk. paper)
1. Emergency management. 2. Crisis management. I. Fagel, Michael J.

HV551.2.P75 2012
363.34--dc23 2011034358

Visit the Taylor & Francis Web site at
http://www.taylorandfrancis.com

and the CRC Press Web site at
http://www.crcpress.com

September 11, 2001. It's been 10 years. Much has changed in the landscape of emergency planning, and a new word has emerged, "Homeland Security," but has it really been that much different? I'd like to take this time to dedicate and rededicate this textbook to my trusted colleagues and friends that I worked alongside with in the aftermath of September 11th and the Oklahoma City bombing in 1995. Those events changed a nation, a world, myself, and my family. The toll those horrific events have taken on the victims, the families, the responders, and the communities is incalculable. I've spent the better part of the last two decades training, teaching, and helping others to prepare for the inevitable. For those that came before me, for those that are with me now, and for those that will come after, I offer this text as a humble tool as you progress.

My family has been my biggest supporter and has suffered greatly from my absences at family gatherings, school plays, and milestones in their lives. To my wife, Patricia, all I can say is, "Thank you all for your unwavering support and dedication to the cause."

Take care of yourself, your family, and the public whom we serve.

Thank you.

Mike

Contents

Foreword

The risk of natural and unnatural disasters has not been greater than it is now and the need for effective emergency management policies and programs has never been more critical. Geophysical and meteorological hazards, pandemic and other medical risks, international and domestic terrorism, industrial accidents, and a host of newly recognized hazards, such as climate change with all of its attendant ramifications, pose growing threats to life, property, and the environment. Increasing community vulnerability, particularly along the nation's coastlines and in seismically active zones, and growing social vulnerability are exacerbating those risks. In short, risk and vulnerability are increasing and the adoption of risk reduction measures is not keeping pace. Families, organizations, communities, and nations cannot afford to ignore the growing potential for catastrophic disaster—it is imperative that they invest in the design, maintenance, and operation of effective emergency management programs.

Unfortunately, finding support for programs to reduce losses of life and property and damage to the environment is complicated by the social, political, and economic milieu. The obstacles have to be overcome. First, the public has little understanding of risks posed by natural and man-made hazards and little comprehension of who is vulnerable. They have to believe that it can happen to them and public education programs have not yet delivered that message. The same is true of elected officials. They too often do not understand the hazards and vulnerabilities in their communities and do not make risk reduction a priority. Second, communities are not always willing to invest in risk reduction programs. Even if risks are understood, selling risk reduction, including emergency management programs, is a challenge. Leadership is lacking and resources are scarce. Risk reduction is a tough sell when there are many other competing needs. Arguments have to be couched in terms of the cost of being unprepared, that is, risk reduction will provide long-term cost savings, and assurances have to be provided that the programs will be effective and efficient. All-hazards emergency management is the most cost-effective approach but it requires attention to the extraordinary requirements of some hazards. In an era in which trust in government is low, the task of selling risk reduction is easiest in the immediate aftermath of disaster when the case has already been made for investments to prevent similar losses. Emergency management officials also have to remind elected leaders and the public at-large that there may be legal costs associated with failing to protect the public. The potential for government officials to be held legally liable for failures to exercise their responsibilities reasonably is increasing. While some states do insulate local officials against lawsuits, there may still be severe political repercussions when risks have been ignored and residents suffer harm.

Developing a strategy to encourage investment in risk reduction is essential. A first step is building public trust that the programs will have a strong foundation built upon what we know about the science of hazards and the measurement of risk and what we know about the design and management of effective programs. Poor management can undermine an otherwise good program.

Good management practice is a focus of the Emergency Management Accreditation Program Standard and the National Fire Protection Association 1600 standard. Without adequate administrative capacity, programs cannot be maintained and operated effectively. A focus on good management can also be found in "The Principles of Emergency Management," the core principles endorsed by the International Association of Emergency Managers, the National Emergency Management Association, and the Federal Emergency Management Agency. Again, the capacity to manage programs effectively is a fundamental requirement. The ability to collaborate effectively with public, private, and nonprofit partners is also fundamental when authority is shared and resources and expertise are dispersed among many stakeholders. Understanding one's legal authority is important. Understanding the legal and practical limits to authority is even more important.

And, it is also imperative that the lessons drawn from past experience be evaluated and used when appropriate. "Best Practices" may not always be transferrable to other organizations or situations, even if they have proven effective in some. Knowing when and how to draw upon past experience is a capability that improves with experience. Distilling generalizations or principles from those lessons is the final step in developing a "science" of emergency management. This book represents just such an integration of science, management expertise, and experience. It brings together a distinguished group of emergency management professionals who have the necessary breadth and depth of experience to identify core functions and values that are not found in the more theoretical literature and lessons that are not always evident in the war stories and case studies. The "nuts and bolts" of emergency management are critical. The chapters in this book give the reader an operational perspective on how to develop planning teams and processes, assure a healthy and safe work environment for responders, develop a workforce with the requisite skills and competencies to maintain and operate effective programs, manage stress in the chaos of disaster operations, manage emergency operations centers, comply with legal requirements and ethical standards, and practice good financial management. They also provide practical guidance on dealing with specific hazards. The issues are seen through the eyes of experienced emergency managers who are grounded in the science and who have managed people and programs.

This is a critical juncture in the development of emergency management. The profession and practice have been transformed over the past three decades. Emergency management has been recognized as an essential function of government and as a critical role in public, private and nonprofit organization. In many jurisdictions, emergency managers have become important participants on the executive decision making team responsible for managing risk. However, history has shown that policymakers may fail to understand and use the expertise that is available. The Katrina disaster demonstrated what happens when emergency management expertise is not brought to bear in the critical first hours and days of a crisis. While there were extraordinary rescues and a remarkable evacuation of much of the population, fundamental errors were made. In fact, officials at all levels were made to pay at the ballot box for their failures. By contrast, the 9/11 disaster illustrated how expertise can be mobilized to manage risk to the community and to support response and recovery operations. Although errors were made, the response and recovery operations were remarkable despite the scale of the event and the loss of New York City's emergency operations center and hundreds of its responders.

Preventing disasters or reducing their effects, preparing communities and responders for catastrophic events, coordinating and supporting response operations, and initiating recovery are part of the larger process of managing risk. Centuries of disaster experience and at least half a century of disaster research provide considerable foundation for that process.

Emergency managers fill a critical role. In large measure, their skill set has been shaped by events. A series of natural disasters in the 1960s and 1970s encouraged the creation of the Federal Emergency Management Agency to concentrate and focus federal efforts to deal with major disasters and to implement an all-hazards approach. A series of billion-dollar disasters in the 1980s and 1990s encouraged states and localities be more proactive and to focus their attention on mitigating known hazards. A series of terrorist incidences, beginning with the first attack on the World Trade Center (WTC) towers in 1993 and the bombing of the Murrah Federal Building in Oklahoma City in 1995, encouraged the expansion of emergency management responsibilities to include unnatural disasters. The practice of emergency management in the United States changed profoundly after the attacks in New York and Washington on 9/11 and changed again with the poor response to Hurricanes Katrina, Rita, and Wilma in 2005. Lessons were learned about the need to share resources. The Emergency Management Assistance Compact was expanded to include all states and territories and more and more states are adopting statewide assistance compacts to facilitate the sharing of personnel and equipment across jurisdictions. Expanding local capacities and the capacities of state offices to provide assistance are the new priorities. Emergency management has evolved from its early cavalry approach with responders riding in to save the day to a mutual aid approach with neighboring jurisdictions providing the extra resources necessary for communities to save themselves.

Clearly, the development of American emergency management has also been stimulated by events elsewhere in the world. The Indian Ocean tsunami of 2004 provided lessons on how to manage mass casualty events and the necessity for early warning systems. The particularly active hurricane season in 2004 that saw four hurricanes hit central Florida in quick succession provided lessons in public warning and recovery. The Pakistan earthquakes of 2005 and the Sichuan earthquake in China in 2008 provided important lessons in seismic safety related to school buildings and other structures. The Haitian and Chilean earthquakes of 2010 illustrated the importance of appropriate building codes and construction standards. The Victoria wildfires in Australia in 2009 stimulated debate over the choice between property owners protecting their homes and businesses and the need to assure safe evacuations of those at risk. Massive flooding in Pakistan in 2010 and in Australia, Brazil, and South Africa in 2011 have been reminders that communities all over the world are vulnerable to natural disasters and that climate change may be increasing the number and severity of weather-related disasters. The Deep Horizon oil spill in the Gulf and the toxic sludge release into the Danube in 2010, as well as numerous lesser hazardous materials spills and chemical fires, have reaffirmed the need to address risks from man-made hazards. Those events are also providing lessons about long-term environmental recovery. The earthquakes in Christchurch, New Zealand, in 2010 and 2011 have also reaffirmed the value of retrofitting structures to withstand seismic events.

Moreover, the bombings in Madrid in 2004 and London in 2005 and the terrorist attack in Mumbai in 2008, as well as the attempted bombings of aircraft in the United States in recent years, have reaffirmed the necessity of focusing on all kinds of hazards, not just natural disasters. Those terrorist attacks are providing crucial lessons about intelligence gathering and information sharing. The point is simply that much is known about hazards and human behavior during disasters and much is still being learned. Putting that knowledge to use is just smart. Finally, the 2011 earthquakes and the tsunami in northern Japan demonstrated that even the best laid plans can fail—preparedness efforts and mitigation measures can be overwhelmed by the scale of events. The unexpected can always happen and the capacity to improvise is essential.

Local capacity is the foundation. In the United States, the Hurricane Katrina disaster, in particular, served as a reminder that emergency management is a specialized area of expertise, not a task that can be addressed in ad hoc fashion by officials without appropriate training and experience. The poor response to Katrina was a national embarrassment and prompted emergency managers to better define their role and function so that the public, the news media, and other public officials can better understand the need for emergency management expertise in large and small disasters. Emergency management is not firefighting or search and rescue or emergency medical services, although emergency managers in some small communities may wear those hats, too. It is more comprehensive and proactive, and it involves the coordination and integration public, private, and nonprofit organization efforts to address risks.

There are also reminders that emergency management is a specialized area of public administration. Coordinating and integrating the efforts of public agencies, private firms, nonprofit organizations, and volunteer groups is a challenge not frequently required of other organizations. Building collaborative relationships is essential in an environment where authority is shared and resources are dispersed. It is a new skill set for many officials. It is also a skill set that runs counter to the training and experience of many. It also requires flexibility. Adaptation, innovation, and improvisation may be required if plans do not fit circumstances. Creativity may also be required in the acquisition and use of resources. The temptation is to bend or break the rules and professional emergency managers have to know where those legal and ethical boundaries are. They have to know the limits beyond which they should not go. This is one of the reasons why emergency management law is becoming more common topic in the textbooks and in training programs.

Emergency management has become more professional through the national Certified Emergency Manager (CEM) program, state certification programs, degree and certificate programs, and so on. In 2010, there were approximately 180 academic programs in emergency management and a variety of related programs focusing on Homeland Security, public health, and humanitarian assistance. There needs to be a firm foundation for those preparing to enter the field, those who come to emergency management from the fire services or law enforcement or other professions to develop the necessary skills and knowledge for their new field, and those who need more management expertise to more into leadership roles in emergency management programs.

Basic skills, however, are paramount. Emergency managers have to work closely with counterparts in other disciplines and with the public. As Thomas Drabek concluded in his classic *The Professional Emergency Manager* in 1987, good interpersonal skills may be the most important attribute of the successful emergency manager. Effective communication and relationship building skills are the keys to success.

The ultimate goal should be emergency management policies and programs that draw upon the physical and social sciences, emergency management experience, and the values of the community. Theoretical insights are necessary to provide structure to the knowledge. Practical lessons are necessary to assure that programs are functional and errors are not repeated. Attention to community values is necessary to assure the fit between public expectations and emergency management practice. The public expects effective and efficient programs. In that regard, emergency managers have to be good managers. However, in the discipline of public administration, the distinction is sometimes made between "doing well" or managing well and "doing good" or helping people. The nature of disaster work is such that "doing good" may well trump "doing well." While effective and efficient programs are usually the stated objective, and it is important that the public get full value for its tax dollars, the task is also to help people in need. This is what draws people to public safety professions and what ultimately is the metric of success. Emergency management is not an easy profession to follow. The pressure of dealing with people in distress is stressful. Long hours and physically and psychologically demanding work is also stressful. In an earlier work, I called emergency management the "quintessential government function" because protecting people is the reason why governments were instituted in the first place. It is why people gathered together in caves, built villages and towns, constructed castles and forts, and created governmental institutions. It is not surprising that emergency responders and emergency managers, and the legions of volunteers, display uncommon altruism and often heroism in crisis. Organizing them around a common purpose is the challenge.

The leadership of professional emergency managers in recent natural and man-made disasters has saved lives and property. In fact, the need for effective leadership has been all too apparent when their expertise has not been brought to bear. The lesson of Katrina was that emergency managers have a distinct and critical skill set that can reduce losses of life and property. The hazard was well known, the problem of evacuating large urban populations was well understood, the need for interoperable communications was well documented, and the necessity for effective intergovernmental and interorganizational cooperation and collaboration was taken as a given within the professional emergency management community. The failure of political leaders to understand those issues and to recognize the need for professional expertise early in the process was difficult to fathom and difficult to excuse. Will officials make the same mistakes again—probably. It is less likely if the profession remains on its current course with the development of benchmarks and standards, appropriate education and training, and the support of a community of practice, a community of academics and practitioners who are committed to risk reduction and community resilience.

This book draws upon the expertise of professional emergency managers who link their experience in the field to the sciences and to their understanding of the communities within which they work. They have taken the practical lessons they have learned into the classroom to share with students. Several of the authors have been very active in developing and promoting national and international standards and in defining the common body of knowledge that all professional emergency managers need to have. This book broadens that body of knowledge.

William L. Waugh Jr., PhD
Andrew Young School of Policy Studies
Georgia State University

Preface

Life is a journey and Emergency Management is no different. Back when I started we called it Civil Defense, and today it is now called Homeland Security. Emergency Management, Homeland Security, Civil Defense, all of these words mean different things to each of us. *Regardless of what it is called in your jurisdiction or organization, at its roots is an All Risk–All Hazards approach to planning, preparedness, mitigation and response.*

I started in this field in the 1960s when we called our work "Civil Defense." Over the past decades, it has grown into Emergency Management, and after the horrific events of September 11, 2001, Homeland Security was created as a new department in 2003. That created the largest reorganization of government in the United States since the National Security Act of 1947.

From my years in fire, rescue, EMS, Cadillac ambulances, "bumper pumpers," and red and yellow hydraulic spreaders

From rambler patrol cars and walkie-talkies that weighed five pounds

To my deployments to the Oklahoma City bombing in 1995

To the World Trade Center attacks in 2001

To my FEMA deployments to floods, hurricanes, ice storms, and tornadoes

From navy bases in the Pacific to army bases in Alaska

From the U.S. Army Solder Biological Chemical Command (SBCCOM) at Aberdeen Proving Grounds

To the Middle East developing FEMA-type organizations and Emergency Operations Centers

Being a part of the FBI National Domestic Preparedness Office four years before DHS was created. I have taught emergency planning at numerous universities, including

Benedictine University
University of Chicago
Eastern Kentucky University
Illinois Institute of Technology
Louisiana State University

Northern Illinois University
Northwestern University
With a teaching start at Waubonsee Community College

My own decades of emergency management and public safety have helped me focus on the future leaders of tomorrow.

When I was deployed to the Oklahoma City bombing in 1995, I was not expecting what I saw, and I had been in fire-rescue, EMS, and law enforcement for more than 25 years at that time. Then in 2001, I was deployed to the WTC attacks for nearly three months in New York City. Again, my training had matured (as we learned from 1995) BUT, again, I was NOT adequately prepared for what I saw and worked on at the WTC.

This book is a companion to *Principles of Emergency Management and Emergency Operations Centers (EOC)*. I have assembled a group of seasoned professionals to help me compile the text. It has expanded on the "All Risk–All Hazard" process by sharing with you some key elements on how to respond to and recover from certain Hazard Specific Issues.

This text will help you expand your understanding of some current events and what is happening around us each and every day. Take the time to surround yourself with current information so that YOU are better prepared to be part of the solution in today's world!

All of these things have changed my life, my career, and my passion. Last, but certainly not least, without the unwavering support of my family, I would not be the responder, the teacher, or the person that I am today. This book is for all of you.

We Will Never Forget!

For those who come behind me, please learn, act, do, and help make this a better place for all!

Mike
Michael J. Fagel, PhD, CEM

Acknowledgments

This is the second book in a series developed with my trusted colleagues to help you to become better prepared for the next event, large or small. Will Rogers was once quoted as saying: “If you want to be successful, it’s just this simple. Know what you are doing. Love what you are doing. And believe in what you are doing.” Never have truer words been spoken than when a group of dedicated professionals work together to create a text to help the next generation, as is evidenced by the many decades of experience represented by the contributors and reviewers of this book. They are all dedicated professionals who have partnered with me to help create an authoritative text for those engaged in emergency planning, preparedness, response, and mitigation. My heartfelt thanks goes to all these trusted colleagues for their support, encouragement, and to our joint success.

I have relied on many discussions throughout my years of training, responding, and teaching in this expanding field. My colleagues at the International Association of Emergency Managers (IAEM) have been relied upon heavily as I researched this text.

Also, I would like to acknowledge that the primary resource and basis for Chapter 9 (The Common-Sense Guide for the CEO) is based on the 1982 version of the International Association of Fire Chiefs (IAFC) document on Disaster Management of a similar name. I have updated the guide here and wish to thank the IAFC for their continued dedication in helping communities plan, prepare for, and recover from events such as these.

To my many colleagues whom I’ve trained with, responded with, and learned from, I owe you all a sincere debt of gratitude for your stewardship, friendship, and guidance.

Please use this book as it is meant to be: one of many tools for you to utilize that may prove valuable throughout your career.

Michael J. Fagel, PhD, CEM

Editor

Michael J. Fagel, PhD, CEM, has more than 40 years of experience in emergency management, fire and emergency medical service, public health, law enforcement, bioterrorism awareness and prevention, as well as corporate safety, security, and threat risk management.

He served as an officer on the North Aurora Fire Department for more than 28 years. He was a Federal Emergency Management Agency (FEMA) reservist and responded to many federally declared emergencies, including deployment to the Oklahoma City bombing in 1995 and the World Trade Center attacks in 2001. He has provided WMD instruction to federal facilities for the U.S. Army, as well as providing technical support to state and county agencies on the Centers for Disease Control and Prevention's (CDC) Strategic National Stockpile Planning. He has served as a subject matter expert to the National Guard Bureau, CERIAC Fusion Center, and provided national EOC planning and CONOPS for U.S. and Middle East venues, including the Salt Lake City Olympics.

He teaches at Northwestern University–Chicago, Northern Illinois University, Illinois Institute of Technology, Benedictine University, Eastern Kentucky University, Louisiana State University's National Center for Biomedical Research and Training (NCBRT), as well as being an SME for Fox Valley Technical College, and the National Center for Security and Preparedness (NCSP) at the University at Albany, State University of New York. Fagel was a technical advisor and taught at the University of Chicago Master's in Threat Response Program. He is now a researcher for NORC at the University of Chicago. Fagel is also currently a Homeland Security Analyst at the Argonne National Laboratory engaged in the protection of critical infrastructure.

He is also a member of the Northern Illinois Critical Incident Stress Debriefing team, the International Association of Fire Chiefs Committee on Safety and Health, and contributed to or published multiple textbooks on emergency management and safety topics.

He can be reached at mjfagel@aol.com.

Contributors

Chad Bowers is vice president of Bold Planning Solutions, a technology company focused on continuity planning, based in Nashville, Tennessee.

Bowers has been actively engaged in the emergency management industry since 2001 and is an expert in the field of Continuity of Operations Planning (COOP) for federal, state, and local government organizations. He has served as senior project manager for dozens of COOP projects, including statewide COOP initiatives for the state of Vermont, the state of South Dakota, and the California Superior Courts. Currently, he serves as senior technical manager for the development and direction of Bold Planning Solutions highly acclaimed web-based Continuity of Operations Planning system, EMplans.com.

He can be contacted at Chad@BoldPlanning.com.

Lucien G. Canton, CEM, is an independent management consultant specializing in strategic planning for crisis. Before starting his own company, he served as the director of Emergency Services for the city and county of San Francisco and as an Emergency Management Programs specialist and chief of the Hazard Mitigation Branch for FEMA Region IX. A popular speaker and lecturer, he is the author of the best-selling *Emergency Management: Concepts and Strategies for Effective Programs*, which is used as a textbook in many higher education courses.

He can be contacted at LCanton@LucienCanton.com.

Adam S. Crowe is the assistant director of Community Preparedness for Johnson County Emergency Management and Homeland Security (JCEMHS). In addition to managing community preparedness activities for JCEMHS, he serves as the cochair of the Emergency Communications Subcommittee of the Regional Association of Public Information Officers, member of the Regional Homeland Security Coordinating Committee, member of Metropolitan Emergency Managers Committee, and past president of the Partnership for Emergency Preparedness in Kansas City.

In 2002, Crowe completed a BS in biochemistry from Clemson University in Clemson, South Carolina, and an MS in public administration from Jacksonville State University.

Crowe is nationally recognized through multiple publications and speaking engagements. He has been published in *Homeland Defense Journal*, *Homeland Security Affairs Journal*, *Crisis Response Journal*, *Disaster Recovery Journal*, and most recently in the *Journal of Homeland Security and Emergency Management*.

He has also provided the keynote address at the Partners in Emergency Preparedness Conference as well as general sessions at the National Severe Weather Workshop, National Preparedness Training and Exercise Conference, CSEEP National Conference, Midwest Consolidated Security Forum, Kansas Emergency Management Association Annual Conference, and International Association of Emergency Managers Annual Conference. He also is scheduled to release *Disasters 2.0: The Application of Social Media in Modern Emergency Management* by mid-2012.

He currently resides in Kansas City, Missouri, with his wife and son.

He can be reached at ACrowe@jocogov.org.

Randall C. Duncan, MPA, CEM, has been a local government emergency manager since 1986, working for both the county and municipal levels of government in Kansas and Oklahoma.

Duncan has been involved in administering nearly a dozen presidential declarations of disaster ranging from floods and tornadoes to ice storms. In addition, he has provided support to FDNY from September 18 to 28, 2001, at the Incident Command Post for the World Trade Center. He has also been involved in other projects ranging from the Biological Warfare–Improved Response Program to serving as an evaluator for the City of Seattle Emergency Operations Center in Topoff II, and the state of Missouri in NLE 11. He is a past president of the USA Council of the International Association of Emergency Managers (IAEM-USA) and currently serves as their Government Affairs Committee Chair. Duncan also serves as adjunct faculty for the Emergency Management Institute in Emmitsburg, Maryland, and Park University in Parkville, Missouri.

He can be contacted at rduncan21@cox.net.

Ron Fisher is the deputy director of the Infrastructure Assurance Center (IAC) at Argonne National Laboratory.

His responsibilities include technical support in many areas of critical infrastructure assurance to the Department of Homeland Security (DHS), Department of Energy, and Department of Defense. Fisher served as a senior consultant to the National Petroleum Council on oil and natural gas infrastructure vulnerabilities and for the President's Commission on Critical Infrastructure Protection. He currently serves as the IAC coordinator for the DHS Office of Infrastructure Protection support activities that include conducting field assessments, vulnerability assessment methodology development, risk analysis, and alignment to the National Infrastructure Protection Plan. He is the author of several interdependencies papers and studies.

He can be contacted at refisher@anl.gov.

Douglas Himberger, PhD, is a senior vice president and the director for Security, Energy, and Environment at NORC at the University of Chicago.

Himberger has more than 35 years of experience in integrating security, survivability, and resiliency best practices across public, private, and civil sectors. He has extensive expertise fostering collaboration between industry, academia, and government to build strong information-sharing networks and advanced technology solutions. He created a Global Risks Initiative Taskforce for analyzing complex all-hazard threats (e.g., pandemics, natural/man-made disasters) and provided multidisciplinary solutions for both government and industry. Himberger serves as a board member, as an advisor, or as a volunteer for numerous resiliency organizations, including the Safe America Foundation (SAF), the Emergency Management Professional Organization for Women's Enrichment (EMPOWER), and others, and he both writes and speaks extensively for professional journals and events.

He can be contacted at himberger-douglas@norc.uchicago.edu.

Patrick J. Jessee, MS, NR-EMTP, CC-EMTP, is an 11-year veteran of the Chicago Fire Department and currently serves as a firefighter/paramedic for the Hazardous Incident Team.

During his service with the fire department, he has functioned in joint operations with the fire department and other agencies and has been intimately involved in writing protocols for response and equipment usage for the team. Additionally, he has previously presented on the Hazardous Incident Team at both the FRI Conference in Chicago, Illinois, and the Midwest Regional Hazardous Materials Conference in Northbrook, Illinois, both in 2010. Previously, he has worked as a paramedic officer for the department and a hazardous materials/WMD instructor for both the department and the Illinois Fire Service Institute. He holds numerous degrees, including a BS in biology (DePaul University) and a BS in chemistry (Southern Illinois University–Edwardsville) and also an MS in biology (University of Illinois at Urbana–Champaign) and Threat and Response Management (University of Chicago).

He may be contacted at pjessee@hotmail.com.

Stephen J. Krill Jr., CEM, PMP, CFCP, is a Senior Associate at Booz Allen Hamilton in McLean, Virginia. He has 24 years of professional experience in emergency management, covering mitigation, preparedness, response, and recovery for government, industry, and nongovernment organizations. He responded to Y2K, 9/11, and Hurricanes Katrina and Rita and served as lead planner for national-level exercises.

Krill is pursuing a PhD in engineering management with a focus in crisis, disaster, and risk management. He is a member of emergency management standards development organizations, such as American Society for Testing and Materials (ASTM), American National Standards Institute (ANSI), and International Organization for Standardization (ISO). He previously served as an adjunct instructor at Northwestern University for transportation terrorism preparedness and presented at emergency management conferences held in both the United States and Europe.

He led the development of three reports to Congress, national guidance on critical infrastructure protection, and published several peer-reviewed journal articles. Krill holds professional certifications in emergency management, business continuity, and project management.

He can be contacted at krill_stephen@bah.com.

Rick C. Mathews, MS, is the director of the National Center for Security and Preparedness (NCSP), an affiliate of the Rockefeller College of Public Affairs and Policy at the University at Albany, State University of New York.

Prior to his current position at the NCSP, he served as the assistant director for research and development at the National Center for Biomedical Research and Training (NCBRT), Academy of Counter-Terrorist Education at Louisiana State University. On April 19, 1995, he participated in the rescue efforts following the explosion at the Alfred P. Murrah Federal Building in Oklahoma City. He served as a technical consultant to the Louisiana Department of Health's Emergency Preparedness Planner during the response and recovery phases of Hurricane Katrina, dealing particularly with issues around the establishment and operation of the PMAC field hospital. In September 2008, he served as a technical consultant to the planning sector of the Louisiana Governor's Office of Homeland Security and Emergency Preparedness following Hurricane Gustav.

His professional experience includes more than 30 years in emergency medical services, hospital administration, emergency preparedness, counterterrorism, and homeland security. He is a member of the International Association of Emergency Managers, the International Medical and Rescue Association, Protect New York, the National Intelligence Education Foundation, InfraGard, and the SESHA Environmental Safety and Health Association of New York.

He can be contacted at rmathews@uamail.albany.edu.

Raymond M. McPartland, MA, MPA, is the founder and chief executive officer of Tier One Associates LLC. Tier One consults and develops training on matters of life safety, emergency preparedness, and homeland security for both the private and public sector and specializes in creating unique, client specific, realistic training and preparedness programs. McPartland is also a subject matter expert with the National Center for Biomedical Research and Training (NCBRT) at Louisiana State University, the National Center for Security and Preparedness (NCSP) at the University of Albany, and the Chemical, Biological, Radiological and Nuclear Defense Information Analysis Center (CBRNIAC). He is also an active trainer and a detective with the New York City Police Department (NYPD).

Currently assigned to the NYPD's Counterterrorism Division's Training Section as a lead instructor and curriculum development specialist, McPartland is a lead instructor with their regional training team responsible for instructing patrol, specialty personnel, and regional partners in various aspects of terrorism and counterterrorism. He is currently the division's subject matter expert on active shooter events and the primary author of the NYPD's published research work—*Active Shooter: Recommendations and Analysis for Risk Mitigation*. Other topics of instruction include active shooter preparedness and response, critical infrastructure protection, maritime terrorism, WMD and radiological awareness, hostile surveillance detection, and behavioral analysis and observation.

McPartland has attended numerous schools and training through the Department of Homeland Security's National Preparedness Consortium and FEMA's Emergency Management Institute.

When not instructing law enforcement or private sector security, he works part time as an adjunct professor at the Metropolitan College of New York in the MPA program in Emergency and Disaster Management and at Mercy College in the undergraduate program for Corporate and Homeland Security.

He can be reached at RMcPartland@tieroneassoc.com.

Kim Morgan is an independent emergency and crisis consultant specializing in preparedness.

She offers more than 22 years of experience in emergency management at the local, state, national, and international levels across private and public sectors. Prior to 9/11, she served as a manager in response and recovery for the Federal Emergency Management Agency (FEMA) during national disasters and for the Environmental Protection Agency (EPA) as project coordinator for hazardous materials response, hazardous materials and transportation studies, and international planning and training. After 9/11, she was asked to relocate to Washington, DC, and lead support in national preparedness operations for various agencies and departments including EPA, FEMA, FDA, USDA, HHS, CDC, State Department, and DHS. Efforts address international, regional, and national level planning and exercises including integration and coordination of emergency preparedness and response operations across multiple agencies, critical infrastructure protection, international preparedness, and specialized preparedness and response tools to maximize available resources and expand capabilities and capacities. Examples of regional, national, and international preparedness include all-hazard, pandemic, crisis and risk management, terrorism, hazardous materials, and natural disasters. Some examples of tools include Emergency Operations Network (EON), Integrated Consortium of Laboratory Networks (ICLN), Computer-Aided Management of Emergency Operations (CAMEO Suite), exercise design on CD, and remote exercise delivery. Morgan is known for creative and innovative solutions and integrated architectural design of programs and processes. Her hands-on experience is supported by a degree and advance studies in emergency management backed by previous undergraduate studies in foreign languages, organizational management, anthropology, finance, and public health.

Kim Morgan can be reached at gokmorgan@gmail.com.

James Peerenboom is the director of the Infrastructure Assurance Center at Argonne National Laboratory.

He has been actively engaged in systems analysis, decision and risk analysis, and advanced modeling and simulation activities for more than 25 years. For the past 15 years, he has focused on critical infrastructure protection and homeland security issues, providing technical support to the Departments of Homeland Security, Energy, and Defense. He is the author of numerous publications on infrastructure interdependencies and received a PhD from the University of Wisconsin–Madison. He currently serves as the director of the Infrastructure Assurance Center and associate director of the Decision and Information Sciences Division at Argonne National Laboratory.

He can be contacted at jpeerenboom@anl.gov.

Nicholas Staikos, AIA, principal of Staikos Associates Architects, has more than 30 years of experience in the design of technologically complex projects including mission critical facilities for emergency management serving government and industry.

Under his leadership, Staikos Associates Architects has completed numerous Mission Critical facilities development projects for all levels of government. The firm has guided the development of EOC design projects beginning with assessment of needs, criteria evaluation, architectural programming and design through to technology deployment and occupancy. Staikos Associates Architects' projects have been recognized for their operational effectiveness and have received regional and national design awards as well as international recognition. Additionally, the firm's work at the Montgomery County (Maryland) Emergency Communications Center and Delaware's State Emergency Operations Center were among the first, if not the first, such centers to integrate emergency management, transportation management, and public safety on a physical as well as functional level. The firm's market segments include public safety, transportation management for projects such as 911 call centers and fusion centers, and transportation management centers.

Staikos has authored articles on integrating security into computer facilities design and site planning.

He can be contacted at NStaikos@Staikos.com.

Michael Steinle is a senior technologist at CH2M HILL, Lawrence, Kansas.

Steinle has more than 20 years of experience in project management, government affairs, emergency response, and environmental, health, and safety stewardship. He served as the manager of Regulatory at Farmland Industries, Inc., in Kansas City, Missouri, and led compliance and government affairs services for this Fortune 500 firm encompassing 31 agricultural chemical plants in 15 states. Steinle also served as the regulatory affairs director at Kansas City International Airport, managing environmental, health, and safety concerns and maintaining compliance with applicable federal, state, and local regulations. He achieved a 58% reduction in employee accidents and exemplary compliance status.

Steinle has served as the project manager on several large-scale state and local emergency preparedness, bioterrorism, and public health planning projects. He has assisted state and local governments in preparing regional pharmaceutical caches and in developing protocols to receive and distribute the Strategic National Stockpile. He served as the project manager for the Mississippi State Department of Health to develop 20 different response plans based on after-action reports from Hurricane Katrina and other requirements. In 2010, he served as the planning coordinator for Emergency Support Function (ESF)-7, Logistics and Resource Management, and ESF-8, Public Health and Medical Services, on the FEMA Region VII New Madrid Seismic Zone Catastrophic Earthquake Plan. Steinle received a commendation from the FEMA Region VII Administrator for his work on this plan.

Steinle has served for five years on the planning board of the Federal Bureau of Investigation's International Symposium on Agroterrorism (ISA) and developed a script for a pandemic influenza video shot in Vietnam for the 2008 ISA. He also served as the chair of the ASIS International Agriculture and Food Security Council from 2006 to 2008 and is an active member of the board of directors for the Safety and Health Council of Western Missouri and Kansas.

He can be reached at msteinle@gmail.com.

S. Shane Stovall graduated from the University of North Texas with a BS in emergency administration and planning. In addition, he received certification as a Certified Emergency Manager (CEM) from the International Association of Emergency Managers in 2004.

He has served as the director for the Department of Emergency Management in Plano, Texas, since October 2006. Before coming to Plano, Stovall began his career in Florida with the Charlotte County Office of Emergency Management, serving as emergency planner, plans and operations supervisor, and then emergency coordinator. Following his eight-year tenure with the Charlotte County Office of Emergency Management, Stovall worked for two years as a project manager with the General Physics Corporation's Homeland Security and Emergency Management Unit in Tampa, Florida. He also serves as cochairman for the Public–Private Partnership Caucus for the International Association of Emergency Managers and serves as chairman for the Public–Private Partnership Committee for the Emergency Management Association of Texas.

Stovall's experience includes developing plans, strategy, and direction of emergency and disaster mitigation, preparedness, response, and recovery projects in multiple jurisdictions. His disaster experience includes response and recovery duties to the Lancaster, Texas, F4 Tornado in 1994, Oklahoma City bombing in 1995, Florida Wildfires of 1998, Tropical Storm Josephine, Tropical Storm Gabrielle, and Hurricane Charley. He also has vast experience in developing and coordinating jurisdictional emergency and disaster response exercises and training for jurisdictional teams in the public, private, and volunteer sectors.

Shane can be contacted at shanes@plano.gov.

Reviewers

Charles R. Blaich, MS, served the New York City Fire Department (FDNY) for 30 years. As deputy chief of the FDNY, he served as the logistics chief of the World Trade Center Task Force from September 2001 to February 2002, responsible for supporting recovery, firefighting, GIS mapping, mortuary activities, care and feeding of the emergency and civilian workers, as well as emergency medical support to the Ground Zero workers. He was responsible for all logistic planning and activities supporting the entire array of emergency and support organizations, including police, EMS, federal, and state. He is a recognized expert on the interaction and management of the various levels of government and agencies necessary to successfully respond to disasters and terrorist incidents. His articles on building construction and collapse as well as emergency operations have appeared in many professional journals. He has lectured on the topics of disaster operations in many conferences and seminars worldwide.

Upon retirement, he accepted the position with the Raytheon Company as the director, Preparedness and Response (2003–2009), in the Raytheon Homeland Security Strategic Business Area (SBA) at its inception. He assumed the position of director, Business Development Integrated Security Systems in 2009 with BAE Systems, Inc., to explore expanding BAE Systems capabilities into the public safety area of Homeland Security. He is currently engaged as a private consultant.

Colonel Blaich served in the active duty and the reserve U.S. Marine Corps from 1968 to 1998. His service included tours of duty in Vietnam and command of two battalions including the 2nd Battalion and 25th Marines Infantry Battalion in support of Operation Desert Storm. Awards include Meritorious Service Medal, Two Navy Commendation Medals with combat "V," Combat Action Ribbon, National Defense Service Medal as well as several campaign and unit citation decorations.

He can be contacted at crblaich@gmail.com.

Edward G. Buikema is a senior consultant at Argonne National Laboratory in Chicago, Illinois. Argonne National Laboratory is a U.S. Department of Energy laboratory managed by the University of Chicago.

Buikema is also the director of Preparedness and Response for Armada Ltd., an emergency management consulting firm based in the Midwest. He is an instructor at the University of Chicago, a subject-matter expert for the Mobile Executive Education Team seminars conducted by the Naval Postgraduate School's Center for Homeland Defense and Security, and a member of the Emergency Management Accreditation Program (EMAP) commission.

He was appointed by President George W. Bush as the administrator of Region V of the Federal Emergency Management Agency (FEMA) (now part of the Department of the Homeland Security) in November 2001, and he served until January 2009. As regional administrator, he coordinated FEMA mitigation, preparedness, and disaster response and recovery activities in six states: Illinois,

Indiana, Michigan, Minnesota, Ohio, and Wisconsin. During his tenure as regional administrator, Buikema oversaw the delivery of disaster relief and assistance in more than sixty-five presidential declared disasters and emergencies. He is a past chair of the Executive Committee of the Chicago Federal Executive Board, a member of the NFPA 1600 Technical Committee, the Executive Board of the Chicago Joint Terrorism Task Force, and past chair of the EMAP Commission. He was also designated as a Principal Federal Official by the secretary of Homeland Security for pandemic influenza for Region C in the United States (combined FEMA Regions V and VIII).

Buikema served as the acting director of FEMA's Response Division from February 2005 through October 2005. In that position, he was responsible for administration and leadership of Response Division programs including the Urban Search and Rescue (US&R) program, the National Disaster Medical System (NDMS) teams, the Mobile Emergency Response System (MERS), logistics, the National Response Coordination Center (NRCC), and the Temporary Disaster Workforce.

Prior to his FEMA position, Buikema was commander of the Emergency Management Division of the Michigan State Police. In this position, he was responsible for leadership and coordination of the state's emergency management program and served as the state coordinating officer and governor's authorized representative for nine presidential disaster declarations. He also served as chair of the Michigan Emergency Planning and Community Right-to-Know Commission, the Michigan Hazard Mitigation Coordinating Council, and Michigan's Anti-Terrorism Task Force. Buikema was an officer with the Michigan State Police for more than 26 years, serving in the Emergency Management Division for 19 of those years.

He was awarded the first ever Eric Tolbert Distinguished Service Award by the EMAP at the 2008 National Emergency Management Association (NEMA) conference "for exemplary service and dedication to EMAP and fostering excellence and accountability in Emergency Management." Recently, he was awarded the Steven R. Cohen Award by the Chicago Federal Executive Board for "giving exemplary service to his own agency and the Federal community at large." A native of Grand Rapids, Michigan, Buikema holds a BA in political science from Calvin College in Grand Rapids and is a graduate of the FBI National Academy at Quantico, Virginia. He is married and has three children and two grandchildren.

Dr. Frank Cartledge served Louisiana State University (LSU) for more than 40 years as professor of chemistry, associate dean of the College of Science, and vice provost for Academic Affairs. His research interests included environmental chemistry of hazardous wastes, particularly in land disposal scenarios, spectroscopic and microscopic characterization methods applied to solids, organosilicon chemistry, and science education at the middle and high school levels. He has been a consultant in these areas to higher education, industry, and government. His work led to more than 100 publications in peer-reviewed journals as well as funding for research and training projects amounting to over eight million dollars. He directed the work of 20 students who received MS or PhD degrees.

As vice provost at LSU, he was a member of the campus executive team responding to Hurricane Katrina. He interacted with both faculty and students during the semester-long return to normalcy after Katrina, organized faculty and staff to provide services to displaced students, and was the principal spokesperson for faculty and staff concerning academic policy changes necessitated by the hurricane. As Chemistry Department chairman, and also the leader of a large research group, he had experience planning for and responding to incidents such as chemical spills, fires, and explosions. Since 2008, he has been certified by the Department of Homeland Security to teach campus emergency management courses for the LSU National Center for Biomedical Research and Training.

Dr. Cartledge's work has been recognized with a National Science Foundation postdoctoral fellowship, the LSU Amoco Foundation Undergraduate Teaching Award, the Coates Award of the Baton Rouge Sections of the American Chemical Society and American Institute of Chemical Engineers, the LSU Alumni Federation Distinguished Faculty Fellowship, and resulted in his appointment as an alumni professor.

He can be contacted at fcartledge@lsu.edu.

Stan W. Casteel, DVM, PhD, is a professor of veterinary pathobiology in the Veterinary Medical Diagnostic Laboratory of the University of Missouri's College of Veterinary Medicine. He is a board-certified veterinary toxicologist and has formal training in the recognition of foreign animal diseases from the Plum Island Animal Disease Center. His expertise spans 25 years as a diagnostician of animal diseases in a laboratory fully accredited by the American Association of Veterinary Laboratory Diagnosticians. His teaching experience spans almost three decades with undergraduate, veterinary, and graduate students in public health and biomedical sciences on topics ranging from foreign animal diseases to zoonotic and vector-borne diseases.

He can be reached at casteels@missouri.edu.

Robert J. Coullahan, CEM, CPP, CBCP, CHS V, is the president and chief operating officer of Readiness Resource Group Incorporated in Las Vegas, Nevada.

He has more than 35 years of professional experience in emergency management and critical infrastructure protection for government and commercial organizations. He is board certified in emergency management, security management, business continuity management, and homeland security. He founded Readiness Resource Group (www.readinessresource.net), which supports national preparedness and public safety programs. He served for 20 years with Science Applications International Corporation including assignment as senior vice president. He served for five years as vice president for Government and International Programs with a university consortium that was named World Data Center for Human Interactions in the Environment, based on U.S. National Research Council peer review. He serves on ANSI and ASIS Business Continuity Management Standards Committees. He holds an MS in telecommunications and an MA in security management–forensic sciences, and is a graduate of the University of California and the George Washington University.

He can be contacted at coullahan@readinessresource.net.

Raymond Crowley is an experienced instructor/trainer in the area of preparedness planning and response to CBRNE (chemical, biological, radiological, nuclear, and high-yield explosives) and other catastrophic domestic events. He has worked for the state of Connecticut's Department of Emergency Management and Homeland Security as a regional trainer. He served 20 years with the New Haven Police Department, 10 of those with the Emergency Services Unit as a bomb technician/explosive detection canine handler and WMD coordinator. There, he was responsible for solutions-oriented training for the department's SWAT, Dive, and Bomb teams. Crowley was a member of the FBI WMD/EOD Task Force in Connecticut and represented law enforcement on the board of directors for the New Haven Area Special Hazards Team (a regional HAZMAT team). In addition to his law enforcement background, Crowley is an adjunct instructor at the Louisiana State University National Center for Biomedical Research and Training Academy of Counter Terrorism Education and is a subject matter expert for the National Center for Security and Preparedness (NCSP) at the University of Albany. He regularly conducts training for the U.S. State Department's Anti-Terrorism Assistance program.

He can be reached at raymondcrowley@sbcglobal.net.

K. R. Juzwin, PsyD, is a clinical psychologist and associate professor of clinical psychology at Argosy University–Schaumburg. She specializes in first responder mental health, high-risk personnel assessment, critical incident response, disaster mental health, and trauma.

Dr. Juzwin is the regional coordinator for mental health response for the Illinois Medical Emergency Response Team (IMERT), a responder on the Illinois Disaster Assistance Team (D-MAT IL-2), and is the current education coordinator and has also been a coordinator on the Northern Illinois Critical Incident Stress Management Team (NICISM). On the IMERT team, she has written the training curriculum for mental health responders and trains personnel related to mental health issues in crisis and disaster response. Clinically, Dr. Juzwin focuses on trauma, forensic trauma, self-injury, eating disorders, and high-risk patients. She is active in education, consultation, writing and research in these areas. She is the director of Self-Injury Recovery Services at Alexian Brothers Behavioral Health Hospital in Hoffman Estates, Illinois, which is the only JCAHO-DSC certified program for self-injury in the country. Dr. Juzwin is the chief psychologist for COPS and FIRE Testing Service and conducts outcome research, assessment, protocol development and training, and supervision in the area of testing and assessment for high-risk hiring for law enforcement, fire/EMS, and emergency dispatch personnel. At Argosy University, Dr. Juzwin teaches testing and assessment, ethics, professional development, police psychology, trauma, forensic trauma, eating disorders, self-injury, and suicide assessment courses.

She can be contacted at kjuzwin@argosy.edu.

Edward J. Krueger is the director for the Criminal Justice Center for Innovation at Fox Valley Technical College. Previously, he served as a police officer and the director of a regional police academy. He has a graduate degree in education and more than 30 years of criminal justice, occupational training, and educational experience. Nationally known for his work in the criminal and juvenile justice fields, he directs numerous education and training programs for criminal justice personnel focusing on criminal–juvenile justice management, tribal community mobilization and community analysis for planning, missing person(s) training, emergency management training for law enforcement and communities, and crime reduction–crime prevention management. Krueger is the recipient of numerous acknowledgments including such distinctions as crime prevention officer, community support officer, and training officer of the year. He also received the Wisconsin Technical College system educator of the year. Most recently, he was recognized by Fox Valley Technical College for his outstanding leadership within the college. He has been the director of hotel security and had the responsibility for special event security for NFL football teams, celebrities, and political dignitaries. In 2004, he served as assistant director of security for the Commission on Presidential Debates. Most recently, he provided services to the U.S. State Department and the Peruvian government regarding the security of unearthed historical communities. He has edited and authored several publications including, most recently, *Accident Prevention Manual for Business and Industry: Security Management*, published by the National Safety Council. His techniques for organizational collaboration as well as his managerial expertise provide a framework for multilevel managerial processes.

He can be reached at kruegere@fvtc.edu.

Howard Murphy, CEM, currently serves on a high-priority Department of Defense civil support mission as the commander of the U.S. Army's first CBRNE Incident Response Force (CIRF). He has 28 years of emergency response, serving as a chief officer and as a commissioner. He is a 23-year veteran of emergency management operations at the local, state, national, and international levels, with extensive experience in disaster operations and development and implementation of policy. He has more than 15 years of national intelligence/security experience, leading and/or participating in the development of numerous national-level programs credited with providing security to the United States and its allies. Murphy is the recipient of numerous civilian and military awards and honors and serves as adjunct faculty with several master's-level education programs.

He can be contacted at jhmurphy@charter.net.

Drew Orsinger is currently the chief of infrastructure analysis at Argonne National Laboratory's Infrastructure Assurance Center. He previously served as the protective security advisor (PSA) for the Chicago, Great Lakes District, for more than four years. Before becoming a PSA, Orsinger worked for Customs and Border Protection (CBP) as a watch commander for the National Targeting Center (NTC) in Reston, Virginia. The NTC is responsible for screening all suspicious cargo and persons prior to entering the United States, and Orsinger resolved complex cases with the Joint Terrorism Task Forces and National Intelligence Community. Before joining CBP, he served as deputy director for current operations at DHS Headquarters, where he coordinated DHS operations with state, local, and private sector partners.

Orsinger served on active duty for nine years in the U.S. Coast Guard and has the distinction of being one of the first Coast Guard liaisons to the DHS Headquarters. In this capacity, Orsinger led a team of Coast Guard officers in the National Operations Center and provided operational and policy expertise to both Secretary Tom Ridge and Secretary Michael Chertoff. Immediately following 9/11, the Coast Guard sent Orsinger to serve as the Coast Guard liaison to the FBI's Strategic Information Operations Center (SIOC) in Washington, DC. Before serving at DHS, Orsinger worked in the Coast Guard Office of Law Enforcement and Policy, the Coast Guard Intelligence Coordination Center, and as operations officer of the Coast Guard cutter *Conifer* stationed in Long Beach, California. Throughout his Coast Guard career, he was a Coast Guard delegate to the Department of State on immigration matters that met regularly in places such as Havana, Cuba; New York City; Miami; and Nassau, Bahamas.

Orsinger earned a BS degree from the U.S. Coast Guard Academy in 1996 and an MA in public policy and administration from Northwestern University in 2010.

He can be reached at dorsinger@anl.gov.

William (Bill) Peterson brings critical emergency management experience spanning more than 40 years in the fire service, emergency management, and homeland security. Most recently, he was appointed regional administrator of the U.S. Department of Homeland Security's Federal Emergency Management Agency (FEMA) Region 6, in Denton, Texas, by President George W. Bush. Peterson was responsible for the delivery of DHS/FEMA Disaster Response and Recovery Operations in Arkansas, Louisiana, New Mexico, Oklahoma, and Texas. During his tenure as regional administrator, he was responsible for handling a total of 83 presidential disaster declarations, emergency declarations, and FMAG incidents within FEMA Region 6, including serving as senior federal official, in theater, for the successful federal response to support the state of Louisiana during Hurricane Gustav and the state of Texas during Hurricane Ike in 2008.

Peterson has an associate degree in fire science, a BA in fire protection administration, and a dual MA in public administration and human relations. He has received numerous awards and recognition including the Benjamin Franklin

Leadership Award by the International Association of Fire Chiefs in 1997; Career Fire Chief of the Year by the International Association of Fire Chiefs and Fire Chief Magazine in 2000; and the Krzysztof Smolarkiewicz Medal (Serial No. 9) for Exemplary Service to the Polish Fire Service in 2005.

Peterson has participated in courses at the National Fire Academy, the Emergency Management Institute, and Harvard University's John F. Kennedy School of Government program for senior executives in state and local government. In addition, he has served as an author of fire service publications, as an instructor of fire science courses, and as a member of national planning and standards committees. He has been a recognized leader during his distinguished career on the local, state, national, and international level. He is a founding member of the United States/United Kingdom Fire Service Symposium. He is a founding member of the U.S.A. Branch of the Institution of Fire Engineers (IFE). He was elected to the International Council (Board of Directors) of the worldwide organization in 1999 and elected to the post of international president of the 11,000-member worldwide organization in 2004. He holds the distinction of being the first individual who is not a subject of the United Kingdom or one of the Commonwealth holdings to be elected to this prestigious office.

Peterson is a Certified Emergency Manager by the International Association of Emergency Managers (IAEM), holds a Chief Fire Officer designation from the Center for Public Safety Excellence (CPSE), and is an elected fellow of the Institution of Fire Engineers.

Edward (Ed) Plaugher, fire chief (retired), is currently the assistant executive director, National Programs and Consulting Services, for the International Association of Fire Chiefs in Fairfax, Virginia. After serving 24 years with the Fairfax County Fire and Rescue Department and retiring as a deputy fire chief, he was appointed to the position of fire chief, Arlington County Fire Department, Arlington, Virginia, a position he held until his retirement in June 2004. In August 2004, at the annual meeting of the International Association of Fire Chiefs, he was named Career Fire Chief of the Year by *Fire Chief Magazine*. Chief Plaugher's career spans more than 40 years as a fire service professional.

In addition, he is a past president of the State Fire Chiefs Association of Virginia. He is a lifetime member of both International Association of Fire Chiefs (IAFC) and the National Fire Protection Association (NFPA). As a member of those associations he served on national fire code and terrorism committees. Chief Plaugher is a member of the Emergency Response Senior Advisory Committee (ERSAC) of the Homeland Security Council (HSC), Department of Homeland Security, and he recently served as a special advisor to the Defense Science Board for the Department of Defense. He holds a BS in fire administration and technology from George Mason University and completed the executive fire officers course at the National Fire Academy.

Chief Plaugher, as chair of the Council of Governments (MWCOG), Fire Chiefs Terrorism Committee, directed the regional terrorism preparedness efforts for the National Capital Region and was the executive agent for the nation's first Metropolitan Medical Strike Team. On September 11, 2001, as chief of the Fire Department, he led Arlington County's response efforts to the terrorist attack at the Pentagon. Arlington's Fire Department coordinated and led the local, regional, state, and federal response to the incident. For both the preparedness and response efforts, Chief Plaugher was awarded the Secretary of the Army Public Service Award in June 2004 and the Secretary of Defense Exceptional Public Service, also in June 2004.

He can be reached at eplaugher@iafc.org.

Thomas D. Schneid, PhD, is a tenured professor in the Department of Safety, Security and Emergency Management (formerly Loss Prevention and Safety) at Eastern Kentucky University and serves as the graduate program director for the online and on-campus master of science degree in safety, security, and emergency management.

He has worked in the safety and human resource fields for more than 30 years at various levels including corporate safety director and industrial relations director. Dr. Schneid has represented numerous corporations in OSHA and labor-related litigations throughout the United States. He holds a BS in education, MS and CAS in safety, an MS in international business, and a PhD in environmental engineering, as well as Juris Doctor (JD in law) from West Virginia University and LLM (graduate law) from the University of San Diego. He is a member of the bar for the U.S. Supreme Court, 6th Circuit Court of Appeals and a number of federal districts as well as the Kentucky and West Virginia Bar.

Dr. Schneid has authored and/or coauthored 15 texts including *Corporate Safety Compliance, Americans with Disabilities Act Handbook, Legal Liabilities for Safety and Loss Prevention Professionals and Fire Law*, as well as more than 100 articles on safety, fire, EMS, law, and related topics.

He can be contacted at Tom.Schneid@eku.edu.

Edward H. Stephenson, DVM, PhD, has a career in veterinary medicine that spans clinical practice to biomedical research to academia to industry to consulting. He completed a 24-year career in the US Army, principally in biomedical R&D. For the last 5 years of his military career, he was the director of the Division of Airborne Diseases, US Army Medical Research Institute for Infectious Diseases (USAMRIID).

Stephenson guided multidisciplinary research programs in aerosol transmission of microbial agents of terrorism and naturally occurring diseases. Subsequently, he was in academia for 12 years, where he developed the Center for Government and Corporate Veterinary Medicine, an internationally recognized program.

Stephenson has been a consultant since 1986 in aerobiology, agriculture and food safety defense, agroterrorism, bioterrorism, human and veterinary biomedical R&D, infectious diseases, airborne transmitted diseases, and vaccine development. Since 2001 he has been an instructor for the National Center for Biomedical

Research and Training, Louisiana State University, as a subject matter expert (SME) in bioterrorism and agroterrorism.

William L. Waugh Jr. is a professor of public management and policy in the Andrew Young School of Policy Studies at Georgia State University and an adjunct faculty member in the executive master of science program in emergency and crisis management at the University of Nevada–Las Vegas. He is the author of more than 100 articles and book chapters on emergency management, terrorism, leadership development, and administrative theory. His books include *International Terrorism: How Nations Respond to Terrorists* (1982), *Terrorism and Emergency Management* (1990), *Cities and Disaster* (1990), *Handbook of Emergency Management* (1990), *Disaster Management in the U.S. and Canada* (1996), *Living with Hazards, Dealing with Disasters* (2000), and *Emergency Management: Principles and Practice for Local Government*, 2nd edition (2007). He is also the editor-in-chief of the *Journal of Emergency Management.*

Dr. Waugh is a member of the Emergency Management Accreditation Program (EMAP) Commission and a former member of the Certified Emergency Manager (CEM) Commission. He has served on the advisory board and as an instructor for the U.S. Department of State Office of Diplomatic Security's Senior Crisis Management Seminar, part of the Anti-Terrorism Assistance Program, and as a consultant to federal, state, and local agencies, nongovernmental organizations, and international organizations. He has participated in workshops on issues ranging from the Partnership for Public Warning's assessment of the Homeland Security Advisory System to Pfizer's assessment of hospital surge capacity issues to the Public Risk Institute's workshop on long-term recovery. In 2007, he testified before Congress on the National Response Framework.

In 1970–1971, he served as an infantryman in South Korea. He received his BA at the University of North Alabama in 1973, his MA from Auburn University in 1976, and his PhD from the University of Mississippi in 1980. He taught at Mississippi State University in 1979, Kansas State University from 1980 to 1985, and Georgia State University since 1985.

He can be reached at wwaugh@gsu.edu.

Chapter 1

Introduction: Why Plan for Disasters?

Michael J. Fagel and Stephen J. Krill Jr.

Contents

Disasters happen. No one is immune to them or the devastation that they can bring to communities of any size. *Disaster* applies to a variety of events, each with varying magnitudes and of varying natures. The Federal Emergency Management Agency (FEMA) defines a disaster as

> an occurrence of a severity and magnitude that normally results in deaths, injuries, and property damage and that cannot be managed through the routine procedures and resources of government

Because of the unpredictable nature of disasters, it is essential to plan for such events. Only through planning is it possible to effectively respond to and mitigate against the potential effects of disaster.

Types of Disasters

Disasters usually fall into one of four categories: natural, man-made, deliberate, or accidental. In the United States, the most frequently occurring disasters are natural. Because natural disasters are not caused by the actions of humans, they are the most difficult to prevent. However, because most, although not all, natural disasters (e.g., weather related) are predictable, they should be the easiest to plan for.

Because it is impossible to stop natural disasters from occurring, emergency planners and responders must increase their awareness of how and why they occur. At the same time, they must analyze the risk in their communities so that they can focus their efforts on the highest risk or highest impact hazards. For example, Florida residents need not prepare for paralyzing blizzards. They must be prepared for hurricanes, however. On the other hand, New England residents need to be prepared for both blizzards and hurricanes as well as a multitude of other hazards.

Man-made disasters fall into two categories: deliberate and accidental. To identify a single cause of any man-made disaster would be nearly impossible. For as many types of man-made disasters that exist, there are a hundred reasons why. Whereas natural disasters are caused by the forces of nature and can only be predicted, not prevented, man-made disasters occur through preventable, sometimes deliberate, and often malicious acts.

Not all man-made disasters are deliberate, however. Many can be categorized as accidental and completely unexpected. As we continue to enter the era of technology, an increasing number of technologically sophisticated incidents will continue to occur, and community disaster plans must take into account the potential for an increasing number of incidents involving toxic spills, radiological releases, and other accidents. The events could be highly disastrous in densely populated areas, and as communities continue to grow, these events could have consequences that expand beyond the realm of anything imaginable in today's society.

Table 1.1 represents the types of disasters that fall into the categories just presented. This list is not intended to be inclusive, and certainly there are other potential disasters that could strike any community. For this reason, when the time comes to create or revise the community's disaster plan, it is critical that planners be prepared to protect their communities from these and other potential disasters that may occur at any time, with or without warning.

Each of these situations poses its own threat to communities and challenges to emergency response personnel. Some are relatively easy to predict, whereas others may occur with virtually no warning. Some will devastate a community for weeks or months, whereas others may pack a powerful punch, then be over, leaving devastation in their wakes.

It can be expected that most disasters, regardless of type or cause, will bring about some degree of death, injury, and property damage. Depending on the circumstances, disruption in communications systems, power, and water supplies; contamination of air, food, and water; and a multitude of other problems may also occur.

Table 1.1 Types of Disasters

Natural	*Man-Made*	*Deliberate*	*Accidental*
Hurricanes	Nuclear accidents	Terrorism	Highway accidents
Tornadoes	Hazmat incidents	Strike violence	Rail accidents
Earthquakes	Explosions	Sabotage	Aircraft accidents
Snow storms		Bombings	Industrial mishaps
Ice storms		Riots and civil disturbances	
Floods			

Phases of Disaster

Each type of disaster and its subsequent response effort—regardless of how, when, or why it takes place—can be expected to consist of a number of phases that are fairly consistent. The success of a community's response plan depends greatly on the planner's understanding of these phases and how they are incorporated into the overall disaster plan.

The initial phase of a disaster is the *warning* phase. It is during this phase that emergency officials have the best opportunity to provide the public with disaster-related information. It is also during this phase that an evacuation or shelter in place, if necessary, will be initiated.

The second phase of a disaster is the *threat and impact* phase. This phase is typically followed by the *inventory*, *rescue*, *remedy*, and *restoration* phases.

Levels of Severity

Within each type of disaster, varying levels of severity will exist, each with its own response requirements. There are three levels of disaster, and even the lowest level can require the involvement of state and federal officials.

> Level I Disaster—A localized multiple-casualty incident wherein local medical resources are available and adequate to provide for field medical treatment and stabilization, including triage. The patients will be transported to the appropriate medical facility for further diagnosis and treatment.
>
> Level II Disaster—A multiple-casualty emergency where the large number of casualties and/or lack of medical care facilities are such as to require multi-jurisdictional medical mutual aid.
>
> Level III Disaster—A mass-casualty emergency wherein local and regional medical resources are exceeded. Deficiencies in medical supplies and personnel are such as to require assistance from state or federal agencies.

Why Plans Fail

Although many jurisdictions have adopted formal disaster plans, local governments often fail to improve their plans, even after a major disaster has occurred. Local governments also fail to adequately and effectively plan for disaster response. These failures can often be attributed to lack of relevant experience with disaster response, failure to study lessons learned, failure to commit to carrying out a disaster planning program, or performing the wrong kind of planning.

Many communities lack the experience to deal with disaster response because their public officials are not involved in enough disasters to gain personal experience. There is often the misconception that responding to a disaster is the same as responding to any other emergency, only on a grander scale.

Another impediment to planning is that although an individual may have been involved in disaster response, it is difficult to view the disaster from the perspective of other organizations. Postdisaster critiques often turn out to be justifications of actions taken, rather than impartial, objective assessments of problems and mistakes.

Disaster planning must have the support of the entire community if it is to be successful. Lack of public awareness can often undercut the community's efforts to plan. Key officials often neglect to read emergency plans. Even after a plan is written, it is often not properly exercised, often resulting in failure of the plan during a true disaster.

Planning as a Blueprint

A disaster plan should serve as a community's blueprint for initiating, managing, and performing operations that will most likely extend beyond the scope of functions carried out in normal day-to-day operations. The disaster plan should serve to coordinate the activities, logistics, and resources involved in disaster response. Plans typically seek to establish the various sectors that need to be implemented in the event of a disaster.

A comprehensive emergency management strategy includes four phases that work in a pattern. The four phases are mitigation, preparedness, response, and recovery. The phases are not linear in nature, but are more cyclical, as illustrated in Figure 1.1.

The goal of prevention is to avert accidents and emergencies. This is not always possible, however, thus creating the need for preparedness, the component of planning in which steps are taken to ensure that all individuals and entities to be involved in a disaster response ready themselves to perform during an emergency.

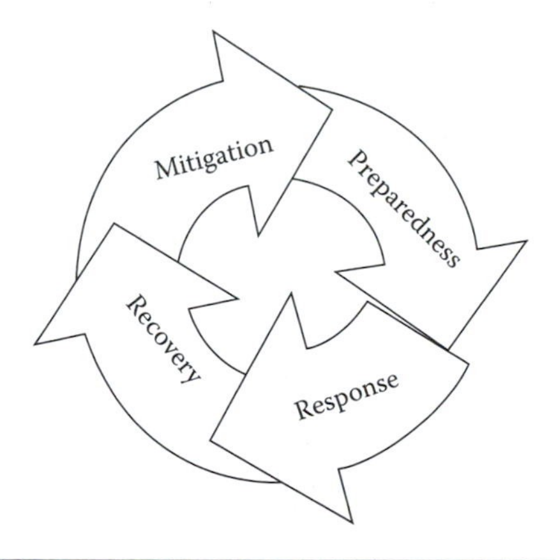

Figure 1.1 The four phases of comprehensive emergency management.

The key to preparedness is to ensure that an adequate level of resources is available to save lives and minimize property damage.

The next stage in disaster planning is response, which is the initiation of activities to save lives and protect property. The final step is recovery. It is in the recovery stage that residents and government try to resume business as usual.

Again, these phases are more cyclic in nature than they are linear as the process is always ongoing and continuous.

The Disaster Cycle and Planning

Without planning, emergency operations can suffer from a multitude of problems, which could lead to serious consequences and even death. Therefore, it is imperative to have a system that enables all participants involved in an emergency response to work together. An integrated emergency management system does just that.

An integrated and comprehensive emergency management system is a conceptual framework that increases emergency management capability by using a structured approach to planning and response. To have this increased capability, it is important to establish networking, coordination, interoperability, partnerships, and creative thinking about resource shortfalls during the planning process before an emergency occurs.

An integrated and comprehensive emergency management system should

- Address all hazards that threaten a community
- Be useful in all phases of emergency management
- Seek resources from any and all sources that are appropriate
- Knit together all partnerships and participants to achieve a mutual goal

The goals of emergency management are to save lives, prevent injuries, and protect property and the environment. Participants in emergency management should include local, state, and federal government; private sector entities (e.g., nonprofit organizations, businesses, and industry); and private citizens.

The role of local government in emergency management is not just limited to response. Local government must ensure the safety of its citizens and acquire knowledge of the threats to the community and the resources required to meet those threats. Some of the measures that local government should take to prepare for an emergency include

- Ensuring that the emergency operations plan (EOP) is developed, trained, exercised, and maintained
- Developing mutual aid agreements, memoranda of understanding (MOUs), and standby contracts for critical emergency resources
- Communicating with the public about potential hazards and how to prepare for them

State government has legal authorities for emergency response and recovery, and serves as the point of contact between local and federal governments. In addition, state governments may require that local jurisdictions develop and submit emergency plans for state review and incorporation into the state plan. State governments also have the authority to activate National Guard resources in response to emergencies and develop emergency mutual aid compacts (EMACs) with other states.

To assist state and local governments in recovering from emergencies, the federal government has legal authorities, fiscal resources, research capabilities, technical information and services, and specialized personnel.

Although government agencies are responsible for protecting the public, they need the help of private citizens in all facets of emergency management to be successful. Nonprofit organizations, such as the American Red Cross, offer critical resources in emergencies.

Private citizens can assist by reducing hazards in and around their communities, preparing disaster supply kits, and monitoring emergency communications carefully. Citizens can also volunteer with established organizations and pursue training in emergency preparedness and response.

Resources for an integrated emergency management system include both equipment resources and personnel resources. Personnel resources include

- Elected and appointed officials
- Emergency program managers
- Emergency operations staff
- Police and fire departments
- Other local service providers, such as public works, transportation, and public health
- Voluntary organizations (e.g., American Red Cross, Salvation Army)

An integrated and comprehensive emergency management system brings all of these personnel resources together through the following:

Planning—Involving all of these key players in the planning process ensures that all roles and responsibilities are clearly defined.

Direction—All parties must have direction clearly defined in the EOP to reduce freelancing and ensure that all response activities are handled according to established policies and procedures.

Coordination—Working together during the planning process develops relationships and promotes teamwork during a response.

Clearly defined roles and responsibilities—Developing and agreeing to roles and responsibilities during the planning process reduces redundancy and ensures the most efficient use of resources during an emergency.

Furthermore, an integrated and comprehensive emergency management program

- Examines potential emergencies and disasters based on the risks posed by likely hazards
- Develops and implements programs aimed at reducing the impact of these events on the community
- Prepares for hazards that cannot be eliminated
- Prescribes actions required to deal with the consequences of actual events

As discussed previously, comprehensive emergency management activities are divided into four phases:

> Mitigation—Taking sustained actions to reduce or eliminate long-term risk to people and property from hazards
> Preparedness—Building the emergency management function to respond effectively to, and recover from, any hazard
> Response—Conducting emergency operations to save lives and property by taking actions to reduce hazards; evacuating potential victims; providing food, water, and medical care to those in need; and completing emergency repairs to restore critical public services
> Recovery—Rebuilding communities so that individuals, businesses, and governments can function on their own, return to normal life, and protect against future hazards

Following an emergency, lessons learned are used to mitigate, prepare, and respond better. As the plan is revised based on lessons learned, the cycle repeats. Each of these phases is described next.

Mitigation

As the costs of disasters continue to rise, it is necessary to take sustained action to reduce or eliminate the long-term risk to people and property from hazards and their effects. These sustained actions are known as mitigation.

Mitigation should be a continued activity that is integrated with each of the other phases of emergency management to use a long-range, community-based approach to disasters. The goals of mitigation activities are to protect people and structures, and reduce the cost of response and recovery.

Mitigation is accomplished in conjunction with a hazard analysis, which helps to identify what events can occur in and around the community; the likelihood that an event will occur; and the consequences of the event in terms of casualties, destruction, disruption to critical services, and costs of recovery. To be successful, mitigation measures must be developed into an overall mitigation strategy that considers ways to reduce hazard losses together with the overall risk from specific hazards and other community goals.

Because it is not possible to mitigate completely against every hazard, preparedness measures can help to reduce the impact of those hazards by taking certain actions before an emergency occurs.

Preparedness

Preparedness involves all of the players in the integrated emergency management system and includes the following activities:

- Developing, training, and exercising the EOP
- Recruiting, assigning, and training staff who can assist in key areas of response operations
- Identifying resources and supplies that might be required in an emergency
- Designating facilities for emergency use

Response

Response begins when an emergency is imminent or immediately after an event occurs. Response encompasses all activities taken to save lives and reduce damage from the event and includes providing emergency assistance to victims, completing emergency repairs critical to infrastructure, and ensuring the continuity of critical services.

One of the first response activities should be to conduct a situation assessment. To fulfill this task, responders must conduct an immediate rapid assessment of the situation. Rapid assessment includes all immediate response activities that are directly related to determining initial lifesaving and life-sustaining needs and identifying imminent hazards. Coordinated and timely assessments enable local government to prioritize response activities, allocate scarce resources, and request additional assistance from mutual aid partners.

Recovery

The final phase in the emergency management cycle is recovery, the goal of which is to return the community's systems and activities to normal. Some recovery operations may be concurrent with response efforts.

Recovery from a disaster is unique to each community and disaster. Short-term recovery is an extension of the response phase, in which basic functions and services are restored. After short-term recovery, the community must rebuild. Considerations for long-term recovery include

- Applying for federal assistance
- Keeping the public informed on the rebuilding process

- Taking mitigation measures to ensure against future disaster damage
- Collecting and distributing donations
- Building partnerships with business and industry for needed resources
- Taking care of environmental concerns
- Meeting the needs of victims
- Taking public health measures to protect against diseases and contamination
- Rebuilding bridges, roads, and other elements of the community's infrastructure

Recovery also involves taking steps necessary to reopen damaged businesses, reemploy workers, and other measures required to return the community to its preemergency status.

A Brief History of Emergency Management

Throughout its history, the United States has faced many disasters—from natural disasters to hazardous materials releases to terrorist attacks. Although all disasters are local, the federal government will provide support—personnel, equipment, resources, and funding—when local capabilities and capacities are exceeded.

Before the 1970s, various federal agencies and programs provided disaster relief services. At one point, more than 100 federal agencies handled disaster and emergencies. However, until four devastating hurricanes struck in the 1970s, the need for improved disaster coordination at the federal level did not happen. Following are the four hurricanes:

- Hurricane Agnes (1972)—This hurricane caused significant East Coast flooding and $2 billion in damages.
- Hurricane Eloise (1975)—This hurricane caused $200 million in damages and 76 fatalities.
- Hurricanes David and Frederick (1979)—These hurricanes were among the deadliest ever seen in the Caribbean, with Frederick causing $2.2 billion in damages.

These disasters and the need for disaster preparedness nationwide pushed the Carter administration to establish the FEMA to coordinate all disaster relief efforts at the federal level.

Authorities and Directives

After creating FEMA, the federal government continued to refine disaster response by creating several acts and directives. Some, such as the Robert T. Stafford Act, were created as general, all-hazards guidelines, whereas others, such as the Post-Katrina Emergency Management Reform Act, were born from a specific event.

Robert T. Stafford Act

To bring a more orderly and systemic means of federal natural disaster assistance for state and local governments in carrying out their responsibilities to aid citizens, the Robert T. Stafford Disaster Relief and Emergency Assistance Act (Public Law 93-288) was passed in 1988. This act describes the programs and processes by which the federal government provides coordination and support to state and local governments, tribal nations, eligible nongovernment organizations (NGOs), and individuals affected by a declared major disaster or emergency. The Stafford Act covers all hazards, including natural disasters and terrorist events.

The Stafford Act was amended several times, most recently with the Post-Katrina Emergency Management Reform Act (PKEMRA) in 2006 (Public Law 109-295) to address the issues and challenges that arose during the response to Hurricane Katrina (see "Post-Katrina Emergency Management Reform Act" section for more information on PKEMRA).

Homeland Security Act of 2002

Title I of the Homeland Security Act of 2002 established the Department of Homeland Security (DHS), with the mission of

- Preventing terrorist attacks within the United States
- Reducing the vulnerability of the United States to terrorism
- Minimizing the damage, and assisting in the recovery, from terrorist attacks that do occur within the United States
- Carrying out all functions of entities transferred to the department, including by acting as a focal point regarding natural and man-made crises and emergency planning

The act organized FEMA into DHS, with a direct line of report between the administrator and the secretary, but it also kept it separate from the new Preparedness Directorate. The Preparedness Directorate consolidated preparedness assets from across the department. It facilitated grants and oversaw nationwide preparedness efforts supporting first-responder training, citizen awareness, public health, infrastructure, and cyber security.

Post-Katrina Emergency Management Reform Act

The Post-Katrina Emergency Management Reform Act of 2006 clarified and modified the Homeland Security Act of 2002 with respect to the organizational structure, authorities, and responsibilities of FEMA and the FEMA administrator. Among other changes, PKEMRA transferred a significant portion of the DHS Preparedness Directorate into FEMA, notably

- United States Fire Administration
- Office of Grants and Training
- Chemical Stockpile Emergency Preparedness Division
- Radiological Emergency Preparedness Program
- Office of National Capital Region Coordination

Presidential Decision Directives

In 1995, the Clinton administration issued Presidential Decision Directive 39 (PDD-39), "U.S. Policy on Counterterrorism" in response to the worst terrorist act on U.S. soil—the bombing of the Alfred P. Murrah Federal Building in Oklahoma City. PDD-39, built on prior directives, outlined three key elements of a national counterterrorism strategy:

1. Reduce vulnerabilities to terrorist attacks and prevent and deter terrorist acts before they occur (threat/vulnerability management)
2. Respond to terrorist acts that occur, end the crisis or deny terrorists their objectives, and apprehend and punish terrorists (crisis management)
3. Manage the consequences of terrorist acts, including restoring essential government services and providing emergency relief, to protect public health and safety (consequence management)

The directive elaborated on specific roles and responsibilities for several federal agencies with respect to each element of the strategy. For example, PDD-39 gave the Federal Bureau of Investigation (FBI) lead agency responsibility for crisis management, and FEMA similar responsibility for consequence management. Reflecting the need for greater interagency coordination, PDD-39 also directed the National Security Council to coordinate interagency terrorism policy issues and to ensure implementation of federal counterterrorism policy and strategy.

Homeland Security Presidential Directives

Following the terrorist attacks of September 11, 2001, the Bush administration issued several Homeland Security Presidential Directives (HSPDs), either augmenting or replacing the PDDs established by the Clinton administration. Among several, the two most notable HSPDs concerning preparedness and response are HSPD-5 and HSPD-8.

HSPD-5, Management of Domestic Incidents

Issued by the White House on February 28, 2003, HSPD-5 established a single, comprehensive national incident management system. It also designated the secretary of Homeland Security as the principal federal official for domestic incident management and recognizes the statutory authorities of the attorney general, secretary of defense, and secretary of state. In addition, HSPD-5 directed the heads of all federal departments and agencies to provide their full and prompt cooperation, resources, and support, as appropriate and consistent with their own responsibilities for protecting national security, to the secretary of Homeland Security, attorney general, secretary of defense, and secretary of state in the exercise of leadership responsibilities and missions assigned.

HSPD-8, National Preparedness

Issued by the White House on December 17, 2003, HSPD-8 established policies to strengthen the preparedness of the United States to prevent and respond to threatened or actual domestic terrorist attacks, major disasters, and other emergencies by requiring a national domestic all-hazards preparedness goal, establishing mechanisms for improved delivery of federal preparedness assistance to state, local, and tribal governments, and outlining actions to strengthen preparedness capabilities of federal, state, local, and tribal entities. Annex 1, Integrated Planning System, published in January 2009, established a standard and comprehensive approach to national planning.

Other Policy References

HSPD-5 and HSPD-8 also helped establish several important policy references around homeland security and emergency management, including the following:

National Incident Management System (NIMS), December 2008, provides a systematic, proactive approach to guide departments and agencies at all levels of government, NGOs, and the private sector to work seamlessly to prevent, protect against, respond to, recover from, and mitigate the effects of incidents, regardless of cause, size, location, or complexity, in order to reduce the loss of life and property and harm to the environment.

National Infrastructure Protection Plan (NIPP), January 2009, establishes a risk management framework for the nation's unified national approach to critical infrastructure and key resource protection.

National Preparedness Guidelines, September 2007, finalize development of the National Preparedness Goal and its related preparedness tools as mandated in HSPD-8. The guidelines consist of four elements: the National

Preparedness Vision, the National Planning Scenarios, the Target Capabilities List, and the Universal Task List.

Response Plans

For nearly a decade, a progression of response plans—Federal Response Plan (FRP), National Response Plan (NRP), and National Response Framework (NRF)—were written to address lessons learned from actual and potential disasters, as well as changes in statutory and policy directives, with the intended aim of improving the nation's preparedness and response coordination.

The creation of the FRP was driven by the PDD-39, which itself was developed following the Oklahoma City bombing of the Murrah Federal Building. The NRP superseded the FRP, following the terrorist attacks of September 11, 2001. Finally, the NRF superseded the NRP after the devastation of Hurricane Katrina. Creation of both the NRP and the NRF was driven by HSPD-5.

These federal plans are supported by emergency support functions (ESFs) annexes that describe the missions, policies, structures, and responsibilities of federal agencies for coordinating resource and programmatic support. The ESFs provide the structure for coordinating federal interagency support for a federal response to an incident. They are mechanisms for grouping functions most frequently used to provide federal support.

The plans are guidelines, not requirements, for the states in the beginning, but states are supposed to follow federal plans to obtain funding. Some states follow the plan, others do not. The development of these plans was based on several premises:

> A basic premise of all the plans is that the state is FEMA's primary client. The response doctrine is rooted in America's federal system and the Constitution's division of responsibilities between federal and state governments.
>
> Another premise is that incidents are handled at the lowest jurisdictional level possible. In the vast majority of incidents, state and local resources and interstate mutual aid will provide the first line of emergency response and incident management support. When state resources and capabilities are overwhelmed, governors may request federal assistance. That strategy provides the framework for federal interaction with local, tribal, state, territory, commonwealth, and private-sector and nongovernmental entities in the context of domestic incident management.
>
> A third premise is that mass care is traditionally a community response. NGOs, such as the American Red Cross and the Salvation Army, and other faith-based and community-based organizations have traditionally provided mass care services to communities during disasters.

All of these plans focus on all-hazard emergencies: natural disasters, technological emergencies (such as hazardous material releases), and acts of terrorism.

Emergency Support Functions

All three plans use ESFs as a means to provide the interagency staff to support federal response operations of the National Response Coordination Center (NRCC), the Regional Response Coordination Center (RRCC), and the Joint Field Office (JFO). Depending on the incident, deployed assets of the ESFs may also participate in the staffing of the Incident Command Post. Under the NRP, each ESF is structured to provide optimal support for evolving incident management requirements.

ESFs may be activated for Stafford Act and non–Stafford Act implementation of the NRP (although some Incidents of National Significance may not require ESF activations). ESF funding for non–Stafford Act situations will be accomplished using NRP Federal-to-Federal support mechanisms and will vary based on the incident.

Within the NRP, each ESF Annex identifies the ESF coordinator and the primary and support agencies pertinent to the ESF. Several ESFs incorporate multiple components, with primary agencies designated for each component to ensure seamless integration of and transition between preparedness, prevention, response, recovery, and mitigation activities. ESFs with multiple primary agencies designate an ESF coordinator for the purposes of preincident planning and coordination.

A federal agency designated as an ESF primary agency serves as a federal executive agent under the Federal Coordinating Officer (FCO) (or Federal Resource Coordinator for non–Stafford Act incidents) to accomplish the ESF mission. When an ESF is activated in response to an Incident of National Significance, the primary agency is responsible for

- Orchestrating federal support within their functional area for an affected state
- Providing staff for the operations functions at fixed and field facilities
- Notifying and requesting assistance from support agencies
- Managing mission assignments and coordinating with support agencies, as well as appropriate state agencies
- Working with appropriate private-sector organizations to maximize use of all available resources
- Supporting and keeping other ESFs and organizational elements informed of ESF operational priorities and activities
- Planning for short-term and long-term incident management and recovery operations

ESF Support Agencies

When an ESF is activated in response to an Incident of National Significance, support agencies are responsible for

- Conducting operations, when requested by DHS or the designated ESF primary agency, using their own authorities, subject-matter experts, capabilities, or resources

- Participating in planning for short-term and long-term incident management and recovery operations and the development of supporting operational plans, procedures, checklists, or other job aids, in concert with existing first-responder standards
- Assisting in the conduct of situational assessments
- Furnishing available personnel, equipment, or other resource support as requested by DHS or the ESF primary agency
- Providing input to periodic readiness assessments
- Identifying new equipment or capabilities required to prevent or respond to new or emerging threats and hazards, or to improve the ability to address existing threats
- Nominating new technologies to DHS for review and evaluation that have the potential to improve performance within or across functional areas
- Providing information or intelligence regarding their agency's area of expertise

Federal Response Plan

The FRP established a new process and structure for the systematic, coordinated, and effective delivery of federal assistance to address the consequences of any major disaster or emergency declared under the Stafford Act. The plan organized the types of federal response assistance that a state is most likely to need less than 12 ESFs. It also described the process and methodology for implementing and managing federal recovery and mitigation programs and support/technical services.

The FRP provided a focus for interagency and intergovernmental emergency preparedness, planning, training, exercising, coordination, and information exchange, serving as the foundation for the development of detailed supplemental plans and procedures to implement federal response and recovery activities rapidly and efficiently.

The FRP applied to a major disaster or emergency as defined under the Stafford Act for which the president determines that federal assistance is needed to supplement state and local efforts and capabilities. The FRP covered the full range of complex and constantly changing requirements following a disaster: saving lives, protecting property, and meeting basic human needs (response); restoring the disaster-affected area (recovery); and reducing vulnerability to future disasters (mitigation). The FRP did not specifically address long-term reconstruction and redevelopment.

The FRP engaged 22 federal agencies, plus the American Red Cross, to assist states with disaster preparedness and response. The FRP augmented other response such as the National Contingency Plan (NCP) for oil and hazardous materials spills and the Federal Radiological Emergency Response Plan (FRERP).

Under the FRP, the federal government and the American Red Cross shared responsibility for sheltering victims, organizing feeding operations, providing emergency first aid at designated sites, collecting and providing information on victims to family members, and coordinating bulk distribution of emergency relief items. As the primary agency for mass care under ESF 6, the American Red Cross coordinated federal mass care assistance in support of state and local mass care efforts. The American Red Cross was the only NGO signatory to the FRP. Other NGOs became formally involved in later plans.

Although the FRP was revised in 2003 in response to the terrorist attacks of September 11, 2001, it was acknowledged that further revision was needed to include long-term recovery activities (e.g., housing) as well as response services.

National Response Plan

In 2003, President George W. Bush directed the DHS, through HSPD-5, to develop a new national response plan to align federal coordination structures, capabilities, and resources to form a unified, all-discipline, and all-hazards approach to domestic incident management. This approach eliminated critical seams and tied together a complete spectrum of incident management activities to include the prevention of, preparedness for, response to, and recovery from terrorism, major natural disasters, and other major emergencies.

In December 2004, Secretary Tom Ridge issued the NRP, which described how to improve coordination among federal, state, local, and tribal organizations to help save lives and protect America's communities by increasing the speed, effectiveness, and efficiency of incident management.

As noted by the preface, Secretary Ridge acknowledged:

> Implementation of the plan and its supporting protocols requires extensive cooperation, collaboration, and information-sharing across jurisdictions, as well as between the government and the private sector at all levels.

The NRP was built on the template of NIMS, which provides a consistent doctrinal framework for incident management at all jurisdictional levels, regardless of the cause, size, or complexity of the incident. It superseded other response plans:

- Federal Response Plan
- United States Government Interagency Domestic Terrorism Concept of Operations Plan
- Federal Radiological Emergency Response Plan

The NRP and its coordinating structures and protocols provided mechanisms for coordination and implementation of a wide variety of incident management and emergency response activities (see Figure 1.2). The NRP was an integration of the state, local, and federal assets. Included in these coordinating activities were federal

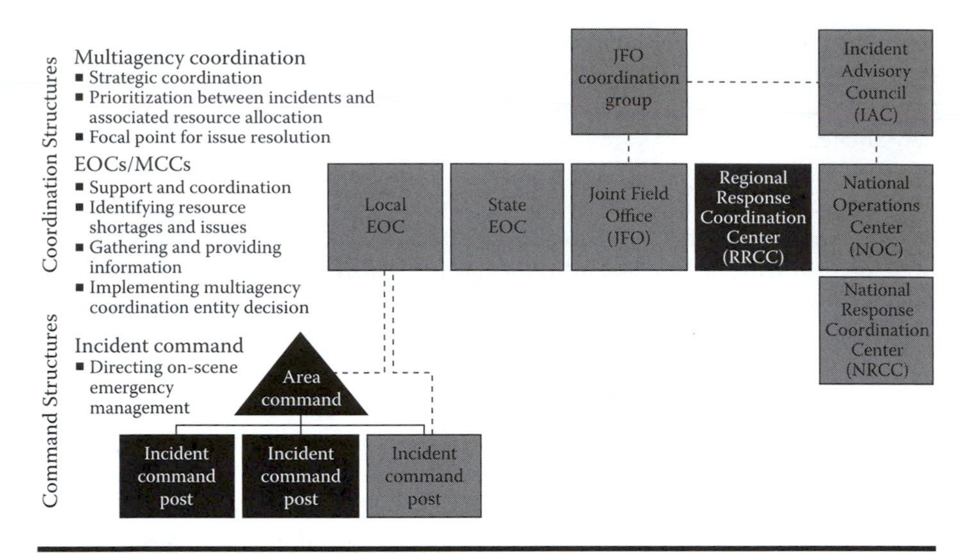

Figure 1.2 Coordination and command structures in the National Response Plan (NRP).

support to local, tribal, and state authorities; interaction with nongovernmental, private-donor, and private-sector organizations; and the coordinated, direct exercise of federal authorities, when appropriate.

Whereas the FRP addressed response activities only, the NRP included response and recovery, as well as a need for long-term recovery activities that was recognized following Hurricane Katrina. The NRP was considered by the emergency management community not to be a plan but rather to set boundaries for hierarchical framework for planning.

The NRP also established a new term of reference for disasters—incidents of national significance—to address potential acts of terrorism. Under the authority of the secretary of Homeland Security, federal response to an incident of national significance could include

- A federal department or agency, responding under its own authorities, requests DHS assistance.
- Resources of state and local authorities are overwhelmed.
    - Stafford Act for major disasters or emergencies
    - Other catastrophic incidents
- More than one federal department or agency is involved.
    - Credible threats or indications of imminent terrorist attack
    - Threats/incidents related to high-profile, large-scale events

Given its sweeping changes, the NRP included its own implementation schedule (see Figure 1.3).

- **Phase I—Transitional Period (0 to 60 days):** This 60-day timeframe is intended to provide a transitional period for departments and agencies and other organizations to modify training, designate staffing of NRP organizational elements, and become familiar with NRP structures, processes, and protocols.
- **Phase II—Plan Modification (60 to 120 days):** This second 60-day timeframe is intended to provide departments and agencies the opportunity to modify existing federal interagency plans to align with the NRP and conduct necessary training.
- **Phase III—Initial Implementation and Testing (120 days to 1 year):** Four months after its issuance, the NRP is to be fully implemented, and the INRP, FRP, CONPLAN, and the FRERP are superseded. Other existing plans remain in effect, modified to align with the NRP. During this timeframe, the Department of Homeland Security (DHS) will conduct systematic assessments of NRP coordinating structures, processes, and protocols implemented for actual incidents of national significance (defined on page 4 of the NRP), national-level homeland security excercises, and national special security events (NSSEs). These assessments gauge the plan's effectiveness in meeting specific objectives outlined in Homeland Security Presidential Directive-5 (HSPD-5). At the end of this period, DHS will conduct a one-year review to assess the implementation on process and make recommendations to the secretary on necessary NRP revisions. Following this initial review, the NRP will begin a deliberate four-year review and reissuance cycle.

Figure 1.3 The three phases of the National Response Plan.

The plan addressed the full spectrum of activities related to domestic incident management, including prevention, preparedness, response, and recovery actions. The NRP focused on those activities that are directly related to an evolving incident or potential incident rather than steady-state preparedness or readiness activities conducted in the absence of a specific threat or hazard.

Additionally, since incidents of national significance typically resulted in impacts far beyond the immediate or initial incident area, the NRP provided a framework to enable the management of cascading impacts and multiple incidents as well as the prevention of and preparation for subsequent events. Examples of incident management actions from a national perspective include

- Increasing nationwide public awareness
- Assessing trends that point to potential terrorist activity
- Elevating the national Homeland Security Advisory System alert condition and coordinating protective measures across jurisdictions
- Increasing countermeasures such as inspections, surveillance, security, counterintelligence, and infrastructure protection
- Conducting public health surveillance and assessment processes and, where appropriate, conducting a wide range of prevention measures to include, but not be limited to, immunizations
- Providing immediate and long-term public health and medical response assets
- Coordinating federal support to state, local, and tribal authorities in the aftermath of an incident

- Providing strategies for coordination of federal resources required to handle subsequent events
- Restoring public confidence after a terrorist attack
- Enabling immediate recovery activities, as well as addressing long-term consequences in the impacted area

On August 30, 2005, Secretary Michael Chertoff invoked the NRP the day after Hurricane Katrina hit the Gulf Coast. By so doing, Secretary Chertoff assumed the leadership role triggered by the law to bear primary responsibility to manage said crisis. Almost a month later, in advance of the landfall of Hurricane Rita, Secretary Chertoff declared the storm an incident of national significance and put preparations in place in the gulf region of Texas.

Because of the lengthy implementation schedule, the increased level of coordination did not sufficiently materialize. This situation became severely problematic when Hurricane Katrina roared into the Gulf of Mexico, then made landfall in Louisiana. Hurricane Katrina caused severe destruction along the Gulf coast. The most severe loss of life and property damage occurred in New Orleans, Louisiana, which flooded as the levee system catastrophically failed, in many cases hours after the storm had moved inland.

Following Hurricane Katrina, the plan was updated on May 25, 2006. The notice of change stated the update "emerged from organizational changes within DHS, as well as the experience of responding to Hurricanes Katrina, Wilma, and Rita in 2005."

National Response Framework

Published in January 2008, the NRF was developed to address the requirements of PKEMRA. It is a framework that guides local, state, and federal entities enabling all response partners to prepare for and provide a unified national response to disasters and emergencies. This framework establishes a comprehensive, national, all-hazards approach to domestic incident approach.

As identified by DHS, the NRF

> presents the guiding principles that enable all response partners to prepare for and provide a unified national response to disasters and emergencies—from the smallest incident to the largest catastrophe. This important document establishes a comprehensive, national, all-hazards approach to domestic incident response. The Framework defines the key principles, roles, and structures that organize the way we respond as a Nation. It describes how communities, tribes, States, the Federal Government, and private-sector and nongovernmental partners apply these principles for a coordinated, effective national response. It also identifies special circumstances where the Federal Government exercises a larger role, including incidents where Federal interests are

> involved and catastrophic incidents where a state would require significant support. The Framework enables first responders, decision-makers, and supporting entities to provide a unified national response.

An underlying basis of the NRF is a set of key principles:

> Engaged partnership—Leaders at all levels must communicate and actively support engaged partnerships by developing shared goals and aligning capabilities so that no one is overwhelmed in times of crisis.
> Tiered response—Incidents must be managed at the lowest possible jurisdictional level and supported by additional capabilities when needed.
> Scalable, flexible, and adaptable operational capabilities—As incidents change in size, scope, and complexity, the response must adapt to meet requirements.
> Unity of effort through unified command—Effective unified command is indispensable to response activities and requires a clear understanding of the roles and responsibilities of each participating organization.
> Readiness to act—Effective response requires readiness to act balanced with an understanding of risk. From individuals, households, and communities to local, tribal, state, and federal governments, national response depends on the instinct and ability to act.

Because of the confusion brought about by the term of reference—incident of national significance—in the NFR, the NRP eliminated this term.

An important concept presented in the NRF included the preparedness life cycle, which represents a systemic approach to build the right capabilities for the nation in response to all hazards. The preparedness life cycle (see Figure 1.4):

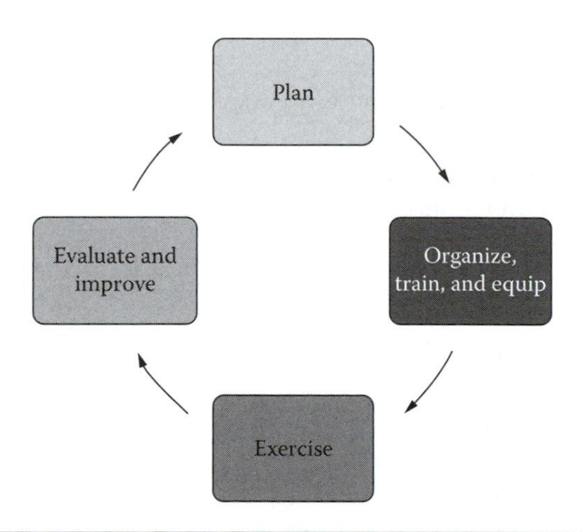

Figure 1.4 Preparedness life cycle of the National Response Framework (NRF).

- Introduces the National Planning System
- Defines response organization
- Requires training
- Advocates interoperability and typing of equipment;
- Emphasizes exercising with broad-based participation
- Describes the process for continuous evaluation and improvement

The NRF establishes 15 ESFs:

- ESF #1—Transportation
- ESF #2—Communications
- ESF #3—Public Works and Engineering
- ESF #4—Firefighting
- ESF #5—Emergency Management
- ESF #6—Mass Care, Emergency Assistance, Housing and Human Services
- ESF #7—Logistics Management and Resource Support
- ESF #8—Public Health and Medical Services
- ESF #9—Search and Rescue
- ESF #10—Oil and Hazardous Materials Response
- ESF #11—Agriculture and Natural Resources
- ESF #12—Energy
- ESF #13—Public Safety and Security
- ESF #14—Long-Term Community Recovery
- ESF #15—External Affairs

It also includes several support annexes and incident annexes to help improve coordination:

- Support Annexes
  - Critical Infrastructure and Key Resources
  - Financial Management
  - International Coordination
  - Private Sector Coordination
  - Public Affairs
  - Tribal Relations
  - Volunteer and Donations Management
  - Worker Safety and Health
- Incident Annexes
  - Biological Incident
  - Catastrophic Incident
  - Cyber Incident
  - Food and Agriculture Incident
  - Mass Evacuation Incident

- Nuclear/Radiological Incident
- Terrorism Incident Law Enforcement and Investigation

To promote awareness and education of the NRF, FEMA developed an independent study training course, IS-800, An Introduction to the NRF, which is available free of charge. FEMA continues to develop other general orientation courses for ESFs and the Support and Incident Annexes through its online study program at the Emergency Management Institute.

Emergency Support Functions

FRP, NRP, and NRF use ESFs as a means to provide the interagency staff to support federal response operations of the NRCC, the RRCC, and the JFO. Depending on the incident, deployed assets of the ESFs may also participate in the staffing of the Incident Command Post. Under the NRP, each ESF is structured to provide optimal support for evolving incident management requirements.

ESFs may be activated for Stafford Act and non–Stafford Act implementation of the NRP (although some incidents of national significance may not require ESF activations). ESF funding for non–Stafford Act situations will be accomplished using NRP federal-to-federal support mechanisms and will vary based on the incident.

Within the NRP, each ESF annex identifies the ESF coordinator and the primary and support agencies pertinent to the ESF. Several ESFs incorporate multiple components, with primary agencies designated for each component to ensure seamless integration of and transition between preparedness, prevention, response, recovery, and mitigation activities. ESFs with multiple primary agencies designate an ESF coordinator for the purposes of preincident planning and coordination.

A federal agency designated as an ESF primary agency serves as a federal executive agent under the federal coordinating officer (FCO; or federal resource coordinator for non–Stafford Act incidents) to accomplish the ESF mission. When an ESF is activated in response to an incident of national significance, the primary agency is responsible for

- Orchestrating federal support within their functional area for an affected state
- Providing staff for the operations functions at fixed and field facilities
- Notifying and requesting assistance from support agencies
- Managing mission assignments and coordinating with support agencies, as well as appropriate state agencies
- Working with appropriate private-sector organizations to maximize use of all available resources
- Supporting and keeping other ESFs and organizational elements informed of ESF operational priorities and activities
- Planning for short-term and long-term incident management and recovery operations

ESF Support Agencies

When an ESF is activated in response to an incident of national significance, support agencies are responsible for

- Conducting operations, when requested by DHS or the designated ESF primary agency, using their own authorities, subject-matter experts, capabilities, or resources
- Participating in planning for short-term and long-term incident management and recovery operations and the development of supporting operational plans, procedures, checklists, or other job aids, in concert with existing first-responder standards
- Assisting in the conduct of situational assessments
- Furnishing available personnel, equipment, or other resource support as requested by DHS or the ESF primary agency
- Providing input to periodic readiness assessments
- Identifying new equipment or capabilities required to prevent or respond to new or emerging threats and hazards, or to improve the ability to address existing threats
- Nominating new technologies to DHS for review and evaluation that have the potential to improve performance within or across functional areas
- Providing information or intelligence regarding their agency's area of expertise

Conclusion

Government at all levels has the basic responsibility for protecting its citizens. In the case of natural, technological, or man-made hazards, the best way to protect the public is by implementing an integrated emergency management system that incorporates all potential players in a response through all phases of emergency management.

As technology continues to advance, man-made disasters will likely reach a magnitude never thought possible. The basic process of disaster planning is only a small part of the process. Equally important is identifying who is responsible for developing or revamping a community's disaster plan. These are the topics to be covered in this book.

Chapter 2

EOC Management and Operations

Lucien G. Canton and Nicholas Staikos

Contents

Introduction

It is a common mistake to confuse the emergency operations center (EOC) with the tasks performed in the EOC and to forget that the EOC is a physical location that generates its own demands. For the EOC team to perform effectively, the physical and organizational demands of the EOC as a facility must be met. This EOC management is distinct from the operational management of the incident.

EOC management can be roughly divided into two main categories: facility management and operational management. Facility management is similar to the activities that take place within any facility. This involves dealing with the physical plant, technology systems, and support services needed to support activities within the EOC. Operational management pertains to the systems and procedures put in place to allow the EOC team to operate efficiently. This pertains to the procedures for performing common tasks and operating EOC equipment. These two components work together to ensure that the EOC team is free to focus on the incident with minimal disruption from the environment in which the team is operating.

Facility Management

EOC facility management shares many commonalities with that found in any major office building. However, there are added complexities that make EOCs unique. Where the typical office building operates for 8 to 12 hours a day, an EOC must be capable of 24-hour operation. This means that there is no downtime for maintenance or support services, and these services must be provided in a way that does not have an impact on operations. In addition, increased security during operations can limit access for support staff or contractors if there is no prior coordination. The EOC also has a considerable number of parallel systems and complex communications systems and may have unique design features such as HEPA filters or overpressure systems to provide security against chemical and biological attack. In some cases, the EOC may also provide living quarters for staff for an extended period.

This level of complexity for EOC systems has implications for day-to-day activities as well. Since an EOC must be ready for activation within a short period, there is no latitude for a lengthy startup period while batteries are charged, software upgraded, or systems checked. This means that a program for ongoing maintenance of EOC systems must be in place.

The EOC as a facility comprises a number of systems. The most obvious are, of course, the environmental, life safety, and utility systems. Even here, though, the EOC is different from a typical office building. Where "emergency power" in an office building means that life safety systems continue to receive power, the EOC's requirement for continuous operation under all conditions may demand alternate commercial power feeds, multiple generators, and the capacity to add external generators.

The same is true for communications and information technology systems. EOCs require multiple parallel systems, each with unique requirements. In some cases, these systems cannot operate simultaneously.

In addition to technical support, the EOC also places demands on support services. For example, 24-hour operation will increase the need for janitorial service and garbage pickup. Additional materials may need to be ordered and delivered. Contractors may be brought in to provide food for EOC staff.

One critical service that must be coordinated is that of security. A fully activated EOC is a focal point for the media and members of the public and could be viewed as a potential target. Although day-to-day security may be adequate, during activation there will be a need for additional security to control access, provide escorts, and conduct patrols. Security responsibilities may include controlling and protecting parking areas or adjacent staging areas.

Unfortunately, facility management is a multidiscipline task performed by many different individuals. The typical jurisdiction will have the physical plant serviced by one department whereas another oversees information technology systems. Radio systems may be handled by a separate communications department, whereas vendor services are overseen by still another department and security by another. It is rare to find a single person charged with looking at the EOC facility systematically. This creates problems during day-to-day operations, but it can prove devastating during actual EOC activation. With no go-to guy for facility issues, problems begin to accumulate and distract the EOC team from its operational responsibilities.

For this reason, emergency managers should consider creating an EOC coordinator position that serves as the central point of coordination for the vendors and teams supporting the EOC facility. This team approach is consistent with how the EOC team routinely operates; the difference is that this team is focused on internal support to the EOC rather than support to the overall operation.

Operational Management

There is often an assumption that when staff report to the EOC, the only thing they really need to function efficiently is the emergency operations plan (EOP). After all, the EOP contains a substantial amount of information, policies, and procedures focused on coordinating disaster operations. Coupled with internal department plans and standard operating procedures (SOPs), the EOP should be sufficient to guide operations.

This may be true in terms of the disaster operations, but is not true for the EOC as a facility. Consultant Art Botterell's Third Law of Emergency Management states that "no matter who you train, someone else will show up." EOC planners need to assume that a significant percentage of the EOC team will be coming to the EOC for the first time. Other members of the team may not have been in the EOC for some time and forgotten how things work.

As an example, consider simple tasks related to information management. How will a team member receive an e-mail? How can he or she direct it to print to the closest printer? What is the EOC fax number? These are simple questions to answer, but if 50 people are asking them at the same time, it can considerably delay operations. Even as simple a question as "How do I turn on the lights?" may create problems if the first person to arrive at the EOC has never been there.

In addition to these more technical issues, there are a number of tasks that need to be performed to support the EOC team. An example of these is a process for activating the EOC. This process should address basic questions such as who has the authority to activate the EOC and under what conditions, but it should also describe in detail the procedure for notifying the EOC team.

Another overlooked process involves the procedure for the initial setup of the EOC upon activation. Something as trivial as not knowing who has the keys or how to turn on a copier can cause considerable problems. Many EOCs are dual-use facilities that require the first arriving staff members to unlock containers, plug in telephones, and set up laptop computers. Assuming that the first person at the EOC will be one of the three or four staff members that know how to set up the facility is a fundamental mistake.

Once operations are concluded, there are things that need to be done to close out the current operation and prepare the facility for future operations. This involves procedural tasks such as archiving files and preparing after-action reports as well as facility-related tasks such as clean up, repairing or performing maintenance on equipment, and reordering supplies. The tendency will be for the last remaining team members to quickly wrap things up and go home, and important tasks can be overlooked without a detailed deactivation process.

This need for structure and organization reinforces the need for standard operating procedures that are focused on the EOC rather than on external operations. The EOC SOP should be written from the perspective of the first-time user and, as much as possible, reduce procedures to short checklists. The checklists should be detailed and task-oriented. Among the items that could be included in the SOP are the following:

- Activation procedures
- Notification procedures
- Setup procedures
- Procedures for using communications and information technology systems
- Procedures for obtaining additional supplies and services
- Security procedures
- Deactivation procedures

One other technique to consider is the posting of key information from the SOP on wall displays in the EOC for ready reference. This could include procedures for transferring calls, fax numbers, printer addresses, and other information that supports operations. Similarly, instructions for use can be posted by copiers, fax machines, and so forth. Remember that staff using the EOC will be under pressure and may be seeing the EOC equipment for the first time. Botterell's First Law of Emergency Management cautions that "stress makes you stupid."

Organizing for EOC Management

Something that is everyone's responsibility is in reality no one's responsibility. An important first step in developing a plan for EOC management is to identify who will be responsible for overall planning and who will serve as the EOC coordinator during activation. These two positions do not necessarily need to be held by the same person.

Planning for EOC management is no different from planning for operations. One begins by identifying key players and stakeholders, and developing a working group to guide the work. As EOC systems and procedures are developed, they should be tested through exercise, preferably by being integrated into regular EOC functional exercises. Although this seems obvious, relatively little planning is traditionally done in the area of EOC management; most organizations focus all their efforts on operations planning. This is the equivalent of assuming that normal departmental functions will be available during a disaster in the absence of a continuity of operations plan. It is a very risky assumption that could result in operational failure.

The following are key points to consider when planning for EOC management:

- Identify a lead person with overall responsibility for EOC management planning.
- Identify key players and stakeholders, and form a working group.
- Develop standard operating procedures for common EOC functions.
- Develop standby contracts for increased or additional services.
- Coordinate supporting operations such as security and support services.
- During activation, consider forming an EOC support team under a single coordinator.
- Use a standard deactivation process that prepares the EOC for immediate reactivation before concluding operations.

Good EOC management can significantly reduce confusion and stress on the EOC team, allowing the team to focus on operations rather than being distracted by their operational environment. It requires preplanning in a manner similar to operations and an ability to forecast the needs of the EOC team. In the end, it may well determine the success or failure of the EOC operation.

Operations Room Design

Origins

The operations room, where internal and external responders report, is the nerve center of today's EOC. Its evolution, like the field of emergency management, is ongoing, driven by the constantly changing technological landscape and by the adoption of new practices derived from lessons learned. It was initially conceived as a space where key public and private agency representatives came together for the collection, evaluation, and dissemination of information. These multipurpose spaces were born out of the context of civil defense. They were fairly similar in layout and typically created with a bunker mentality as survivability was paramount.

The accommodations were spartan in nature, and the facility was used for other purposes until escalating threat levels warranted activation. Although their principal function was to provide the responding agencies a seemingly protected place to maintain communications with their respective organizations' operational structures, they were also viewed as a place of refuge for governing authorities, thus ensuring the continuity of government. It was not unusual to find these centers located in the lower basement levels of a municipal building or at times as a stand-alone underground facility. Interestingly, these presumed-to-be-well-protected locations were held hostage by external events outside of their control such as plumbing failures from above or flooding via backed-up drainage systems from below. Furthermore, by being collocated with many other users, any form of building evacuation such as a fire alarm or other such alert would mandate exit from the occupied space not to mention the impact of sprinkler discharge seeking the lowest level.

As the Cold War tensions defused, officials soon came to the realization that the threats from natural disasters would be more likely to occur and could have significant consequences on the day-to-day function of government as well as affecting the lives and welfare of a large proportion of the jurisdictions' population. This growing awareness produced the need for enlarged staffs to administer recovery programs as well as focused support during the crisis and drove the need for more capable facilities.

The September 11 attacks shifted emergency management's focus from an all-hazards approach to one that was biased toward homeland security and counterterrorism. The concern over the short-sightedness in this shift in emphasis was raised by the emergency management community. Then, the impact of Hurricane Katrina and the other storms of the period reminded everyone that emergency management and response demands a broad base of preparedness found in an all-hazards approach.

This transitional awareness is helping drive a new mindset among community leaders that EOCs and their operations centers should not only ensure survivability of government but facilitate continuity of operations for the private sector by striving to quickly return to normalcy. Our leaders recognize that this is even more important now that globalization and interdependency are no longer academic concepts but a reality. The linkage between suppliers, manufacturers, producers, and consumers is one of the key threads that bind a nation together. Having a robust and strong economic base has come to represent a key component of a country's strength. As a result, the EOC and its operation's room play a pivotal role in managing a crisis, and the EOC has had to become a sophisticated communications hub to fulfill its mission during the cycle of mitigation, preparation, response, and recovery rather than just a command-and-control center during a crisis.

Today's Focus

Regardless of the scale of an emergency, the success of a coordinated response almost always depends on several key factors:

1. Redundant and interoperable communications systems, which are balanced vertically and horizontally
2. Comprehensive ability to quickly determine and coordinate asset utilization
3. Organizational flexibility to accommodate a variety of responding entities, which are driven by event type
4. In-depth situational awareness
5. Access to all supporting resources to formulate alternative response scenarios

One must also recognize that because of the speed at which events can unfold and the need to engage a myriad of supporting players, a well-crafted plan should be in place. To be effective, it should target the most likely types of events for the locale and be tuned as a result of multiagency exercises to be fully effective. The plan—the tools—are the baseline of preparedness, for as we have often seen, events never follow the script and scenario adaptation will be necessary.

Another reality that our political establishment can sometimes forget is that all disasters are local. Even during a widening crisis, organizational effectiveness starts with the local responders and then gradually draws upon the next level of support. Having said this, we have all seen situations where the local entity failed to quickly recognize that events were spiraling beyond their capabilities and failed to request support quickly enough. This is where the value of the EOC's operations room is leveraged. For while incident command is focusing on the immediate issues on the ground, the professionals in the EOC with a theater-wide view can implement and guide a strategic response measured by the needs coming from the field.

Additionally, we must further recognize that staffing realities dictate that regardless of the size of the jurisdiction, support during the initial stages of an event will come from a 24/7 watch component. This on-duty team acts as the trip wire providing the vital linkages in the response chain. They will do so until such time as the facility reaches full staffing wherein interface with each supporting agency's organization will be handled by their responding entity. This watch staff, therefore, will need to have all the skills and tools to capably manage the initial response until such time as the appropriate representatives of the activated agencies are dispatched to the EOC to provide real-time coordination within their respective infrastructure.

Can Organizational Structure Impact Room Design?

The short answer to this question is a definite yes. Our experience has shown that the management structure of the responsible agency most definitely influences the style of layout for the OPS room. Predictably, it will tilt either toward a C2 (command and control) or to a C4 (which I prefer to refer to as communicate, collaborate, coordinate, and, to a more limited degree, control) setup. The determining factor for this bias is quite often dependent on whether the agency has a public safety or emergency management heritage. In public safety environments where there is a clear chain of command as in the military, room layout will focus its attention on either a common information display wall or command structure. Those agencies from an emergency management lineage will tend to tilt toward a collaborative or clustered environment. In many jurisdictions, hybrid models have begun to be implemented.

Regardless of the layout, the reality is that a fundamental shift has occurred and the modern EOC's success will be dependent on its ability to foster the coordination and dissemination of information to the appropriate consumers. These activities will be coordinated through the event managers and the emergency support function (ESF) personnel who are positioned in the operations room. And even as advancements in technology lend credibility to the notion that the future will be in the creation of a virtual EOC, recent events continue to suggest that face-to-face collaboration is a more efficient and effective form of problem solving. This reality, I believe, extends to all levels of government.

Design's Role in Supporting the Evolving Mission

One must keep in mind that the development of a comprehensive design for an EOC and its operations room must take into account a wide range of considerations that will affect internal and external features. These items encompass issues such as responder accessibility, hazard zone proximity, availability of redundant services, maintenance of secure operations, and leveraging natural hardening through siting to physical hardening of both structures and systems. All of these factors as well as others must be considered when developing the design criteria and programmatic requirements for the facility. The degree to which a facility is hardened is largely influenced by the risk analysis developed from the threat and hazard assessment. The focus of this discussion will be on the various design options for the internal organization of the operations room of the modern EOC to create a well-conceived and functional center.

Management studies have shown that design does indeed have a significant impact on workplace effectiveness. Repeatedly, environmental comfort for both the physical as well as psychological needs of the occupants has been shown to play a significant role in mitigating the detrimental affects of high stress, which accompany crisis situations. Of the many factors that form the basis of an integrated design, the following are a few key components:

Ergonomics
- Console design
- Visual display design
- Seating comfort
- Technology integration
- Adaptability to a diversity of user body types

Environmental comfort
- Variable glare-free lighting control
- Acoustical control
- Thermal comfort and control

Space allocation
- Operator positions
- Supporting services
- Breakout areas
- Policy room
- Strategic response planning
- Quiet rooms
- Resource management
- Extended stay accommodation
- Self-sufficiency

Circulation and access control
- Hierarchical circulation system
- Electronics used to augment physical security design
- Layout supports work flow

Relationship to support spaces
- Ease of accessibility
- Ability to be serviced while in operation

Sustainable utility systems
- Redundant services
- Diverse routing
- Resupply capability
- Flexible cable management system

By properly considering these as well as other requirements, the designers of the workplace environment will have a significant impact on the operational effectiveness, thus shaping the quality of an entity's response. When implemented in the appropriate manner, these features and concepts will become transparent to the user as they will not be a source of discomfort. This will dramatically improve the level of performance as fatigue and frustration play a prominent role in degrading operational effectiveness.

The impact of design becomes even more apparent as the scale of an event escalates as these challenges become more complex due to the impact of their potential consequences. Time and time again, we read of the failings of response efforts, such as with Hurricane Katrina and in the earthquake in Haiti, when events of a catastrophic nature can overwhelm the system. Logistical entanglements, lack of communication, or conflicting requests produce chaos and unacceptable results. Make no mistake, facilities, whether physical or virtual, without a well-coordinated plan along with effective communications and logistical support will not produce the needed result.

Fundamentals

Each jurisdiction needs to have a solution tailored to its needs. The following narrative and accompanying diagrams illustrate several of the fundamental ways in which the focal point of an EOC, the operations room, can be configured to optimize the effectiveness of a jurisdiction's response. As the reader reviews the concepts, it is important to keep in mind that the efficiency of operation improves when the functional space is purposefully built yet affords the flexibility to adjust for refinement of operation. At times this may seem an unreachable goal, but it is achievable. This discussion deals primarily with the space and big-picture issues of technology integration and not its deployment nor optimal position assignment for the responding entities. Additionally, each of the plans presented can support the requirements and goals of the National Incident Management System (NIMS), some better than others.

One should also remember that many of these requirements, including situational awareness, asset control, and collaborative problem solving, are scalable to all jurisdictional levels. Furthermore, the need for this capability and attendant sophistication increases as the jurisdictional landscape and physical area encompassed grow due to increased political complexity. Regardless of the many variations of layout currently in use, the design of the operations room can be characterized by six basic configurations, each of which can be applied to all levels of response. For the purposes of this analysis, we will describe them as follows:

- Traditional multipurpose
- Cubicle cluster
- Horseshoe
- Stadium/theater
- Collaboration pods—theater style
- Iris

Even virtual centers will utilize similar organizational structures for the network control center.

Although they may be referred to by other names, these are the basic configurations for today's modern operations room. As previously mentioned, the preference of one design form over another is often influenced by the branch of government in which the agency finds itself. This, in turn, influences the management style of the leadership and whether the organization's roots are from public safety, which favors a C2 arrangement, or from emergency management, which tilts toward a C4. Of course, this does not mean that you cannot have a collaborative environment with a central focus context.

In addition to the design of the internal arrangement of the operations room, it is essential that a well-thought-out concept of the supporting spaces for functions such as policy making, strategic planning, breakout rooms, quiet rooms, sleeping areas, equipment rooms, locker rooms, and break rooms be implemented. The interrelationship of these components plays a significant role in establishing an effective center and has great influence on the success of the center during periods of activation.

Review of Layouts

Multipurpose

The multipurpose layout was traditionally used in smaller jurisdictions where dedicated space for response activities was not available. This template featured a simple room with a flat floor, which could serve a multiplicity of uses from conference space to community meetings (Figure 2.1). Because of its multipurpose nature, conversion to a full-fledged OPS room/EOC required setup time for furniture configuration, technology deployment, communications installation, and so forth.

Positives

1. Multipurposed
2. Flexible configuration

Negatives

1. Time required for physical setup.
2. Technology deployment can be challenging; use of floor boxes can reduce setup time.
3. Need storage space to store supporting equipment.
4. Challenges in lighting control when layout changes.
5. With flexibility comes lack of focus.
6. Acoustics generally are substandard.

Cluster/Pod Example

Figure 2.2 shows a cluster/pod example.

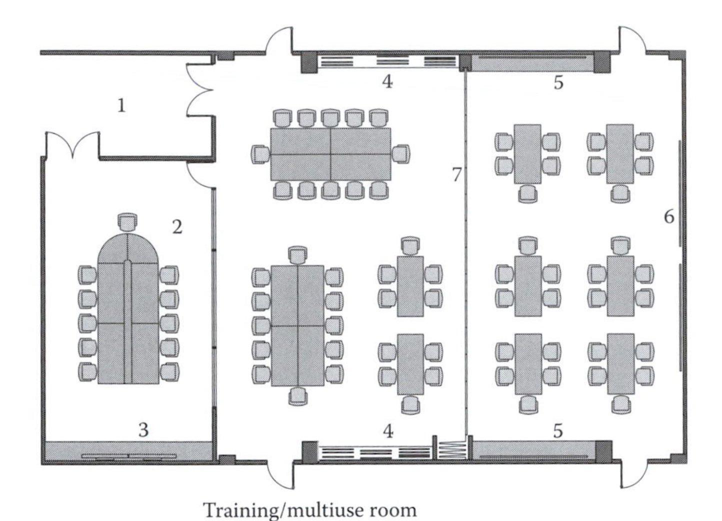

Key plan

1. Suite entry
2. Conference room
3. Displays
4. Map boards
5. Tack/white boards
6. Drop-down screens
7. Folding partition

© 2010

Figure 2.1 Traditional multipurpose.

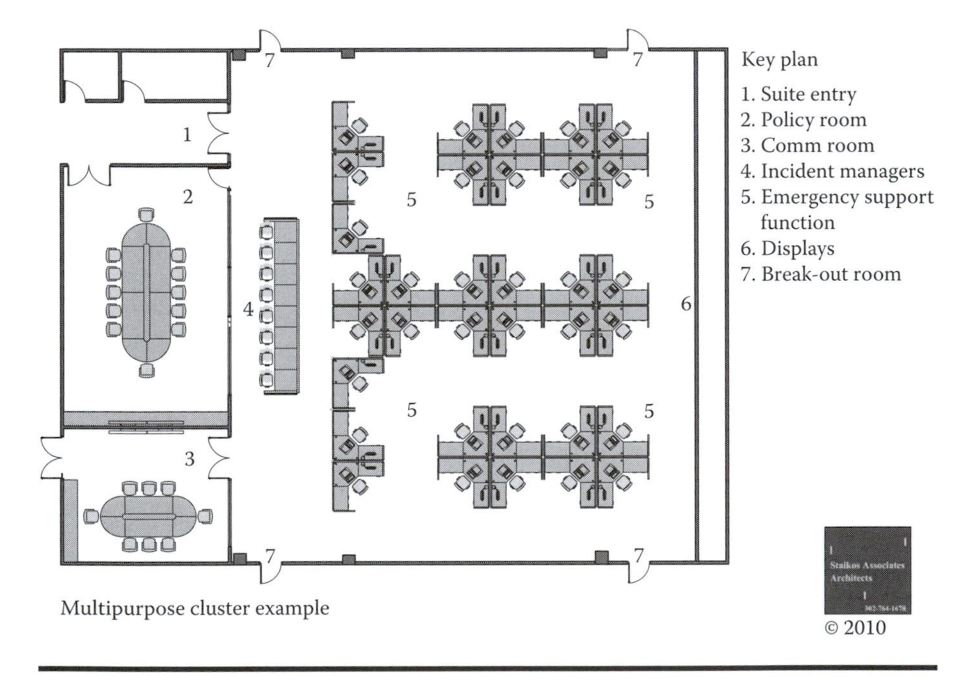

Figure 2.2 Cluster/pod example.

Positives

1. Allows for reconfiguration to enhance collaboration of responding entities.
2. Allows for reassignment of cubes to entities having a higher level of interaction.
3. Demands uniformity of technology to be conveniently flexible.
4. Allows adoption of a hoteling concept with user logins to redefine occupant.
5. Can adjust to variable numbers of related responder personnel.

Negatives

1. Does not provide a central focus for disseminated or displayed information.
2. On-the-fly reconfiguration is not as simple as it would seem and therefore not typically done until after an event, on lessons learned basis before the next event.
3. Reconfiguration based on specific event experience may not apply to a different type of activation.
4. Use of cube positions even with seated height walls are more difficult to reorient than simple desking.
5. Conflict between dedicated services and storage capability versus flexibility.

Horseshoe

The development of the horseshoe layout allows the participants to view commonly displayed information yet maintain eye contact with their fellow responders (Figure 2.3). This arrangement is typically geared to a smaller room so that some direct conversation can occur across the room. When arranged on a stepped floor, supporting associates of the principals seated at the primary table can be positioned in close proximity.

Positives

1. In a smaller room environment, eye and voice contact can be maintained so that each speaker can follow the body language of the other principal.
2. Opportunity for tiered or level OPS floor.
3. Opportunity for centrally focused display wall.
4. Allows for OPS floor breakout area or miniconference room, close at hand supervision.
5. Provides room for supporting staff providing back up to the principals.

Negatives

1. Has a tendency to encourage and generate increase noise levels if multiple conversations are occurring simultaneously.
2. Does not effectively allow for cross-agency collaboration or subgroup collaboration as it must occur outside the room.
3. Shape limits optimal or preferred sight lines.
4. Limited number of ESF positions.
5. Not ideal for small-group collaboration.
6. OPS room becomes longer, narrower as positions increase.

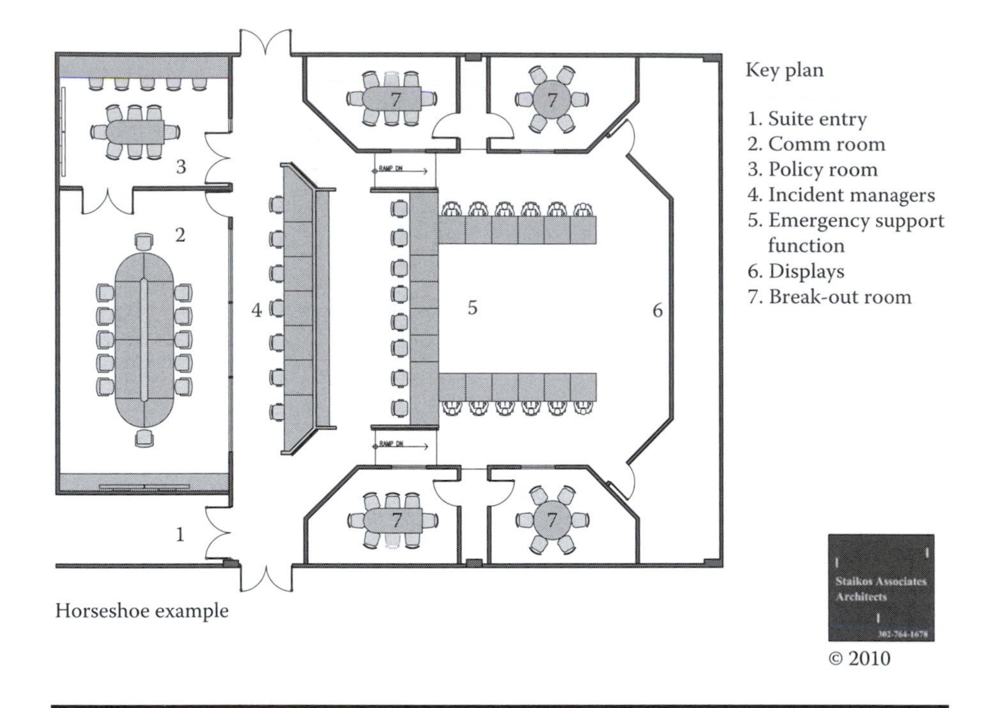

Figure 2.3 Horseshoe.

Stadium/Theater

The stadium/theater layout became popular with the advent of space missions wherein multiple activities could be commonly displayed for use by many participants. The rooms are generally stepped with tiers of consoles to allow unobstructed sight lines to displayed information (Figure 2.4). It fits a command-and-control model influenced by military models.

Positives

1. Visual focus of room allows all participants to share in tangential information used on an as-needed basis; messaging and pictorial information displays are easily viewed.
2. Minimizes excess noise generation due to people asking what is going on.
3. Allows attention to be focused on the speaker who may address the group.
4. Incident managers are positioned to oversee activity on OPS floor.
5. Technology is fully deployed and tested so that startup is almost immediate.
6. Can be effectively used for multipurpose training without reorganization.
7. Adjacent or on OPS floor breakout area or miniconference room.
8. Tiered concept allows for executive management to have overlook but be off floor.
9. Allows for designed acoustical and light control methods to be in place.

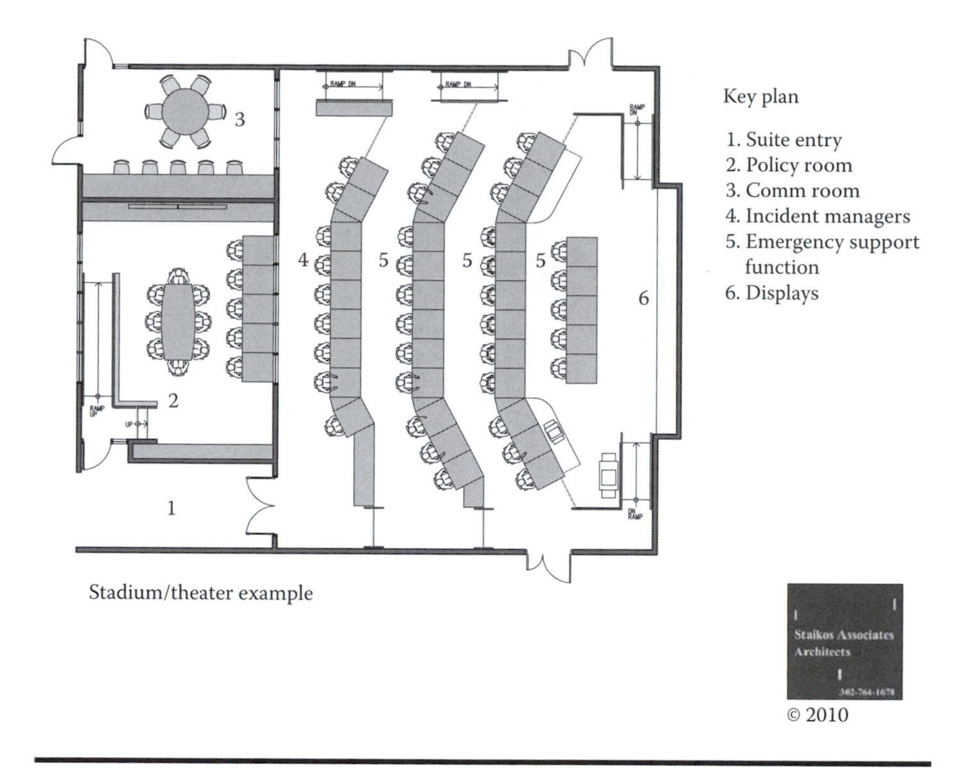

Figure 2.4 Stadium.

Negatives

1. Dedicated space with limited multipurpose use.
2. Although positions can be reassigned easily, it does not readily lend itself to podlike clustering or grouping of responding agencies.
3. Display size must increase with overall depth of room.
4. Requires buffer area at front of room to maintain optimum viewing angles.
5. Room proportions dictated by display characteristics.
6. Large group collaboration must occur in break rooms.

Collaboration Pods/Theater Variation

The pods/theater layout allows for the establishment clustering of ESF functions while providing common focus to displayed information. Configuration establishes subgroup principals allowing direct voice and visual interaction of the principals while still at their positions, allowing assimilation of ongoing event data (Figure 2.5).

Positives

1. Subgrouping or ESF clusters work well for seven to 11 participants as speaking levels can be held in check.

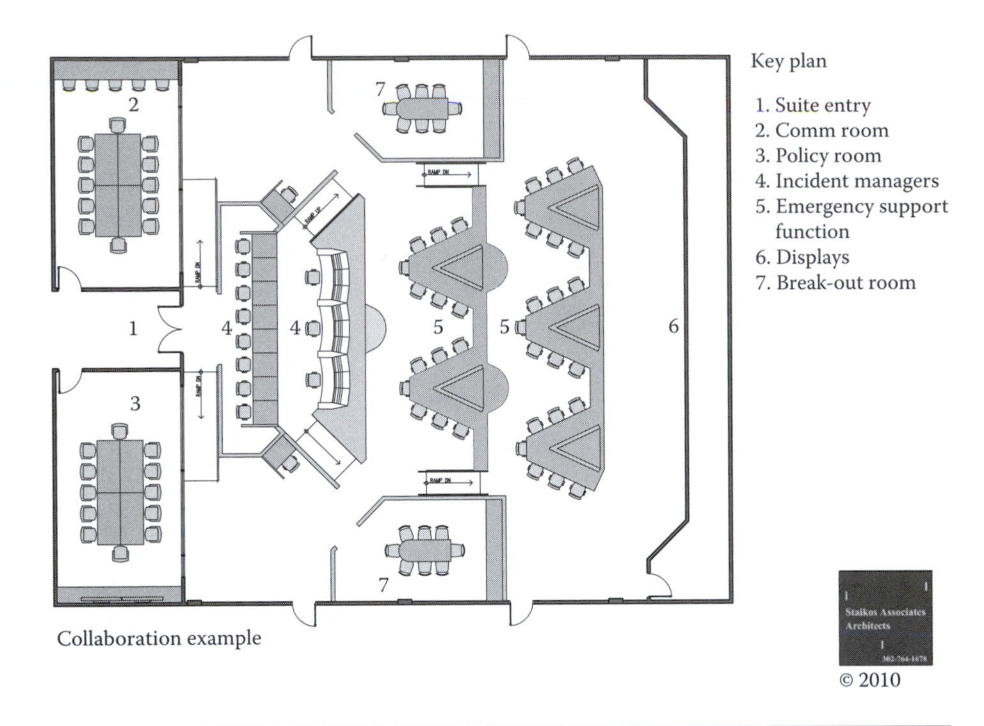

Figure 2.5 Collaborative.

2. Provides all participants with front of the room focus to view displayed information.
3. Stepped configuration allows for clear sight lines.
4. Subgroup principals have direct eye contact with team members and information displays.
5. Room for on-the-floor huddles between ESF subgroups.
6. Breakout rooms can be located on either side as in theater/stadium layout.

Negatives

1. Viewing angles are compromised for a portion of the group.
2. Does not easily lend itself to changing ESF populations without migrating across POD structure.
3. As depth of room increases, display surfaces need to enlarge, affecting optimal sight lines for those closer to display.

Iris Example

The iris layout is patterned after a nonhierarchical structure where there is no self-evident position of room leadership (Figure 2.6). This deference to equality can be seen in the UN Security Council layout.

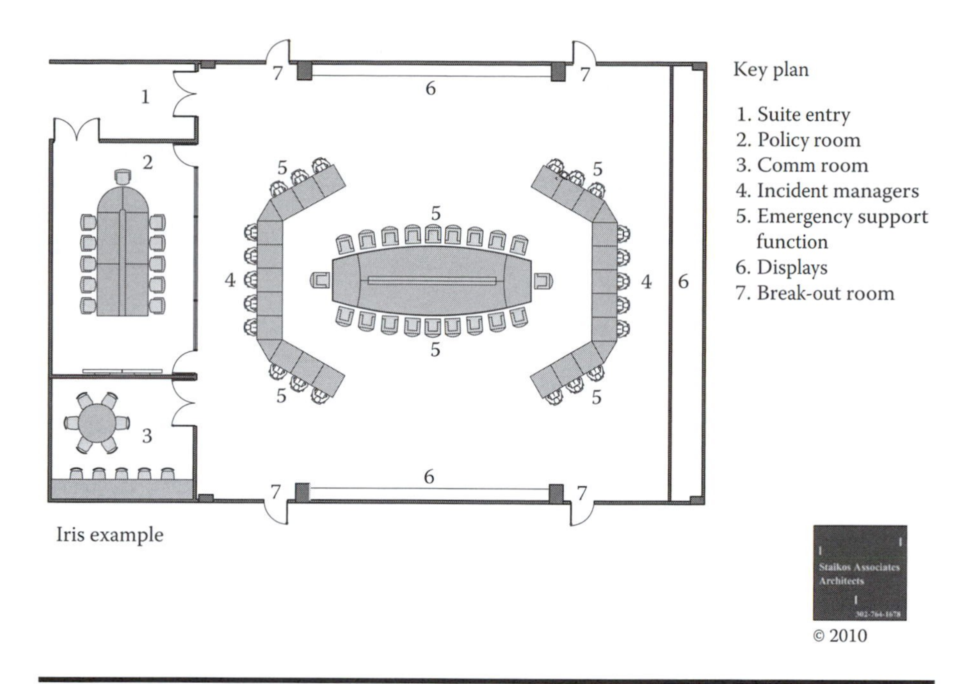

Figure 2.6 Iris.

Positives

1. Inward focus not hierarchical
2. Acceptable sight lines but for a limited number of participants
3. Breakout clusters/teaming outside OPS room

Negatives

1. Poor sight-line angles due to need to place display walls at an elevated height
2. Limited single point overview/observation opportunities caused by multi-display walls
3. Single level/flat OPS floor degrading sight lines
4. Less flexibility for routine (nonactivation time) activities

Summary

As the reader well understands, the design of today's EOC is a complex problem requiring an understanding of functional requirements, operational methodologies, and systems technologies. Regardless of the direction taken, it is a workplace that must be adaptable to meet ever-changing systems, management styles, and mission objectives. This can be accomplished through a concerted team-planning effort by fully engaging the emergency manager, architectural and engineering design professional, and facilities management team to produce a flexible, reliable, and maintainable complex. When this collaboration is implemented at the outset with a broad-based needs assessment, programming, team-oriented design, and continuing through to live performance evaluation, one can be assured that everyone's voice has been heard. Additionally, if compromise is dictated, the opportunity for buy-in is facilitated as all will be aware of the constraints, which will enable acceptance and produce a successful result.

Chapter 3

Continuity of Operations Planning

Chad Bowers

Contents

Continuity of Operations Planning

Continuity of operations planning (COOP) facilitates the performance of essential functions during all-hazards emergencies or other situations that may disrupt normal operations. Continuity planning is a fundamental responsibility of public institutions and private entities to our nation's citizens. Continuity planning facilitates the performance of essential functions during an emergency situation that disrupts normal operations and the timely resumption of normal operations once the emergency has ended. A strong continuity plan provides the organization with the means to address the numerous issues involved in performing essential functions and services during an emergency. Without detailed and coordinated continuity plans, and effective continuity programs to implement these plans, jurisdictions risk leaving our nation's citizens without vital services in what could be their time of greatest need.

It is the policy of the United States to maintain a comprehensive and effective continuity capability composed of continuity of operations and continuity of government programs to ensure the preservation of our form of government under the Constitution and the continuing performance of government operations and functions under all conditions as outlined in the following:

- National Security Presidential Directive 51 (NSPD-51)
- Homeland Security Presidential Directive 20 (HSPD-20)
- Federal Continuity Directive 1 (FCD1)
- Federal Continuity Directive 2 (FCD2)
- Continuity Guidance Circular 1 (CGC1)

Continuity requirements must be incorporated into the daily operations of all agencies to ensure seamless and immediate continuation of mission essential function (MEF)/primary mission essential function (PMEF) capabilities so that critical government functions and services remain available to the nation's citizens.

Continuity planning is the good business practice of ensuring the execution of essential functions under all circumstances. Continuity includes all activities conducted by jurisdictions to ensure that their essential functions can be performed. This includes plans and procedures that delineate essential functions; specify succession to office and emergency delegation of authority; provide for the safekeeping of vital records and databases; identify alternate operating strategies; provide for continuity communications; and validate these capabilities through test, training, and exercise (TT&E) programs. Today's changing threat environment and the potential for no-notice emergencies, including localized acts of nature, accidents, technological system failures, and military or terrorist attack-related incidents, have increased the need for continuity capabilities and planning across all levels of government and the private sector.

The goal of continuity planning is to reduce the consequence of any disruptive event to a manageable level. The specific objectives of a particular organization's continuity plan may vary, depending on its mission and functions, its capabilities, and its overall continuity strategy. In general, continuity plans are designed to

- Minimize loss of life, injury, and property damage.
- Mitigate the duration, severity, or pervasiveness of disruptions that occur.
- Achieve the timely and orderly resumption of essential functions and the return to normal operations.
- Protect essential facilities, equipment, records, and assets.
- Be executable with or without warning.
- Meet the operational requirements of the respective organization. Continuity plans may need to be operational within minutes of activation, depending on the essential function or service, but certainly should be operational no later than 12 hours after activation.
- Meet the sustainment needs of the respective organization. An organization may need to plan for sustained continuity operations for up to 30 days or longer, depending on resources, support relationships, and the respective continuity strategy adopted.
- Ensure the continuous performance of essential functions and operations during an emergency, such as pandemic influenza that require additional considerations beyond traditional continuity planning.
- Provide an integrated and coordinated continuity framework that takes into consideration other relevant organizational, governmental, and private sector continuity plans and procedures.

Responsibility for continuity planning resides with the highest level of management of the organization involved. The senior elected official or the administrative head of a state or local organization is ultimately responsible for the continuation of essential services during an emergency and for the related planning. Organizational responsibilities typically include the development of the strategic continuity vision and overarching policy, the appointment of key continuity personnel, and the development of a program budget that provides for adequate facilities, equipment, and training.

Organizational continuity planning cannot be approached in isolation. The effectiveness of one continuity plan is often dependent on the execution of another organization's continuity plan as many agency functions rely on the availability of resources or functions controlled by another organization. Such interdependencies routinely occur between government and private sector organizations. Likewise, many government continuity plans are dependent on private sector resources, especially in the area of critical infrastructure and key resources support.

Effective implementation of continuity plans and programs requires the support of senior leaders and decision makers who have the authority to commit the organization and the necessary resources to support the programs. Emergency management officials are often responsible for developing or assisting in the development of continuity plans and programs for their jurisdictions. They are also available to assist in reestablishing essential functions and services during emergencies and disasters.

An organization's resiliency is directly related to the effectiveness of its continuity capability. An organization's continuity capability—its ability to perform its essential functions continuously—rests on key components and pillars, which are in turn built on the foundation of continuity planning and program management. These pillars are leadership, staff, communications, and facilities. The continuity program staff within an organization should coordinate and oversee the development and implementation of continuity plans and supporting procedures.

Pillars 1 and 2: People/Leadership and Staff

Continuity of leadership is critical to ensure continuity of essential functions. Organizations should provide for a clear line of succession in the absence of existing leadership and the necessary delegations of authority to ensure that succeeding leadership has the legal and other authorities to carry out their duties. Continuity of leadership during crisis, especially in the case of senior positions, is important to reassure the nation and give confidence to its citizens that the principal or appropriate successor is managing the crisis and ensuring the performance of essential functions. Leaders need to set priorities and keep focus.

Leaders and staff should be sufficiently trained to be able to perform their duties in a continuity environment. To ensure that required skill sets are available, personnel should be both cross-trained and vertically trained to be able to perform the functions of their peers and the persons above and below them in an emergency.

Pillar 3: Communications and Technology

The ability to communicate is critical to daily operations and absolutely essential in a crisis. The nation's domestic and international telecommunications resources, including commercial, private, and government-owned services and facilities, are essential to support continuity plans and programs. All organizations should identify the communication requirements needed to perform their essential functions during both routine and continuity conditions. Communication systems and technology should be interoperable, robust, and reliable. Planners should consider the resilience of their systems to operate in disaster scenarios that may include power and other infrastructure problems.

Organizations should use technology to perform essential functions as an intrinsic part of daily operations, utilizing voice, data, and video solutions as appropriate. Communications and business systems, including hardware and software for continuity operations, should mirror those used in day-to-day business to assist continuity leadership and staff in a seamless transition to crisis operations.

Pillar 4: Facilities

Facilities are the locations where essential functions are performed by leadership and staff. Organizations should have adequate, separate locations to ensure execution of their functions. Physical dispersion should allow for easy transfer of function responsibility in the event of a problem in one location.

The Foundation: Continuity Planning and Program Management

Although an organization needs leaders, staff, communications, and facilities to perform its essential functions, it also needs well-thought-out and detailed plans for what to do with those key resources. Planning should include all of the requirements and procedures needed to perform essential functions.

Other key continuity concepts include geographic dispersion, risk management, security, readiness, and preparedness. Geographic dispersion of an organization's normal daily operations can significantly enhance the organization's resilience and reduce the risk of losing the capability to perform essential functions. Geographic dispersion of leadership, data storage, personnel, and other capabilities may be essential to the performance of essential functions following a catastrophic event and will enable operational continuity during an event that requires social distancing (e.g., pandemic influenza and other biological events).

Risk management is the process to identify, control, and minimize the impact of uncertain events. Security is a key element to any continuity program to protect plans, personnel, facilities, and capabilities to prevent adversaries from interfering with continuity plans and operations. To ensure the safety and success of continuity operations, an effective security strategy should address personnel, physical, and information security.

Continuity Program Management Cycle

A standardized continuity program management cycle ensures consistency across all continuity programs and supports the foundation and pillars that comprise the nation's continuity capability. It establishes consistent performance metrics, prioritizes implementation plans, promulgates best practices, and facilitates consistent cross-agency continuity evaluations. Such a cyclic-based model that incorporates planning, training, evaluating, and the implementation of corrective actions gives key leaders and essential personnel the baseline information, awareness, and experience necessary to fulfill their continuity program management responsibilities. The continuity program management cycle consists not only of its programmatic elements, but also should include the plans and procedures that support implementation of the continuity program. These plans and procedures should also be evaluated pre- and postevent, tested or exercised, and assessed during the development of corrective action plans. Objective evaluations and assessments, developed from tests and exercises, provide feedback on continuity planning, procedures, and training. This feedback in turn supports a corrective action process that helps to establish priorities, informs budget decision making, and drives improvements in plans and procedures. This continuity program management cycle, as illustrated in Figure 3.1, should be used by all organizations as they develop and implement their continuity programs.

To support the continuity program management cycle, organizations should develop a continuity multiyear strategy and program management plan that provides for the development, maintenance, and annual review of continuity capabilities, requiring an organization to

- Designate and review MEFs and PMEFs, as applicable.
- Define both short-term and long-term goals and objectives for plans and procedures.

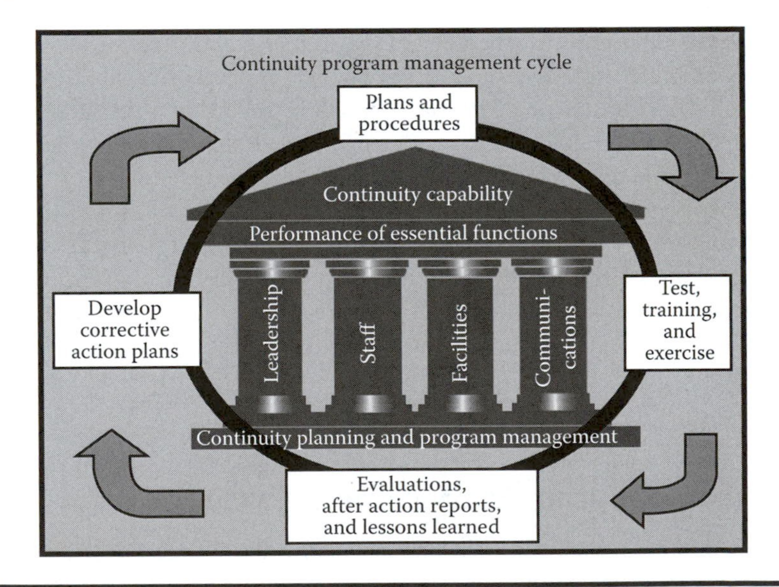

Figure 3.1 Continuity program management cycle.

- Identify issues, concerns, and potential obstacles to implementing the program, as well as a strategy for addressing these, as appropriate.
- Establish planning, training, and exercise activities, as well as milestones for accomplishing these activities.
- Identify the people, infrastructure, communications, transportation, and other resources needed to support the program.
- Forecast and establish budgetary requirements to support the program.
- Apply risk management principles to ensure that appropriate operational readiness decisions are based on the probability of an attack or other incident and its consequences.
- Incorporate geographic dispersion into the organization's normal daily operations, as appropriate.
- Integrate the organization's security strategies that address personal, physical, and information security to protect plans, personnel, facilities, and capabilities, to prevent adversaries from disrupting continuity plans and operations.

Each organization should develop a corrective action program (CAP) to assist in documenting, prioritizing, and resourcing continuity issues identified during TT&E, assessments, and emergency operations.

Essential Functions

All organizations should identify and prioritize their essential functions as the foundation for continuity planning. Essential functions, broadly speaking, are those functions that enable an organization to provide vital services, exercise civil authority, maintain the safety of the general public, and sustain the industrial/economic base during an emergency.

The identification and prioritization of essential functions are a prerequisite for continuity planning, because they establish the planning parameters that drive an organization's efforts in all other planning and preparedness areas. Resources and staff will likely be limited during an event that disrupts or has the potential to disrupt normal activities and that necessitates the activation of continuity plans, preventing the organization from performing all of its normal functions or services. Therefore, a subset of those functions determined to be critical activities are defined as the organization's essential functions. These essential functions are then used to identify supporting tasks and resources that should be included in the organization's continuity planning process.

The National Continuity Policy Implementation Plan has established three categories of essential functions: national essential functions (NEFs), PMEFs, and MEFs. The ultimate goal of continuity in the federal executive branch is the continuation of NEFs. To achieve that goal, the objective for nonfederal entities is to identify their MEFs and PMEFs, as appropriate, and ensure that those functions can be continued throughout, or resumed rapidly after, a disruption of normal activities.

The eight NEFs represent the overarching responsibilities of the federal government to lead and sustain the nation and will be the primary focus of the federal government's leadership during and in the aftermath of an emergency.

PMEFs are MEFs that must be performed in order to support the performance of NEFs before, during, and in the aftermath of an emergency. PMEFs need to be continuous or resumed within 12 hours after an event and maintained for up to 30 days or until normal operations can be resumed.

MEFs are a broader set of essential functions that includes not only an organization's PMEFs, but also all other organization functions that must be continued throughout or resumed rapidly after a disruption of normal activities, but that do not rise to the level of being PMEFs. MEFs are those functions that enable an organization to provide vital services, exercise civil authority, maintain the safety of the public, and sustain the industrial/economic base during disruption of normal operations.

When identifying an organization's essential functions and categorizing them as MEFs or PMEFs, organizations with incident management responsibilities must incorporate them into their continuity planning requirements for performing these functions. Integration of continuity planning with incident management planning and operations includes responsibilities delineated in the National Response Framework (NRF) and is linked to an organization's ability to conduct its essential functions.

In short, MEFs are the responsibilities/tasks an organization is required to complete to be considered "operational." Each organization has its own distinct operational responsibilities. Therefore, each organization will then have to have its own unique list of MEFs it is required to conduct, and all of these functions serve a distinct purpose in ensuring continuity of government. Following is an example list of MEFs from various organizations:

- Transport inmates to/from court proceedings (Police)
- Book and process incoming offenders (Police)
- Conduct daily audit of accounts payable log (Accounting)
- Detect and suppress urban, rural, and wildland fires (Fire)
- Respond to 9-1-1 calls and vehicular injuries [emergency medical service (EMS)]
- Issue medications to tuberculosis/HIV patients (Health)
- Inspect and maintain water/wastewater system (Public works)

When listing MEFs, organizations first need to decide the frequency of how often the function must be conducted (daily, weekly, monthly). Next, organizations need to identify how many personnel it requires to complete the function as well as any specialized resources it may require to complete (vehicles, tools, software, etc.).

Table 3.1 Examples of Essential Functions by Priority

Priority	*Function*	*Division*
Tier 1 – Functions to be performed, given a *One Day* disruption (highest priority to lowest)		
#1	Record and index land records, maps, trade names, armed forces discharges	Land Records
#2	Receive and index birth, marriage, death and burial records	Vital Statistics
#3	Issue marriage and civil union licenses	Vital Statistics
#4	Collect fees on all transactions, print reports, transmit receipts to Accounting	Accounting
#5	Provide access to public records (land records, maps, trade names, minutes, etc.). For attorneys, title searchers, genealogists, and public; provide copies of above as required	Administrative Services
Tier 2 – Functions to be performed, given a *One Day–One Week* disruption (highest priority to lowest)		
#1	Scan land records, trade names, armed forces discharges	Land Records
#2	Send biweekly payroll records to Accounting	Accounting
#3	Register notary certificates, name changes; perform certifications	Vital Statistics
#4	Track vacancies in public office and notify state as required	Administrative Services

Once an organization has added its MEFs to the list, it then needs to prioritize the functions based on their importance of being completed. Organizations must also keep in mind that there may also be laws, ordinances, and regulations that stipulate functions the organization must conduct. An example of how an organization should create and display a list of its essential functions is provided in Table 3.1; first by length of disruption and then by priority of carrying out the function.

Human Capital

People are critical to the operations of any organization. Choosing the right people for an organization's staff is vitally important, particularly in a crisis situation.

Leaders need to set priorities and keep focus. During a continuity event, emergency employees and other special categories of employees will be activated by an organization to perform assigned response duties. One of these categories is continuity personnel, referred to as the emergency relocation group, relocation team, or similarly named group of designated personnel.

An organization's continuity of operations program, plans, and procedures should incorporate existing organization-specific guidance and direction for human capital management. These can include guidance on pay, leave, work scheduling, benefits, telework, hiring, authorities, and flexibilities. An organization's continuity coordinator (or continuity manager) should work closely with the organization's chief human capital officer or director of human resources to resolve human capital issues related to a continuity event. Human capital issues can be solved using available laws, regulations and guidance, as well as organization implementing instructions.

The planning and preparedness related to leadership, staff, and human capital considerations for a continuity of operations situation encompass the following six activities:

- Organizations should develop and implement a process to identify, document, communicate with, and train continuity personnel.
- Organizations should provide guidance to continuity personnel on individual preparedness measures they should take to ensure a coordinated response to a continuity event.
- Organizations should implement a process to communicate the organization's operating status with all staff.
- Organizations should implement a process to contact and account for all staff in the event of an emergency.
- Organizations should identify a human capital liaison—a continuity coordinator or a continuity manager—to work with the organization's human resources and emergency planning staff when developing the organization's emergency plans.
- Organizations should implement a process to communicate their human capital guidance for emergencies (pay, leave, staffing, and other human resources flexibilities) to managers and make staff aware of that guidance in an effort to help organizations continue essential functions during an emergency.

Continuity Teams and Leadership

The following responsibilities are assigned to the leadership of designated entities. Government organizations play an integral role in determining and supporting the needs of the general public and ensuring the continuation of essential services on a daily basis (e.g., police and fire services, road construction, and public education).

These organizations should work with their tribal, local, state, and federal partners and the private sector in developing and coordinating continuity plans. This coordination helps facilitate the resourcing and allocation of resources for the development of continuity plans and the procurement of emergency response equipment, as appropriate.

Elected Officials/Executive Continuity Team

At all jurisdictional levels, elected officials are responsible for ensuring that continuity programs are appropriately resourced, and that responsible and effective continuity leaders and managers are appointed or hired to direct those programs. Elected officials should develop an executive continuity team for the jurisdiction that encompasses all of the departments, divisions, or other offices within the jurisdiction. The elected officials should sign off on the final plans and policies developed by the executive continuity team and each of the participating organizations' continuity of operations plans within the jurisdiction.

Planning Team

The continuity planning team coordinates continuity planning and duties for the entire organization. These duties include

- Overall continuity coordination for the organization.
- Providing guidance and support for the development of the organization's continuity plan.
- Establishing designated members who serve on the planning team for their organization or office, and will serve as members of the principal continuity coordinating organization and forum for exchanging ideas and information regarding continuity planning, procedures, and resources for that organization.
- Coordinating continuity exercises, documenting postexercise lessons learned, and conducting periodic evaluations of organizational continuity capabilities.
- Understanding the role that adjacent jurisdictions and organizations might be expected to play in certain types of emergency conditions and what support those adjacent organizations might provide.
- Understanding the limits of their continuity resources and support capabilities.
- Anticipating the point at which adjacent organizational or mutual aid resources will be required.

When developing continuity teams, it is essential that planners include contact information for each member, as well as an individual description of each member's role for serving on the team. An example layout of a continuity planning team is shown in Table 3.2.

Table 3.2 Example of Continuity Team Members, Contact Information, and Team Roles

COOP Planning Team	
Title	*COOP Planning Team Role*
Planning Chief	As Planning Chief, will review all changes to the COOP plan before submission for final approval.
EMS Dept Manager Emergency Services	Will serve as the main coordinator to the planning team. Responsibilities include scheduling meetings, notifying team members of meetings.
Staff Coordinator Ambulance Services	Will review and update COOP plan documentation on a quarterly basis.
Operations Manager Emergency Services	Responsible for the training of agency personnel in the actions and responsibilities contained within the plan in preparation of a COOP event.
Training Coordinator Fire and Rescue	Will document all developments and changes for the COOP.

Team members are responsible for the following:

- Understanding their continuity roles and responsibilities within their respective organizations.
- Knowing and being committed to their duties in a continuity environment.
- Understanding and being willing to perform in continuity situations to ensure an organization can continue its essential functions.
- Ensuring that family members are prepared for and taken care of in an emergency situation.

Orders of Succession

All organizations are responsible for establishing, promulgating, and maintaining orders of succession to key positions. Simply stated, orders of succession can be summed up by the statement "Who comes next?" It is critical to have a clear line of succession to office established in the event leadership becomes debilitated or incapable of performing its legal and authorized duties, roles, and responsibilities. The designation as a successor enables that individual to act for and exercise the powers of a principal in the event of that principal's death, incapacity, or resignation. Orders of succession enable an orderly and predefined transition of leadership within the organization. Orders of succession are an essential part of a continuity plan and should reach to a sufficient depth and have sufficient breadth—at least three positions deep and geographically dispersed where feasible—to ensure that essential functions continue during the course of any emergency. An example of how an organization's order of succession should be arranged is provided in Table 3.3.

As a minimum, orders of succession should do the following:

- Establish an order of succession for leadership. There should be a designated official available to serve as acting head until that official is appointed by appropriate authority, replaced by the permanently appointed official, or otherwise relieved.
  - Geographical dispersion to include, if applicable, regional, field, or satellite leadership in the line of succession, is encouraged and ensures roles and responsibilities can transfer in all contingencies.
  - Where a suitable field structure exists, appropriate personnel located outside of the subject region should be considered in the order of succession.
- Establish orders of succession for other key leadership positions, including administrators, key managers, and other key mission essential personnel or equivalent positions. Order of succession should also be established for devolution counterparts in these positions.
- Describe orders of succession by positions or titles, rather than by the names of the individuals holding those offices. To ensure their legal sufficiency, coordinate the development of orders of succession with the general counsel or other comparable legal authority.

Table 3.3 Example List of an Order of Succession

Leadership Succession for Health Director		
Position	*Title*	*Agency*
Primary	Health Director	Monroe County Health Department
#1 Alternate	Assistant Director	Monroe County Health Department
#2 Alternate	Operations Chief	Monroe County Health Department
#3 Alternate	Manager of Research	Monroe County Health Department

- Establish the rules and procedures designated officials should follow when facing the issues of succession to office.
- Include in the succession procedures the conditions under which succession will take place in accordance with applicable laws and procedures; the method of notification; and any temporal, geographical, or organizational limitations to the authorities granted by the orders of succession.
- Include orders of succession in the vital records and ensure that they are available at the continuity facilities or other continuity of operations locations in the event the continuity plan is activated.
- Revise orders of succession, as necessary, and distribute the revisions promptly as changes occur.

Delegations of Authority

To ensure a rapid response to any emergency requiring the implementation of its continuity plan, an organization should delegate authorities for making policy determinations and other decisions, at the field, satellite, and other organizational levels, as appropriate. It is vital to clearly establish delegations of authority, so that all organization personnel know who has the right to make key decisions during a continuity situation. Generally, a predetermined delegation of authority will take effect when normal channels of direction and control are disrupted and will lapse when those channels are reestablished.

Primary Facilities

When creating a continuity plan, it is important to first identify the locations where an organization operates. This information is used to identify the types of amenities and specific requirements currently in place to support the operations of an organization under normal operations.

Alternate Facilities

As part of the continuity planning process, all organizations should identify continuity facilities; alternate uses for existing facilities; and, as appropriate, virtual office options including telework. Risk assessments should be conducted on these facilities to provide reliable and comprehensive data to inform risk mitigation decisions that will allow nonfederal entities to protect assets, systems, networks, and functions while determining the likely causes and impacts of any disruption. All personnel should be briefed on organization continuity plans that involve using or relocating personnel to continuity facilities, existing facilities, or virtual offices. Continuity personnel should be provided supplemental training and guidance on relocation procedures.

A major section of your plan revolves around the concept of identifying an alternate facility or backup location for each of your primary facilities. Imagine if your workspace or building was unusable due to an event. Your organization would be confronted with many immediate questions, including

- Where do we relocate our operations; what facility?
- What do we need in place at that alternate facility?
- What items are already at the alternate facility?
- What other items do we need to bring, and in what quantities?

Every organization has different needs and requirements to operate; therefore, they need to think these questions through for themselves. An organization should discuss and decide upon as many of these questions as possible ahead of an event to ensure that relocation to a new facility is accomplished as efficiently as possible.

An organization should try to identify at least two alternate facility choices for each its primary facilities. The first choice facility should be a facility close to the primary facility and easily accessible assuming a small-scale disruption that only impacts the single building or work area (fire, pipe burst, mold in the walls, etc.). The second choice facility should be on a regional level and should assume a large-scale event has impacted the surrounding area (tornado, hurricane, earthquake, etc.).

For each alternate facility, it is important to provide specific details about the facility, including the resources that are located at this location versus the resources that would need to be transported to this location to continue operations. Resources to identify might include computers, communication equipment, office furniture, emergency supplies, and any other amenities the organization relies upon to operate. Table 3.4 is an example of how to identify and list the requirements of an alternate facility, including a detailed list of items that are prepositioned at the alternate facility in addition to the extra items that need to be transported to the facility for operations.

In many instances, organizations have a difficult time identifying a specific facility ahead of time that would be available for relocation. The fact is that there usually is no empty working space just waiting around for people to move into. This fact should not deter an organization from moving forward with its planning efforts. In this case, it is recommended to address a facility as "to be determined." With this approach, an organization does not immediately have to be able to identify a specific location to relocate, but can at least begin the discussion about the specific types and number of resources needed for operations. Once a list of resources is developed, the planning team should then have a better understanding of the type and size of location its organization would require for operations.

Table 3.4 Example of an Alternate Facility and List of Required Resources

Name Location (Physical Address)	*Resources Required to Perform Mission-Essential Functions*
Alternate Facility Monroe County Complex 1220 Greeley Street Monroe, FL 45245	**Transported** 12 — Desks — *Furniture* 1 — Generators — *Emergency Equipment* 10 — Desktop/laptop computers — *Computer Hardware* 4 — Walkie talkies — *Communications* 1 — VHF Base Station USMS frequencies — *Communications* 2 — Satellite Phones — *Communications*
Alternate For Monroe County FD Headquarters — *Primary Facility* 270 Main Street Monroe City, FL 52545	**Prepositioned** 3 — Fire cabinets (Tall/3 Drawer) — *Furniture* 1 — Copy machine — *Computer Hardware* 1 — Projector/screen — *Computer Hardware* 16 — Phones — *Communications* 1 — Internet/Intranet connection — *Communications*

Vital Records/Vital Resources

Another critical element of a viable continuity plan and program includes the identification, protection, and availability of electronic and hardcopy documents, references, records, information systems, and data management software and equipment (including classified and other sensitive data) needed to support essential functions during a continuity situation. Personnel should have access to and be able to use these records and systems to perform essential functions and to reconstitute back to normal organization operations. Organizations should compile a complete list of vital records and resources used for their day-to-day operations. Vital records and resources used by an organization should then be prepositioned at an alternate facility or stored at a backup location to ensure performance of essential functions upon COOP activation. Table 3.5 shows an example of vital records for an organization, including the name and description of each vital record shown in a list in order of priority.

Each organization has different functional responsibilities and business needs. An organization should decide which records are vital to its operations and then

Table 3.5 Example of Vital Records and Resources Listed by Priority

Vital Records/Resources		
	Record Name	*Data*
#1	Rapid Fire Software	Rapid Fire Software is used for continual training and education of our responders to enhance response times and safety measures.
#2	FARSITE	FARSITE is a fire behavior and growth simulator for use on Windows computers. It is used by Fire Behavior Analysts from the USDA and is taught in the S493 course. FARSITE is designed for use by trained, professional wildland fire planners and managers familiar with fuels, weather, topography, wildfire situations, and the associated concepts and terminology.
#3	Genesis Software	Genesis Software manages the historical calls and prior Responses Database.
#4	SMART System Software	The SMART system is the state-operated database that is used for ...
#5	Emergency Call-Out List	Emergency call-out list consisting of personnel contact information

should assign responsibility for those records to the appropriate personnel, who may be a combination of continuity personnel, personnel in the chief information officer's department, and records management personnel. An effective vital records program should have the following characteristics.

1. An official vital records program:
   - Identifies and protects those records that specify how an organization will operate in an emergency or disaster
   - Identifies those records necessary to the organization's continuing operations
   - Identifies those records needed to protect the legal and financial rights of the organization and citizens
2. A vital records program should be incorporated into the overall continuity of operations plan, and it needs a clear authority to include:
   - Policies
   - Authorities

 - Procedures
 - The written designation of a vital records manager
3. As soon as possible after continuity of operations activation, but recommended within 12 hours of such activation, continuity personnel at the continuity facility should have access to the appropriate media for accessing vital records, such as
 - A local area network
 - Electronic versions of vital records
 - Supporting information systems and data
 - Internal and external e-mail and e-mail archives
 - Hard copies of vital records
4. Organizations should strongly consider multiple redundant media for storing their vital records.
5. Organizations should maintain a complete inventory of records along with the locations of and instructions on accessing those records. This inventory should be maintained at a backup/offsite location to ensure continuity if the primary site is damaged, destroyed, or unavailable. Organizations should consider maintaining these inventories at a number of different sites to support continuity operations.
6. Organizations should conduct vital records and database risk assessment to
 - Identify the risks involved if vital records are retained in their current locations and media, and the difficulty of reconstituting those records if they are destroyed.
 - Identify offsite storage locations and requirements.
 - Determine if alternative storage media is available.
 - Determine requirements to duplicate records and provide alternate storage locations to provide readily available vital records under all conditions.
7. Appropriate protections for vital records will include dispersing those records to other organization locations or storing those records offsite. When determining and selecting protection methods, it is important to take into account the special protections needed by the different kinds of storage media. Microforms, paper photographs, and computer disks, tapes, and drives, all require different methods of protection. Some of these media may also require equipment to facilitate access.
8. At a minimum, vital records should be annually reviewed, rotated, or cycled so that the latest versions will be available.
9. A vital records plan packet should be developed and maintained. The packet should include:
 - A hard copy or electronic list of key organization personnel and disaster staff with up-to-date telephone numbers
 - A vital records inventory with the precise locations of vital records
 - Updates to the vital records
 - Necessary keys or access codes

- Continuity-facility locations
- Access requirements and lists of sources of equipment necessary to access the records (this may include hardware and software, microfilm readers, Internet access, and dedicated telephone lines)
- Lists of records-recovery experts and vendors
- A copy of the organization's continuity of operations plan

This packet should be annually reviewed with the date and names of the personnel conducting the review documented in writing to ensure that the information is current. A copy should be securely maintained at the organization's continuity facilities and other locations where it is easily accessible to appropriate personnel when needed.

10. The development of an annual training program for all staff should include periodic briefings to managers about the vital records program and its relationship to the organization's vital records and business needs. Staff training should focus on identifying, inventorying, protecting, storing, accessing, and updating the vital records.
11. There should be an annual review of the vital records program to address new security issues, identify problem areas, update information, and incorporate any additional vital records generated by new organization programs or functions, or by organizational changes to existing programs or functions. The review will provide an opportunity to familiarize staff with all aspects of the vital records program. It is appropriate to conduct a review of the vital records program in conjunction with continuity exercises.
12. There should be annual testing of the capabilities for protecting classified and unclassified vital records and for providing access to them from the continuity facility.

Devolution

Devolution is the capability to transfer statutory authority and responsibility for essential functions from an organization's primary operating staff and facilities to other organization employees and facilities, and to sustain that operational capability for an extended period.

Devolution planning supports overall continuity planning and addresses the full spectrum of threats and all-hazards emergency events that may render an organization's leadership or staff unavailable to support, or incapable of supporting, the execution of the organization's essential functions from either its primary location or its alternate location(s). Organizations should develop a devolution option for continuity, to address how the organization will identify and conduct its essential functions during an increased threat situation or in the aftermath of a catastrophic emergency.

Reconstitution

Reconstitution is the process by which surviving or replacement organization personnel resume normal operations from the original or replacement primary operating facility. Reconstitution embodies the ability of an organization to recover from an event that disrupts normal operations and consolidates the necessary resources so that the organization can resume its operations as a fully functional entity. In some cases, extensive coordination may be necessary to procure a new operating facility if an organization suffers the complete loss of a facility or in the event that collateral damage from a disaster renders a facility structure unsafe for reoccupation.

Testing, Training, and Exercising

A well-defined TT&E program is necessary to assist organizations to prepare and validate their organization's continuity capabilities and program to perform essential functions during any emergency. This requires the identification, training, and preparedness of personnel capable of performing their continuity responsibilities and implementing procedures to support the continuation of organization essential functions.

Training provides the skills and familiarizes leadership and staff with the procedures and tasks they should perform in executing continuity plans. Tests and exercises serve to assess and validate all the components of continuity plans, policies, procedures, systems, and facilities used to respond to and recover from an emergency situation and identify issues for subsequent improvement. All organizations should plan, conduct, and document periodic tests, training, and exercises to prepare for all-hazards continuity emergencies and disasters, identify deficiencies, and demonstrate the viability of their continuity plans and programs. Deficiencies, actions to correct them, and a timeline for remedy should be documented within an organization's COOP.

Testing

Testing ensures that equipment and procedures are maintained in a constant state of readiness to support continuity activation and operations. An organization's test program should include

1. Annual testing (at a minimum) of alert, notification, and activation procedures for continuity personnel, with recommended quarterly testing of such procedures for continuity personnel.
2. Annual testing of plans for recovering vital records (both classified and unclassified), critical information systems, services, and data.
3. Annual testing of primary and backup infrastructure systems and services (e.g., for power, water, fuel) at continuity facilities.

4. Annual testing and exercising of required physical security capabilities.
5. Testing and validating equipment to ensure the internal and external interoperability and viability of communications systems, through quarterly testing of the continuity communications capabilities (e.g., secure and nonsecure voice and data communications).
6. Annual testing of the capabilities required to perform an organization's essential functions, as identified in the BPA.
7. A process for formally documenting and reporting tests and their results.
8. Conducting annual testing of internal and external interdependencies identified in the organization's continuity plan, with respect to performance of an organization's and other organization's essential functions.

Training

Training familiarizes continuity personnel with their procedures, tasks, roles, and responsibilities in executing an organization's essential functions in a continuity environment. An organization's training program should include

1. Annual continuity awareness briefings (or other means of orientation) for the entire workforce.
2. Annual training for personnel (including host or contractor personnel) who are assigned to activate, support, and sustain continuity operations.
3. Annual training for the organization's leadership on that organization's essential functions, including training on individual position responsibilities.
4. Annual training for all organization personnel who assume the authority and responsibility of the organization's leadership if that leadership is incapacitated or becomes otherwise unavailable during a continuity situation.
5. Annual training for all predelegated authorities for making policy determinations and other decisions, at the field, satellite, and other organizational levels, as appropriate.
6. Personnel briefings on organization continuity plans that involve using or relocating to continuity facilities, existing facilities, or virtual offices.
7. Annual training on the capabilities of communications and information technology (IT) systems to be used during an incident.
8. Annual training regarding identification, protection, and ready availability of electronic and hardcopy documents, references, records, information systems, and data management software and equipment (including sensitive data) needed to support essential functions during a continuity situation.
9. Annual training on an organization's devolution option for continuity, to address how each organization will identify and conduct its essential functions during an increased threat situation or in the aftermath of a catastrophic emergency.

10. Annual training for all reconstitution plans and procedures to resume normal organization operations from the original or replacement primary operating facility.

Training should prepare continuity personnel to respond to all emergencies and disasters and ensure performance of the organization's essential functions. These include interdependencies both within and external to the organization. As part of its training program, the organization should document the training conducted, the date of training, those completing the training, and by whom.

Exercises

An organization's continuity exercise program focuses primarily on evaluating capabilities or an element of a capability, such as a plan or policy, in a simulated situation. Organizations should refer to the Homeland Security Exercise and Evaluation Program for additional exercise and evaluation guidance. An organization's exercise program should include

1. An annual opportunity for continuity personnel to demonstrate their familiarity with continuity plans and procedures and to demonstrate the organization's capability to continue its essential functions.
2. An annual exercise that incorporates the deliberate and preplanned movement of continuity personnel to an alternative facility or other continuity location.
3. Communications capabilities and both inter- and intraorganization dependencies.
4. An opportunity to demonstrate that backup data and records required to support essential functions at continuity facilities or locations are sufficient, complete, and current.
5. An opportunity for continuity personnel to demonstrate their familiarity with the reconstitution procedures to transition from a continuity environment to normal activities when appropriate.
6. An opportunity for continuity personnel to demonstrate their familiarity with the devolution procedures to reconstitute from a continuity environment to normal activities when appropriate.
7. A comprehensive debriefing after each exercise, which allows participants to identify systemic weakness in plans and procedures, and to recommend revisions to the organization's continuity plan.
8. A cycle of events that incorporates evaluations, after-action reports, and lessons learned into the development and implementation of a CAP, to include an improvement plan.
9. Organizational participation: conducting and documenting annual assessments of their continuity TT&E programs, and continuity plans and programs.

Each organization should develop a CAP to assist in documenting, prioritizing, and resourcing continuity issues identified during TT&E, assessments, and emergency operations. The purpose of CAP is to accomplish the following:

- Identify continuity deficiencies and other areas requiring improvement and provide responsibilities and a timeline for corrective action.
- Identify program and other continuity funding requirements for submission to the organization leadership.
- Identify and incorporate efficient acquisition processes, and where appropriate, collect all interorganization requirements into one action.
- Identify continuity personnel requirements for an organization's leadership and their supporting human resource offices.

Phases and Implementation of a Continuity Plan

A continuity plan is implemented to ensure the continuation or rapid resumption of essential functions during a continuity event. Organizations should develop an executive decision-making process that allows for a review of the emergency situation and a determination of the best course of action based on the organization's readiness posture. An organization's continuity implementation process should include the following four phases: readiness and preparedness, activation and relocation, continuity operations, and reconstitution. The four phases are implemented as illustrated in Figure 3.2.

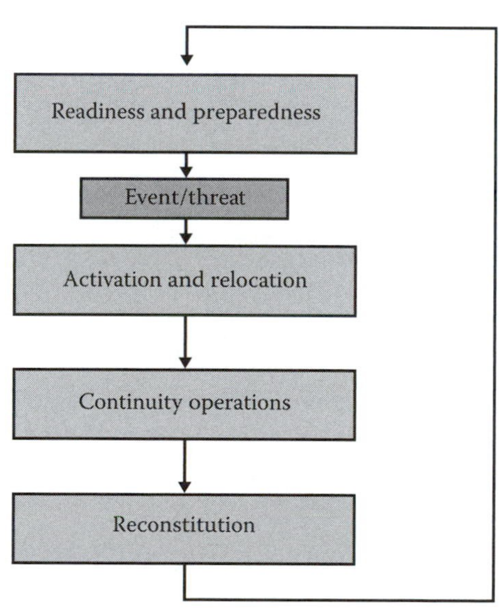

Figure 3.2 Phases and implementation of a continuity plan.

Chapter 4

Strategizing Emergency Management Programs

S. Shane Stovall

Contents

Introduction

As the profession of Emergency Management continues to advance itself, it is important that Emergency Management agencies and organizations have a clear programmatic direction. In order to accomplish this, it is critical that the Emergency Manager establish a solid mission, set of objectives, and initiatives to carry out these objectives. Each of these elements are components that make up a complete strategy that will provide direction and guidance to the Emergency Manager as they progress with their agency or organization. With the litany of mandates, laws, rules, regulations, ordinances, guidance, and other internal and external influences, managing an Emergency Management program can be daunting at best. No matter what the staffing is or what the resources are for an Emergency Management program, it is critical that a solid departmental or organizational strategy be developed. A strategy can serve as the foundation for an Emergency Management program. It is also considered good business practice. In this chapter, we will look deeper into what strategies are, why departmental and organization strategies are necessary, and the elements of an Emergency Management program strategy.

Strategy Defined

In order for an Emergency Manager to develop a strategy, it is important to know what one actually is. To achieve this, here are a couple of definitions that can help to clarify the term:

> *Merriam-Webster's Dictionary*: (a) A careful plan or method. (b) The art of devising or using plans of strategems toward a goal.*
> *BusinessDictionary.com:* (1) Alternative chosen to make happen a desired future, such as achievement of a goal or solution to a problem. (2) Art and science of planning and marshalling resources for their most efficient and effective use. The term is derived from the Greek word for generalship or leading an army. See also tactics.†

* "strategy." Merriam-Webster Online Dictionary. 2010. Merriam-Webster Online. http://www.merriam-webster.com/dictionary/strategy. Accessed 14 December 2010.

† "strategy." BusinessDictionary.com. 2010. Web Finance Inc. http://www.businessdictionary.com/definition/strategy.html. Accessed 6 January 2010.

In these definitions, some common words appear, such as "plans/planning" and "goals." For the purposes of this chapter, a hybrid of these definitions is going to be used—a strategy is a plan or method used to provide efficient and effective direction for use of resources in meeting organizational goals. In this chapter, the purposes for developing an Emergency Management program strategy, and the methodologies that can be used in their development, will be discussed.

Why Develop a Strategy?

There are various reasons for why a strategy should be developed. In this section, reasons for strategy development will be outlined. These reasons can be used to "sell" the idea for developing a strategy to other staff members (strategy development is not a solo project—this will be discussed later in this chapter). Also, whether they realize it or not, Emergency Management departments and organizations are businesses. They can be a division (department) of an overall business (jurisdiction or organization). They usually have a budget (no matter how large or small), and they hopefully are providing a service to their constituents (whether internally or externally). If the Emergency Manager begins to think from a business standpoint, the development of a strategy (business plan) begins to make sense. It does not matter if the Emergency Manager works in the public sector, the private sector, or the volunteer sector—the development of a strategy for the Emergency Management department, program, or organization is a necessity. Let us look at why.

Justification of Program and Projects

As the economy fluctuates, situations arise that may bring people to question the justification for particular programs and functions. Emergency Management is not immune from this. Many have had the unfortunate experience of cutbacks and reductions in funding that have left many Emergency Management programs with minimal staff and resources to deal with an ever-increasing workload. During such economic times, Emergency Managers often find themselves answering the question: "So, what are you working on when there aren't any disasters?" or "What is it that you do?" Emergency Management in its purest sense, as a profession, is largely misunderstood by the public and sometimes even by those that hired them. Emergency Management is a profession that usually is not on the "front lines" in the public eye on a day-to-day basis. Therefore, what the Emergency Manager does is typically not very obvious to the casual observer. For this reason, it is important that Emergency Managers have a written document showing what it is that their department or organization has done, is doing, and plans to do in the future. The strategy document helps to further legitimize programs and efforts put forth by the Emergency Management department or organization.

An Emergency Manager armed with a well-written strategy can provide this document to anyone who may be looking to justify further cuts in personnel, funding, or other resources. A properly written and updated strategy should be able to answer many questions as to what each person (by position) is working on and plans to work on, as well as what current and future programs will require funding. Obviously, a strategy does not prevent or deter potential questions, scrutiny, or even cuts to programs, personnel, and funding. However, it does provide a plan and, if written correctly, a justification for personnel, funding, and other resources that help to support the Emergency Management programs.

Mandated Goals and Objectives

Emergency Manager often find themselves inundated with projects that they have to do in order to meet local, state, or federal requirements. These requirements can come in the form of plan review and acceptance, grant deadlines, and compliance with the National Incident Management System (NIMS), just to name a few. The development of a strategy will assist the Emergency Manager in organizing resources and determining tasks required to meet these mandates. This is critical in order for the Emergency Manager to be proactive rather than reactive with their workload.

Program Development and Direction

Development of a departmental or organizational strategy can give the Emergency Manager a solid idea of what has been accomplished, what is currently being completed, and what needs to be done. The development of a strategy (as will be seen later in this chapter) also allows the Emergency Manager to take a look at overall goals for the department or organization and what steps or tasks are needed to be completed in order to reach that goal. The thought process in developing these tasks and subtasks, or initiatives, is typically very intense and time consuming. However, the end product gives a clear look at what needs to be accomplished in the present and in the future.

Work Plans and Assignments

Once goals and initiatives to meet each goal are set forth, personnel and resources can be assigned to each one of them. The Emergency Manager, at this point, can begin to get a clear picture as to what resources and personnel may or may not be necessary as the organization or department moves forth. This is very important in staffing and budget planning. During this point, a Work Plan can also be developed that outlines each task and subtask within an initiative and assigns personnel and other resources to it. This can be a separate document that can also be used to drive the Performance Objectives of staff (to be discussed later in this chapter).

The preceding should provide the Emergency Manager with ample reasons for developing their own departmental or organization strategy. Not only is it a good business practice, but it also can provide justification and help to outline details of the direction that the department or organization will be taking in the future.

Elements of the Strategy

The strategy document should be separated into sections and easy to read. One should feel free to use pictures, charts, and other graphics to illustrate and emphasize any points that need to be made, particularly those that quantify any accomplishments (i.e., amount of grant funds brought in, number of plans written or reviewed, number of exercises held). The end user/reader should certainly be kept in mind as the strategy document is developed. The following outlines sections that the Emergency Manager many want to include in their overall strategy document.

Introduction

This section gives the Emergency Manager an opportunity to explain the departmental or organizational responsibilities. This will be the first page that the reader will see, other than possibly a table of contents. The Emergency Manager must ensure that they say everything that they want to demonstrate about their department on the first page. Some readers may not read past the first page, and this will be the only impression of your department or agency that they get.

Mission Statement

The Mission Statement of any organization is essentially a concise definition of what the Emergency Manager organization or department does. The Mission Statement should be no more than two sentences long. There should be no extraneous wording. Just a simple answer to "What do you do?" Here is an example:

> The City of Plano Department of Emergency Management will serve the citizens of the City of Plano by directing and coordinating emergency management programs to prevent/mitigate, prepare for, respond to, and recover from emergencies and disasters.*

This statement is to the point, and clearly defines the role of the organization.

---

* Stovall, Shane. City of Plano, Texas. City of Plano Department of Emergency Management Strategic Plan 2011–2016, p. 3, Plano: 2011. Print.

Vision Statement

The vision statement differs from the mission statement in that it gives readers an overall view of the direction of the department or organization. This statement can be a little bit longer than the mission statement, but should be short in length yet encompass the future of the department or organization in general terms. Here is an example:

> The City of Plano Department of Emergency Management shall continue to develop and maintain a leading edge, all-hazards emergency management and homeland security program that encompasses all organizations in the public and private sectors. This will include citizens; government agencies at the city, county, regional, state, and federal levels; school boards, businesses (small and corporate), faith-based, and volunteer agencies. The program will coordinate the comprehensive community planning, training, and exercises needed to ensure maximum efficiency and benefit from hazard prevention/mitigation, preparedness, response, and recovery in order to protect lives and property in the City of Plano. The program will be professional, responsive, and shall strive to serve as a model municipal Emergency Management agency.*

Organizational Values

This section does not necessarily need to be placed in any particular part of the strategy. However, when stating the department or organization's mission and vision statements, it may make more sense to list the qualities for which the Emergency Management agency is trying to achieve, as shown in Table 4.1.

Following each term, the organization should add a statement such as "We will...." This helps to demonstrate the departmental or organizational understanding of the term. This section also helps to serve as a reminder of the qualities for which the agency wants to portray.

Organizational Chart

In this section, the Emergency Manager can insert the organizational chart for their department or organization. This will depict the staffing structure for the organization, and hopefully show the areas for which employees are responsible (Figure 4.1).

---

* Stovall, Shane. City of Plano, Texas. City of Plano Department of Emergency Management Strategic Plan 2011–2016, p. 3, Plano: 2011. Print.

Table 4.1 Organizational Qualities

• Excellence	• Teamwork
• Dependable	• Adaptable
• Commitment	• Respect
• Ethics and Integrity	• Loyalty
• Empowerment	• Safety
• Education	• Communication

Source: Stovall, Shane. City of Plano, Texas. City of Plano Department of Emergency Management Strategic Plan 2011–2016. p. 3, Plano: 2011. Print. With permission.

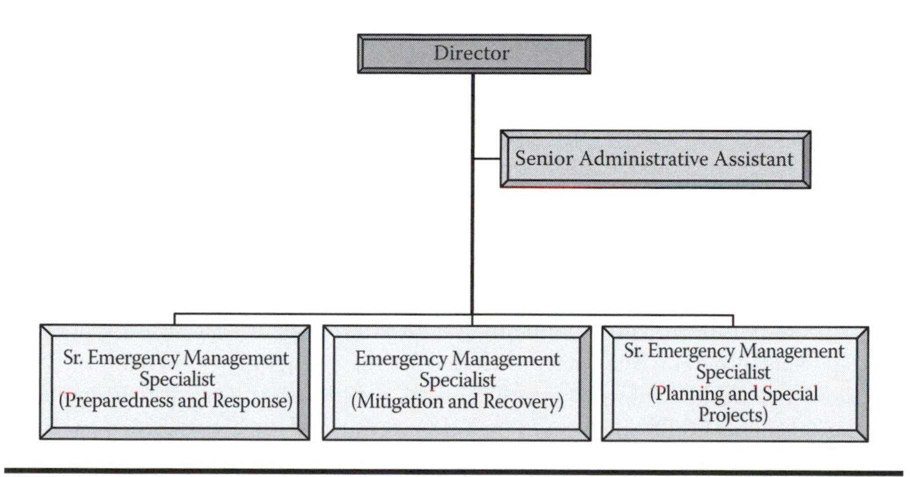

Figure 4.1 Sample organizational chart. (From Stovall, S., City of Plano, Texas. City of Plano Department of Emergency Management Strategic Plan 2011–2016, p. 3, Plano: 2011, Print. With permission.)

The preceding elements of the strategy can be considered preliminary elements that set the stage for the rest of the strategy. The following elements can be considered the body, or core, of the strategy document.

Executive Summary

This section should provide an explanation of the contents and structure of the strategy. The intent of this part of the strategy should also be to provide the readers with an opportunity to understand what they are getting ready to read in the rest of the strategy document. The Executive Summary should contain the following information:

- **Who** is covered in the strategy—what department, organization, or program.
- **What** the department does and what agency has generally accomplished in the past.
- **Why** the strategy is being developed—internal requirement, to provide a foundation for future direction of the agency, other? All of the above?
- **When** the strategy will be reviewed and updated.
- **How** the strategy is laid out structurally in order to give the reader an idea of the layout of the strategy document (section names, chapters, etc.).

In general, the Executive Summary should be no more than one page in length. It is important to keep the word "summary" in mind for this section.

Standards

Emergency Management professionals are faced with many different types of standards that should be figured in when developing a program strategy. This section will take a look at a few of these standards, and why they should be considered when developing an Emergency Management program strategy.

State Emergency Management Standards

Many states put forth requirements that dictate the elements that local Emergency Management programs shall have included in their programs. Oftentimes, these requirements are in the form of requirements for inclusion in a local Emergency Operations Plan (EOP). Some states include training and exercise standards as well.

For instance, the "State of Texas Preparedness Standards" focuses on three activities that contribute to the overall readiness of a community. These three activities are planning, training, and exercises. The Texas Division of Emergency Management (TDEM) uses a series of collective assessments of local emergency preparedness programs to measure their effectiveness and determine where areas of additional emphasis are needed. The "State of Texas Preparedness Standards" set forth criteria for local jurisdictions to meet in order to meet certain levels of preparedness. These levels include Basic, Intermediate, and Advanced levels of preparedness. Each level of preparedness contains a specific set of criteria prescribed by TDEM with respect to planning, training, and exercise programs.

Compliance with these criteria does not imply that a jurisdiction's efforts in emergency management should be limited only to the criteria outlined in the "State of Texas Preparedness Standards." The standards that are set forth do not assess staffing levels, funding for emergency programs, the level of training provided for emergency responders, or the availability of response equipment or emergency facilities. These are seen to be local responsibilities, and are taken into consideration by the local jurisdictions.

Another example can be found with the Michigan Department of State Police Emergency Management Division. They have issued a "Local Emergency Management Standards"* document. This listing of standards includes many required and recommended elements that local jurisdictions should include in their Emergency Management programs. The elements in the Michigan document include requirements and recommendations for planning, training, exercise, and administration for the four phases of Emergency Management—mitigation, preparedness, response, and recovery.

Many times these state standards and requirements can be tied to funding streams, and compliance with these standards can affect the eligibility to access these funds. The funds can come in the form of grants, technical assistance, and disaster assistance. Therefore, it is important that the Emergency Manager become familiar with any State Emergency Management standards that may be set forth for their organization, and take them into consideration when constructing their Emergency Management program strategy.

NIMS Requirements

Following the attacks on the World Trade Center and the Pentagon on September 11, 2001, it was determined that a national approach to incident management would further improve the effectiveness of emergency response providers† and incident management organizations when dealing with any hazard, whether it be natural, man-made, or technological. This national approach would be applicable to all jurisdictional levels and functional disciplines in order to allow for consistency in emergency and disaster preparedness, response, and recovery efforts.

On February 28, 2003, the President issued Homeland Security Presidential Directive (HSPD)-5, which directed the Secretary of Homeland Security to develop and administer a NIMS. According to HSPD-5:

> This system will provide a consistent nationwide approach for Federal, State, and local governments to work effectively and efficiently together to prepare for, respond to, and recover from domestic incidents, regardless of cause, size, or complexity. To provide for interoperability and

---

* State of Michigan. Local Emergency Management Standards. Lansing: State of Michigan, 1998. Print.

† As defined in the Homeland Security Act of 2002, Section 2(6), "The term 'emergency response providers' includes Federal, State, and local emergency public safety, law enforcement, emergency response, emergency medical (including hospital emergency facilities), and related personnel, agencies, and authorities." 6 U.S.C. 101 (6). This definition includes all City of Plano agencies and personnel, as those outside of public safety are categorized in the "related personnel, agencies, and authorities."

> compatibility among Federal, State, and local capabilities, the NIMS will include a core set of concepts, principles, terminology, and technologies covering the incident command system; multiagency coordination systems; unified command; training; identification and management of resources (including systems for classifying types of resources); qualifications and certification; and the collection, tracking and reporting of incident information and incident resources.*

To provide the framework for interoperability and compatibility, the NIMS is based on the appropriate balance of flexibility and standardization in order to allow for consistent integration of multiple internal and external agencies during incident management. The major components of NIMS are

- **Command and Management**—Mandates consistent use of the Incident Command System (ICS), Multiagency Coordination Systems, and Public Information Systems.
- **Preparedness**—Requires standardized planning, training, and exercises; consistent methods for qualification and certification of emergency personnel; uniform response and recovery equipment acquisition and certification; and publication management.
- **Resource Management**—Defines uniform mechanisms for inventorying, mobilizing, dispatching, tracking, and recovering resources over the life cycle of an incident.
- **Communications and Information Management**—Identifies the requirement for standardized communications, information management (collection, analysis, and dissemination), and information sharing at all levels of incident management.
- **Supporting Technologies**—Includes identification and acquisition of technology and technological systems that support capabilities that are essential to implementing and continuously refining the NIMS. These include voice and data communications systems, information management systems, and data display systems.
- **Ongoing Management and Maintenance**—Establishes activities to provide strategic direction for oversight of the NIMS. This includes routine review and refinement of the system.

Since the establishment of the NIMS, the U.S. Department of Homeland Security has issued requirements that local, state, and federal governments are to meet annually in order to be compliant with the NIMS for each respective year. These requirements cover each of the elements described above. Many of the

* United States. Homeland Security Presidential Directive 5: Management of Domestic Incidents. Washington, DC: White House, 2003. Print.

requirements placed upon state and local governments typically require many hours of work in order to be fully compliant. In addition, there is a continual refinement of each of the standards in order to clarify the "spirit and intent" of each requirement. Because of these factors, full implementation of the NIMS at the federal, state, and local levels of government is a phased process, and is expected to take several more years for all requirements to be fully mandated.

There has been some debate regarding the life expectancy of the NIMS requirements. However, these requirements and standards should still be considered as the Emergency Management professional constructs their program strategy. This will assist in long-term NIMS compliance for the organization.

Emergency Management Accreditation Program

The Emergency Management Accreditation Program (EMAP) is a voluntary, non-governmental process of self-assessment, documentation, and independent review designed to evaluate, enhance, and recognize quality in emergency management programs. The accreditation process is intended to improve emergency management program capabilities and increase professionalism at the federal, state, and local levels of government, thus benefiting the communities that these programs serve. This process has been used internationally, and is also being used at colleges and universities. The goal of the accreditation is to evaluate an emergency management program's organization, resources, plans, and capabilities against current standards to increase effectiveness in protecting the lives and properties of residents.

The EMAP has been designed to facilitate compliance with a set of standards called the "EMAP Standard." The *EMAP Standard* is now a stand-alone standard. The *EMAP Standard* was built upon the *NFPA 1600 Standard on Disaster/Emergency Management and Business Continuity Programs* adopted by the National Fire Protection Association (NFPA). The *NFPA 1600* earlier adopted a portion of its program element framework from the Capability Assessment for Readiness created by the Federal Emergency Management Agency (FEMA).

The *EMAP Standard* contains 64 standards that are intended to indicate the components that a quality emergency management program should have in place. These standards are often difficult to meet and prove to be challenging for most emergency management agencies. The standards describe "what" a program should accomplish, but not necessarily "how" compliance with a standard should be achieved. This provides flexibility to the local governments in developing emergency management programs based around the *EMAP Standard.*

Although meeting this set of standards is voluntary, it provides a good "bar" or quality level for the Emergency Management professional to strive for. Therefore, strong consideration should be given to using this standard when developing an Emergency Management program strategy.

The standards mentioned above are a few examples that Emergency Management professionals must consider during the development of their overall strategy. The

Emergency Manager must stay apprised of local, state, and federal rules, regulations, ordinances, statutes, and laws that may govern the way in which they conduct business. Some standards are voluntary standards, whereas others are mandatory. The Emergency Manager must be able to make a determination as to what is in the best interest of their organization, and include possibly a mixture of standards into their strategy planning.

Goal and Initiative Development

Planning Team

Developing the department or organizational goals for an Emergency Management program strategy should involve all members of the Emergency Management staff, and in some cases, any supervisors of the Emergency Management Coordinator or Director. The Emergency Manager may also decide that it is relevant to invite other Emergency Management stakeholders (i.e., Public Works, Public Information, Schools, etc.) to participate in the strategy building process. Having multiple inputs into what goals and initiatives shall be set forth is critical in gaining holistic understanding and buy-in to the overall strategy. Having multiple people involved in the strategy building process also ensures that the strategy is comprehensive and that all elements in the Emergency Management program are addressed.

Format and Structure of Strategy Document

Before goals can be established, it is wise to develop an outline or format of how the goals and initiative are to be laid out. There are several formats that can be used for structure. The decision as to what format should be chosen is a matter of preference of the Emergency Manager. However, the format should reflect the general foundational structure of the department or organization (if one has been developed). Some examples of formats for Emergency Management program strategies can include:

Four Phases Format

As discussed previously in Chapter 1, in this approach the four phases (some now use five to include "prevention") of Emergency Management are used to outlined program areas. The four phases of Emergency Management include prevention/mitigation, preparedness, response, and recovery. Under each area, a list of pertinent goals and initiatives can be developed (Figure 4.2). Discussion of goal and initiative development will be explained later in this chapter.

Incident Command System Format

This approach can be used by Emergency Management agencies that have set themselves up in accordance with the ICS Structure. This format splits goals

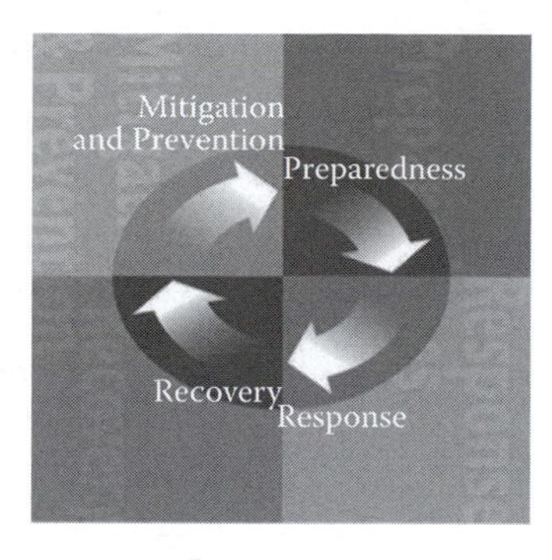

Figure 4.2 Four phases of emergency management. (Idaho Department of Education.)

and initiatives into the four areas defined as General Staff in the ICS structure: Finance and Administration, Logistics, Operations, and Planning. This format is not, contrary to many beliefs, necessarily "consistent" with the NIMS requirements. The reason for this is that ICS is only one component of the overall NIMS structure and mandates. This format is merely a structure that can be used to organize a department, and therefore the strategy (Figure 4.3).

Emergency Support Function Format

Some Emergency Managers may decide to format their strategy in accordance with the Federal Emergency Support Functions (ESFs) outlined in the Department of Homeland Security's National Response Framework. Some states have also set forth ESF criteria for local Emergency Management Plans that may include additional ESFs. The ESF format is an option, although is can be a labor-intensive format (Table 4.2).

Operational Area Format

Some Emergency Managers may decide to organize their agency, and thus their program, in accordance with specific operational areas. These operational areas can be varied and can include areas listed in previous discussed format, or any hybrid thereof, including

- Administrative
- Natural hazards

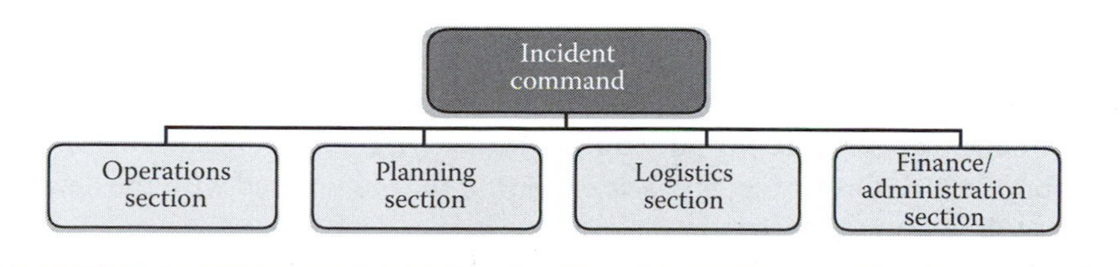

Figure 4.3 Incident command system general staff (FEMA).

Table 4.2 Federal Emergency Support Functions—National Response Framework

ESF-1 Transportation	ESF-9 Search and Rescue
ESF-2 Communications	ESF-10 Oil and Hazardous Materials Response
ESF-3 Public Works and Engineering	ESF-11 Agriculture and Natural Resources
ESF-4 Firefighting	ESF-12 Energy
ESF-5 Emergency Management	ESF-13 Public Safety and Security
ESF-6 Mass Care, Emergency Assistance, Housing, and Human Services	ESF-14 Long-Term Community Recovery
ESF-7 Logistics Management and Resource Support	ESF-15 External Affairs
ESF-8 Public Health and Medical Services	

Source: Department of Homeland Security, 2010. With permission.

- Man-made hazards
- Financial
- Technological hazards
- Public education
- Risk and hazards analysis
- Direction and control
- National Incident Management System
- Community Emergency Response Teams
- Public–private partnerships
- Hazard-specific sections
- Evacuation
- Multiagency coordination
- Drills and exercises
- Training

As can be seen, there are a variety of different formats that can be used, as well as sections, to make up an Emergency Management program strategy. There is no "silver bullet" or "one-size-fits-all" strategy format. The Emergency Manager must determine which format best fits their organization, and proceed with the development of their strategy document.

Once the Emergency Manager decides on a format, it is important to lay out the structure within the overall format. Figure 4.4 shows an example of a structure that

Chapter Title (Choose Phase of Emergency Management)

I. Definitions
II. Past and Current Department Activities
III. Summary of Overall Goals for This Chapter
  a. Goal 1
  b. Goal 2
  c. Goal 3
IV. Individual Goal #1 (taken in order from goals in Section III above)
  a. Initiative 1
  b. Initiative 2
  c. Initiative 3
V. Individual Goal #2
  a. Initiative 1
  b. Initiative 2
  c. Initiative 3
VI. Individual Goal #3
  a. Initiative 1
  b. Initiative 2
  c. Initiative 3
VII. Summary

Figure 4.4 Four phases format for emergency management program strategies. (From Stovall, S., City of Plano, Texas. City of Plano Department of Emergency Management Strategic Plan 2011–2016, p. 3, Plano: 2011, Print. With permission.)

can be used when laying out the content of the overall strategy. This example can be used in each section under the strategy format to provide for a consistent layout that will make the strategy easier to read and follow.

Once the format and structure of the Emergency Management program strategy document are complete, the Emergency Manager must begin working with the planning team to fill out the format and structure that have been set forth. This is done with the development of goals and initiatives to meet these goals.

Developing Strategic Goals

Before continuing with the development of the agency strategy, the Emergency Manager must understand how to develop a goal. This is one of the most misunderstood elements when building a strategy or a business plan. Many do not develop their goals correctly. As a result, the goals are misunderstood and many times never reached. A "goal" is defined as follows:

> *Merriam-Webster's Dictionary:* "2. : the end toward which effort is directed : (Aim)..."*

* "goal." Merriam-Webster Online Dictionary. 2010. Merriam-Webster Online. 14 December 2010 http://www.merriam-webster.com/dictionary/goal.

> *BusinessDictionary.com:* "... a goal is an observable and measurable end result having one or more objectives to be achieved within a more or less fixed timeframe."*

As can be seen, a "goal" is considered an end result of some sort of effort or action, and typically is completed within a particular timeframe. Goals that are contained in the Emergency Management program strategy should include both short-term and long-term goals. The strategy is a living document, and the goals will drive the future of the Emergency Management agency.

As an Emergency Manager and their planning team begin constructing the goals for the strategy, there are some things to consider in order to ensure that the goal is written effectively. One of the best tools in developing goals and initiatives is to use the SMART acronym.

Specific—A specific goal is much easier to understand than a general or a vague goal. To set a specific goal, it is important to include answers to as many of the following questions as possible:

- Who is to achieve this goal (can be individuals or a team)?
- What needs to be accomplished?
- Where is this goal to be reached or accomplished?
- Why does this goal need to be accomplished?
- Which way is the goal to be attained, taking into account legal and other requirements?
- When should the goal be completed by (this is also covered under the "T" part of the SMART acronym)?

Measurable—Establish milestones for accomplishing the goal. Determine how the goal can be measured quantitatively so that successful completion can be measured.

Attainable—The planning team has to agree that the goal is attainable. Sometimes it is a good exercise to "reverse engineer," or start with the result and work backward to examine what personnel, funding, and other resources will be needed to attain your goal.

Realistic—Be realistic. Just as with being "attainable," a "realistic" goal is also critical (although many may see how these go hand in hand). A goal that is too lofty or that is not realistic or attainable is not worth developing. Unmet goals can lead to lower morale and unclear direction.

Timely—Time constraints or definitions are probably some of the most difficult attributes of a goal to establish. Emergency Management professionals are often inundated with projects that are influenced by entities external to their departments (i.e., grant deadlines, plan reviews, budgets). However, when

* "goal." BusinessDictionary.com. 2010. Web Finance Inc. http://www.businessdictionary.com/definition/goal.html. Accessed 6 January 2010.

possible, it is necessary to tie a timeframe to a goal. Without a timeframe, the goal loses its sense of urgency and may or may not ever be accomplished. Even if it is necessary to move a self-imposed deadline due to unexpected increases in workload or other factors, it is still important to add a time element to goals when possible.

Examples of goals that do not institute the SMART methodology include

- Complete EOP
- Ensure readiness of the Emergency Operations Center
- Establish pre-event contract for debris management.

Taking these same non-SMART goals, here are examples of SMART goals:

- Establish a planning team to complete revisions to the EOP no later than July 1 (can add year).
- Develop a schedule for Operations Chief to test and complete operational readiness checks before the beginning of hurricane season each year.
- The director will work with Purchasing personnel to develop specifications and bid package for debris management contractor no later than April 10 (can add year).

Using the SMART methodology helps the Emergency Management strategy planning team by forcing a thought process that makes the goals clearer, manageable, and attainable. SMART goals will also be clearer to whoever reads the strategy document, and will help to develop a clearer direction for the Emergency Management agency.

Constructing Strategic Initiatives

Once the Emergency Management program strategy goals are established, there needs to be initiatives to help meet those goals. In some situations, these initiatives may be called "objectives." However, the word "initiative" implies that an action will be taken. The word "initiative" indicates that a leading action will be taken to meet an end result. For this purpose, we will use the term "initiative" to define those tasks that will help the Emergency Manager and their staff meet a goal.

Initiatives, much like goals, can follow the SMART acronym for their development. Developing initiatives that are specific, measurable, attainable, realistic, and timely can help to clearly outline tasks that are designed to meet an overall goal in the Emergency Management program. When constructing the initiatives, the planning team must think about action words. After all, the tasks required to meet a goal always require some sort of action. The following list suggests some of the action verbs that can be used:

- Develop
- Construct
- Build
- Complete
- Update
- Enhance
- Review
- Change
- Exercise
- Evaluate
- Assess
- Perform
- Justify
- Propose
- Compose
- Generate
- Operate
- Check
- Speak
- Execute
- Identify
- Evaluate
- Monitor
- Test
- Modify
- Mobilize
- Classify
- Support
- Negotiate
- Hire
- Identify
- Determine
- Operate
- Collect

The list of action verbs is extensive. However, they are necessary in order to show some sort of action taken toward achieving a particular goal. Figure 4.5 shows what a set of goals and initiatives can look like.

Defining Challenges

Following the development of goals and initiatives, it is important that the strategy planning team define any challenges that could arise when working toward

Strategic Goal 4

By May 2011, the Department of Emergency Management staff (all staff members) shall complete a review of and update the Comprehensive Emergency Management Plan as needed based on lessons learned and new requirements.

- Emergency Management staff, as assigned, shall continue identification of planning gaps and work with emergency management stakeholders to attain information to address gaps. This shall be completed no later than the end of February 2011.
- By the end of April 2011, Emergency Management staff, as assigned, will complete initial training agencies on CEMP and specific roles outlined in the 22 Emergency Support Functions (ESFs).
- By the end of March 2011, Emergency Management staff, as assigned, shall perform Tabletop Exercise (TTX) that tests out the new CEMP and make corrections based on After Action Review (AAR) and Improvement Plan.

Figure 4.5 Example of SMART goal and strategy. (From Stovall, S., City of Plano, Texas Department of Emergency Management Strategic Plan 2011–2016. With permission.)

completion of the goals and initiatives. Emergency Managers are faced with a litany of challenges that can include apathy, lack of funding, lack of personnel, external deadlines, politics, and having to deal with actual emergencies and disasters. Identification of these challenges allows the Emergency Manager to determine ways that these influences can be controlled (when possible). Once the challenges have been identified, it is important to make sure that the challenges are thoroughly explained. This is important so that readers can understand that there may be certain influences that can affect the timely completion of goals and/or initiatives.

Capabilities and Future Needs

In this section, the Emergency Manager has the ability to describe what their agency's capabilities are in terms of completing the listed initiatives and reaching the strategic goals. Areas that can be covered in this section include budget, facilities, equipment, personnel, and any other area that influences the capabilities of the Emergency Management agency. The Emergency Manager should point out any shortfalls that may exist. Although the Emergency Manager may be hesitant to point out any shortfalls, it is important to indicate them, as they may be obstacles in the completion of initiatives and goals. There is not one Emergency Management agency in existence that has all of their administrative and operational needs fulfilled.

Once shortfalls are identified, it is important that the Emergency Managers describe how they plan to fill any gaps. Solutions to filling gaps can include budget supplements, personnel hiring, acquisition of resources, attainment of grants, further coordination, and development of plans (just to name a few). Once this process is complete, the Emergency Manager should ensure that these action items are included in the Goals and Initiatives set forth in the Emergency Management program strategy.

Summary/Conclusion

This section is the wrap-up section where the Emergency Manager can finalize the strategy and make any last statements that they want the reader to know. The Emergency Manager may also want to summarize all strategic goals that have been outlined in the strategy (sans initiatives). The Emergency Manager needs to realize that this section will be the place in the strategy that should leave a final impression about the Emergency Management agency, its current status, and what direction it will be taking in the future.

This chapter has been designed to give Emergency Managers and their planning team steps that can be taken when developing an Emergency Management program strategy. As mentioned previously, a departmental or organizational strategy should be treated much like a business plan for the agency. This strategy should remain a living document that is reviewed and refined over time. Without a solid strategy document, it is difficult for the Emergency Manager to explain what their agency does, and the vision for their agency's future. The strategy document can be a tool used to develop task lists for staff, and to build performance measures for employee evaluation. The preceding steps are not part of an "exact science." As discussed, there are many different formats and structures for Emergency Management program strategies. However, the intent was to help Emergency Manager in examining their program and developing a roadmap of sorts for their agency's successful future.

Chapter 5

The Hazards among Us

S. Shane Stovall

Contents

Introduction

In order for a community to be able to understand the risks and vulnerabilities for which it is to plan, it is critical that the person charged with emergency management duties perform a hazard vulnerability analysis. To complete this analysis, it is necessary for the Emergency Manager to understand the different hazards and vulnerabilities for the community that they serve. Vulnerabilities can include special populations, critical facilities, and environmental concerns. To clarify what a hazard vulnerability analysis is, it is important that the Emergency Manager understands a few definitions. This includes

> **Hazard**—An occurrence, whether accidental or intentional, that can threaten or actually cause damage or destruction to lives or property. Hazards can be natural, man-made, or technological.
> **Vulnerability**—A site or a community's exposure to loss from a hazardous incident. This can be influenced by demographics, economic considerations, locations of political significance, geologic factors, or other environmental concerns.
> **Risk**—Probability, or likelihood, of a hazardous occurrence.

In this chapter, we will look at a multitude of hazards that could potentially affect a community. This listing is not entirely all-inclusive due to the ever-changing variables associated with hazards. This is also not an attempt to state that these hazards will all occur in one community. It is the responsibility of the Emergency Manager to complete a hazard vulnerability analysis to determine the hazard scenarios for which the community should plan, train, and exercise.

Hazards

Hazards come in many different forms. For the purposes of this chapter, hazards will be divided into three categories: Natural, Man-Made, and Technological. It is important that the Emergency Manager become familiar with each of these hazards, and understand the implications of each of these hazards on their community, as well as other communities. Disasters caused by these hazards do not know jurisdictional boundaries. Therefore, it must be understood that the effect(s) of a disaster in a neighboring jurisdiction, or even within the region, could have long-reaching effects into many communities.

Natural Hazards

Flooding

Description and History

Floods cause more deaths each year than any other hazard produced from severe weather events. Although the number of fatalities can vary dramatically with weather conditions from year to year, the national 30-year average (1977–2006) for flood deaths is 99. That compares with a 30-year average of 58 deaths for lightning, 54 for tornadoes, and 49 for hurricanes.[1] There are different types of flooding that can occur. It is important to discern between the different types of flooding for emergency planning purposes. The different types of flooding that can occur include

Riverine Flooding—Typically caused by overbanking of water in rivers, creeks, or streams. This is the most common type of flood event. Typically, these types of floods are a result of large-scale weather systems that generate prolonged rainfall over wide geographic areas. This type of flooding can also be caused by upstream snow melts, alluvial fan floods, and ice jam floods.

In 1993, the Midwestern United States experienced extensive rainfall for four months. This rainfall caused the Mississippi River and many of its tributaries to flood. Flooding was experienced in North Dakota, South Dakota, Nebraska, Kansas, Minnesota, Iowa, Missouri, Wisconsin, and Illinois. In this area, hundreds of levees that were built to protect agricultural interests and towns were breached due to the excessive amounts of rainfall. Damages for this event were estimated at approximately $15 billion. Approximately 10,000 homes were destroyed, and 50 people lost their lives in this event.[2]

Alluvial Fan Flooding—Alluvial fans are fan-shaped deposits where a fast moving stream flattens out and spreads over a flatter plain. Typically, this is seen from a stream moving from a mountainside into a canyon or a plain. Risks associated with these floods include fast, widespread flooding since there are no banks to control the stream flow. Also, due to the velocity of the water, large debris can be carried in the floodwaters.

In December 1999, debris flows and flash floods on alluvial fans inundated coastal communities in Cordillera de la Costa and Vargas, Venezuela. These floods caused severe property destruction, and resulted in a death toll estimated at 19,000 people. Because most of the coastal zone in Vargas consists of steep mountain fronts that rise abruptly from the Caribbean Sea, the alluvial fans are the only areas that have slopes that can be built upon.[3]

Ice Jam Flooding—Caused typically when a frozen river or stream begins to melt and break up. The masses of ice then block normal downstream drainage paths, forming an ice dam and causing a backup of water. Secondarily, once the ice dams break, the velocity of the water rushing downstream can cause flash flooding. Ice jams on the Yukon River in Alaska contributed to severe flooding during the spring breakup of 1992.

Flash Flooding—Can occur within seconds, or may occur over a period of a few hours. These floods typically cause a rapid rise in water flow and velocity. Several factors contribute to flash flooding. These include surface conditions, rainfall duration, rainfall intensity, and topography.

On June 14, 1990, a storm system dropped 4 inches of rain in less than 2 hours. This produced a 30-foot-high wall of water and killed 26 people. Damage estimates range anywhere from $6 to $8 million.[4]

Dam and Levee Failure Flooding—Dams and levees are built to hold back a computed amount of water. However, if the amount of water behind the dam or levee exceeds the computed amount that the dam can hold, overtopping occurs. Another type of failure can occur in these systems when the dam or levee structure fails either due to it being washed out or possibly a mechanical or maintenance failure. The amount of water released can cause catastrophic flash floods due to the high velocity of the moving water.

One of the most historic disasters involving a dam break occurred on May 31, 1889. The dam broke, resulting in a 36-foot to 40-foot wall of water washing through the town of Johnstown, Pennsylvania (Figure 5.1). A total of 2200 people lost their lives in this event, making it the worst flood in the history of the United States.[4]

Storm Surge Flooding—Storm surge is water pushed up onto dry land by onshore winds. This phenomenon is primarily caused by tropical storms, hurricanes, and other intense low-pressure systems. Storm surge is the most dangerous and most destructive part of most hurricanes. The waves and the velocity of the storms surge often causes great destruction and is known to be the highest cause of deaths during hurricanes. At times, storm surge can come during high tide, which causes a storm tide. The height of the waves associated with this type of flooding can reach more than 25 feet in worst-case scenarios.

Hurricane Katrina in August 2005 was one of the most devastating hurricanes in the history of the United States. It produced storm surge flooding of 25 feet to 28 feet above normal tide level along portions of the Mississippi coast and storm surge flooding of 10 feet to 20 feet above normal tide levels along the southeastern Louisiana coast. Hurricane Katrina produced

Figure 5.1 Damage from Johnstown Flood, May 1, 1889. (National Park Service.)

catastrophic damage—estimated at $75 billion in the New Orleans area and along the Mississippi coast. The death toll is estimated at 1200, including about 1000 in Louisiana and 200 in Mississippi. Seven additional deaths occurred in southern Florida.[5]

Debris Flooding—These types of floods are caused by an accumulation of debris, such as rocks, sediment, logs, or other debris in a stream or creek channel. The accumulated debris causes a temporary dam, and as water builds up behind the temporary dam, the dam gives way and the release of water becomes a flash flood.

Mudflow Flooding—Also known as lahars, these occur when volcanic activity melts snow and glaciers on the mountain. The result is a large amount of water mixed with debris and mud that rapidly moves down the mountain slope.

This type of flooding occurred on May 18, 1980, with the eruption of Mount St. Helens. The eruption melted snow and ice atop Mount St. Helens, which in turn caused large amounts of mud to stream down the side of the mountain. This contributed to the 57 lives that were lost.

Consequences of Flooding

Mass amounts of water from flood events can cause extensive damage throughout a community (Figure 5.2). Like many disasters, flooding can cause many different types of secondary disasters that must be considered and planned for. These include power outages from destroyed electrical grid infrastructure, road washouts, sewer infiltration, septic system destruction, erosion, mass evacuations of people and animals (leading to temporary housing issues), crop damage, contamination of water supplies, public and private property damage, large amounts of debris, contamination of homes due to mold and other contaminants, and mass casualties.

Figure 5.2 Flooding from the Tar River in Greenville, North Carolina (1999). (Dave Gately, Federal Emergency Management Agency.)

Preparation and Planning

Emergency managers must know the different watches and warnings issued for flooding. These will assist in prompting actions needed to protect lives and the public before, during, and after a flood event.

Flood Watch—Flooding is possible. Tune in to the National Oceanic and Atmospheric Administration (NOAA) Weather Radio, commercial radio, or television for information.

Flash Flood Watch—Flash flooding is possible. Be prepared to move to higher ground; listen to the NOAA Weather Radio, commercial radio, or television for information.

Flood Warning—Flooding is occurring or will occur soon; if advised to evacuate, do so immediately.

Flash Flood Warning—A flash flood is occurring; seek higher ground on foot immediately.

Another consideration that Emergency Managers must look into is getting their communities involved in the National Flood Insurance Program (NFIP). The NFIP is administered by the Flood Insurance and Mitigation Administration, which is a component of the Federal Emergency management Agency (FEMA). The NFIP is a voluntary program that encourages communities to adopt and enforce floodplain management ordinances to reduce future flood damage. In exchange for participating in the program, the NFIP makes federally backed flood insurance available to businesses, homeowners, and renters in the Community. *Damage due to flooding is not covered in homeowners' policies. Flood insurance is a separate policy.*

Communities that are involved in the NFIP can also enroll in the Community Rating System (CRS). The CRS is a voluntary incentive program that promotes and recognizes community floodplain management and flood mitigation activities that exceed the minimum NFIP standards. Communities exceeding the minimum NFIP regulations can receive points that can equate out to reduced flood insurance rates for the community. More information on the CRS can be found at http://www.fema.gov/business/nfip/crs.shtm, or you can find the information via search engine on the Internet.

Lightning

Description and History

Lightning is another one of the deadliest natural hazards in the United States. Although lightning and the associated thunderstorms are most prevalent during the summer months, they can strike year round. During the past 30 years, an average of 58 people have been killed annually by lightning strikes.[6] Hundreds more are injured. Although this natural hazard is more likely to affect a local area, it still must be planned for. Special events such as concerts, fairs, carnivals, and other outdoor recreational and sports activities are especially vulnerable to lightning strikes during thunderstorms. Thunderstorms do not have to be in the immediate area for lightning to strike. Oftentimes, lightning strikes well away from the actual storm. Lightning has also been witnessed in large forest fires, nuclear detonations, heavy snowstorms, and volcanic eruptions.

Notable lightning strikes that have occurred worldwide include a strike on November 2, 1994 in Dronka, Egypt. This strike hit an aviation fuel storage facility and ignited three of the eight 5000-gallon fuel tanks. The resulting explosion sent blazing fuel into the village of Dronka, where a majority of the 469 fatalities from this incident occurred. Airplanes are also vulnerable to lightning strikes. On December 8, 1963, Pan Am Flight 214 crashed in Elkton, Maryland, after it was struck by lightning while in a holding pattern. All 81 people on board were killed. Modern protection systems now protect aircraft from lightning strikes.

Consequences of Lightning

Lightning strikes can cause severe damages in a localized setting. From time to time, it is possible for lightning to trigger a secondary event. Such secondary events include power outages, structure fires, wildfires, mass casualties, and infrastructure damage.

Preparation and Planning

Lightning strikes are localized events. Primary planning should involve monitoring watches and warnings from the National Weather Service. Although lightning watches and warnings are not specifically issued, they can be imbedded in the text of severe weather watches and warnings. If thunderstorms are forecast, then lightning should be planned for.

Emergency managers can also coordinate with the organizers of outdoor events to ensure that lightning detection equipment is available. In some cases, it may also be possible to utilize equipment that will divert lightning away from populated parts of the event, such as lightning rods. This equipment should *never* be a substitute for good emergency planning. The equipment is only a tool to protect just in case a storm rapidly approached. All special events should have a way to monitor the weather and have a plan for delaying activities and evacuating attendees and participants to a safe shelter.

Hurricanes and Tropical Storms

Description and History

Each year, an average of 11 tropical storms develop over the Atlantic Ocean, Caribbean Sea, and Gulf of Mexico. Many of these remain over the ocean and never impact the U.S. coastline. Six of these storms become hurricanes each year. In an average three-year period, roughly five hurricanes strike the U.S. coastline, killing approximately 50 to 100 people anywhere from Texas to Maine. Of these, two are typically "major" or "intense" hurricanes (a Category 3 or higher storm on the Saffir–Simpson Hurricane Scale).[7]

Hurricanes and tropical storms are categorized as tropical cyclones. Tropical cyclones are generally low pressure systems that, for the most part, form in the tropics. The tropics are defined as the region along the equator with the Tropic of Cancer in the Northern Hemisphere and the Tropic of Capricorn in the Southern Hemisphere. On occasion, tropical cyclones can form outside of this area. Due to the Earth's rotation, tropical cyclones in the Northern Hemisphere rotate counterclockwise. Tropical cyclones in the Southern Hemisphere rotate clockwise. Several elements must come together to form a tropical cyclone. These include preexisting weather disturbances (such as thunderstorms), moisture, warm ocean waters (at least 80°F), and winds in the upper atmosphere. If these conditions combine and remain consistent and undisturbed, a tropical cyclone can form. These storms can cause widespread damages and loss of life with their high winds, storm surge, torrential rains, and potential tornadoes. Although these storms directly impact coastal areas, effects from tropical cyclones can be felt for hundreds of miles inland.

The Atlantic Hurricane Season runs from June 1 to November 30 each year. The Pacific Basin Hurricane season runs from May 15 to November 30 each year. Emergency management professionals should be familiar with the following definitions:

> Tropical Depression—An organized tropical system, with a well-defined center of circulation that has sustained winds of 38 miles per hour or less.
> Tropical Storm—An organized tropical system, with a well-defined center of circulation that has sustained winds of 39–74 miles per hour.
> Hurricane—An organized tropical system, with a well-defined center of circulation that has sustained winds of 75 miles per hour or higher.

Hurricanes are categorized by the Saffir–Simpson Hurricane Scale, which categorizes hurricanes from Category 1 (lowest strength) to Category 5 (highest strength). Table 5.1 shows the Saffir–Simpson Scale.

Notable hurricanes include the 1900 Hurricane (the United States did not name storms until 1951), which struck Galveston, Texas. This was a Category 4 hurricane. Estimates for this storm includes 8000 deaths, with some estimates being as high as 12,000.[8] The 1900 Hurricane brought a storm surge that inundated most of Galveston Island and the city of Galveston. More recently, in 2005, Hurricane Katrina, a Category 3 hurricane, caused an estimated $80 billion in damages and caused an estimated 1500 deaths in Louisiana, Mississippi, and Alabama.

Table 5.1 Saffir–Simpson Scale

Saffir–Simpson Hurricane Scale			
Scale Number (Category)	*Sustained Winds (mph)*	*Damage*	*Storm Surge (ft)*
1	74–95	Minimal: Unanchored mobile homes, vegetation, and signs.	4–5
2	96–110	Moderate: All mobile homes, roofs, small crafts, flooding.	6–8
3	111–130	Extensive: Small buildings, low-lying roads cut off.	9–12
4	131–155	Extreme: Roofs destroyed, trees down, roads cut off, mobile homes destroyed. Beach homes flooded.	13–18
5	>155	Catastrophic: Most buildings destroyed. Vegetation destroyed. Major roads cut off. Homes flooded.	>18

Source: Saffir–Simpson Scale, Federal Emergency Management Agency.

Consequences of Tropical Storms and Hurricanes

Tropical storms and hurricanes can cause widespread damage over large areas. Size, strength, and forward speed of tropical systems contribute to what damages could occur. Initial effects of a tropical storm or hurricane can include flash flooding, storm surge inundation, structural damage to residences and businesses (leading to temporary housing and economic loss issues), infrastructure damage, and large amounts of debris. Some of the secondary effects of tropical cyclones include long-term power outages, mass evacuations of people and animals, mass casualties, infrastructure failure, agricultural damage or destruction, contamination of water supplies (through contaminated fresh water or salt water inundation), and hazardous materials incidents. As with other hazards, the cascading effects of disasters depend on multiple variables, so this list is by no means comprehensive.

Preparation and Planning

The Emergency Manager that works in a hurricane-prone area must know the different watches and warnings that are issued for tropical depressions, tropical storms, and hurricanes. These will assist in prompting actions needed to protect lives and the public before, during, and after tropical cyclone events.

Tropical Storm Watch—Issued when tropical storm conditions (sustained winds of 39 mph to 73 mph) are *possible* within the specified coastal area within 48 hours.
Tropical Storm Warning—Issued when tropical storm conditions (sustained winds of 39 mph to 73 mph) are *expected* somewhere within the specified coastal area within 36 hours.
Hurricane Watch—Issued when hurricane conditions (sustained winds of 74 mph or higher) are *possible* within the specified coastal area. Because hurricane preparedness activities become difficult once winds reach tropical storm force, the hurricane watch is issued 48 hours in advance of the anticipated onset of tropical-storm-force winds.
Hurricane Warning—An announcement that hurricane conditions (sustained winds of 74 mph or higher) are *expected* somewhere within the specified coastal area. Because hurricane preparedness activities become difficult once winds reach tropical storm force, the hurricane warning is issued 36 hours in advance of the anticipated onset of tropical-storm-force winds.

Unlike many hazards, hurricanes typically have a warning period that can last several days. However, it is imperative that the Emergency Manager not wait until there is a threat of a landfalling tropical storm or hurricane before preparing his or her organization and the public. Tropical storms and hurricanes have two primary facets (although not the only facets), both of which can contribute to extensive damage and loss of life:

- Storm Surge and Storm Tide

  Storm surge is the rise of water that results from the wind pushing water ashore. This surge of water can be much larger than the predicted astronomical tides. Storm tide is the combination of the astronomical tide with the storm surge (Figure 5.3). These phenomena are typically coupled with battering waves, which exacerbate damages onset by the storm surge or storm tide.

  There are several variables that influence the maximum potential storm surge that can be received at a particular location. Slight deviations in these variables can vastly change the amount of storm surge received onshore. These variables include storm intensity, storm size (radius of winds), forward speed, angle of approach to the coast, the shape of the coastal geography (bays and estuaries can cause a buildup of water), and the width and depth of the continental shelf (shallow areas in the continental shelf can cause quicker build up of storm surge).

  To determine a location's vulnerability, the Emergency Manager can refer to the Sea Lake and Overland Surges from Hurricanes (SLOSH) model. This

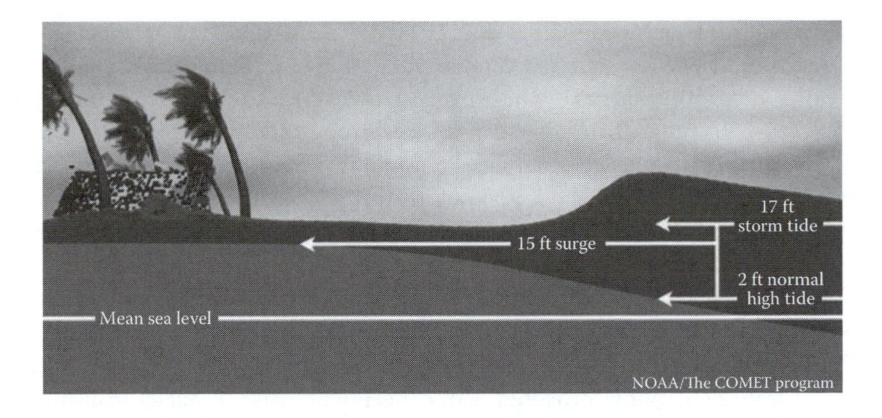

Figure 5.3 Storm surge versus storm tide. (National Oceanic and Atmospheric Administration.)

model gives the Emergency Manager an idea of what size tropical cyclone will be needed in order to inundate a particular location with storm surge. The SLOSH model uses the Maximum of Maximums or the Maximum Envelope of Water to illustrate the highest potential storm surge at a particular location within the SLOSH model (Figure 5.3).

- High Winds

  Strong sustained winds are what most people associate with tropical cyclones. These high winds can reach upward of more than 155 mph and turn loose objects into missiles. Tornadoes are also common with tropical cyclones. Although typically short-lived, they can contribute to the extensive damage caused by tropical storm-force and hurricane-force winds. Another element that is associated with tropical cyclones is the "eye," or the center of the rotating storm. Oftentimes, the weather in the eye is calm and can often mislead people into thinking that the tropical storm or hurricane may have passed. However, this is the time where people should remain in a protected shelter—away from additional high winds, torrential rain, and flooding that may occur. The highest winds on record from a hurricane are an estimated 190 miles per hour[9] with Hurricane Camille in 1969. It is estimated that the winds were higher, but the instruments taking the readings were damaged in the high winds.

Tornadoes

Description and History

Tornadoes are one of the most destructive natural hazards that are faced today. Tornadoes are defined as violently rotating columns of air that reach from a thunderstorm and make contact with the ground. Tornadoes occur worldwide, but the highest concentration of tornadoes occurs in the United States. On average each year, more than 800 tornadoes are reported nationwide, resulting in more than 80 deaths and 1500 injuries.[10] The strongest tornadoes can have rotating winds that exceed 250 mph. Their paths can be up to one mile wide and several miles long. These extreme tornadoes only make up about 2% of all reported tornadoes. However, they account for about 70% of all tornado deaths. Tornadoes are associated with severe thunderstorms, called supercells, which can occur across the United States. Most times tornadoes come with very little notice, although advances in Doppler radar technologies have the ability to recognize developing tornadoes before they form on the ground.

In February 2007, the National Weather Service Storm Prediction Center switched from classifying tornadoes using the Fujita Scale (F-Scale) to using the Enhanced Fujita Scale (EF Scale). The reasons for the switch was that the F-Scale did not take into account differences in construction of structures, the categories were only based on damage in worst-case location, and it was felt that the wind speeds were overestimated in F-3 tornadoes and stronger. Most of the time, the classification of a tornado is done after the fact, and after damages have been assessed. The Enhanced Fujita scale uses 28 damage factors that take into account building construction and object (tree, flagpole, etc.) type, and allows for more accurate estimation of winds (Table 5.2).

One tornado of historical record is the Oklahoma/Kansas Tornado Outbreak on May 3, 1999 (Figure 5.4). During a period of 21 hours, 74 tornadoes touched down across the two states. At one point, there were four tornadoes reported to be on the ground at the same time. The strongest tornado was an F-5 on the Fujita Tornado Scale (the EF scale had not yet been adopted). This F5 tornado tracked for

Table 5.2 Fujita versus Enhanced Fujita Scale

Fujita Scale			*Operational EF Scale*	
F Number	*Fastest 1/4 Mile (mph)*	*3-s Gust (mph)*	*EF Number*	*3-s Gust (mph)*
0	40–72	45–78	0	65–85
1	73–112	79–117	1	86–110
2	113–157	118–161	2	111–135
3	158–207	162–209	3	136–165
4	208–260	210–261	4	166–200
5	261–318	262–317	5	Over 200

Source: National Weather Service, Storm Prediction Center.

Figure 5.4 F3 tornado in Oklahoma, May 3, 1999. (National Oceanic and Atmospheric Administration/National Severe Storms Laboratory.)

nearly an hour and a half along a 38-mile path. The death toll was 46, with 800 injured, and more than 8000 homes destroyed. Property damage was estimated at nearly $1.5 billion.[11]

Consequences of Tornadoes

The damages caused by a tornado are dependent on several factors including the size of the tornado and the structural integrity of the object being impacted. An EF-0 tornado can cause damage to mobile homes, carports, and canopies, while not causing much (if any) to a reinforced brick structure. Also, it must be noted that tornadoes can be preceded by hail, which can cause damage as well. Some of the secondary effects of tornadoes can include infrastructure damage, mass casualties, damage or destruction of businesses and residences (leading to economic recovery and temporary housing needs), and hazardous material incidents.

Preparation and Planning

The Emergency Manager that works in a location prone to tornadoes must know the different definitions for tornadoes, and their associated watches and warnings. These will assist in prompting actions needed to protect lives and the public before, during, and after tornadic event.

> Funnel Cloud—This is a sign of a developing tornado, and consists of a violent rotating column of air that *does not* touch the ground. People commonly mistakenly call these tornadoes.
> Waterspout—There are wo types: (1) Nontornadic, which are not associated with supercell thunderstorms, are the most common type, and usually rate no higher than an EF-0 on the Enhance Fujita Scale. (2) Tornadic are tornadoes over water and usually form in connection with severe thunderstorms.

Tornado—Violently rotating columns of air that reach from a thunderstorm and make contact with the ground.
Tornado Watch—Conditions are favorable for the formation of tornadoes. No tornado has been sighted or detected at this point.
Tornado Warning—A tornado has been actually sighted or is being detected on radar covering your area. Take shelter immediately in a safe and sturdy structure.

Tornadoes often form without much notice. However, there are a couple of tools that the Emergency Manager can use that could give an idea if there will be a threat of thunderstorms that could produce tornadoes. The National Weather Service provides a Hazardous Weather Outlook product that can provide information on weather that may threaten the community. Another product that can be used, which is issued by the Storm Prediction Center, is the Mesoscale Discussion. This Discussion is typically issued when atmospheric conditions look like they may produce severe weather. The Storm Prediction Center often issues their Mesoscale Discussion anywhere from half an hour to several hours before issuing a weather watch. As with any hazard, it is critical that the Emergency Manager take actions to prepare their organization, as well as the public, well before a hazardous incident occurs.

Wildfires

Description and History

Wildfires can be caused naturally (i.e., lightning) as well as by man (i.e., accidental or arson). Wildfires are considered to be any uncontrolled fire in combustible vegetation that occurs in a forest, countryside, or other open wilderness area (Figure 5.5). Wildfires are often given other names such as grass fire, brush fire, bush fire, forest fire, hill fire, muck fire, peat fire, and wildland fire (typically associated with

Figure 5.5 2003 Cedar Fire. (San Diego Fire-Rescue Department.)

the type of vegetation being burned). Wildfires are typically large, can spread at rapid speeds, and can change direction unexpectedly. Wildfires are also known to jump fire breaks, roads, and rivers. Some wildfires even get large enough to create their own weather, which can further exacerbate the spread of the wildfire.

In 1991, the Oakland Hills Fire in Oakland, California, burned 1800 acres, killed 25 people, injured 150 people, and destroyed 3810 living units. The economic loss from this fire has been estimated at over $1 billion.[12] This fire was a reignition of a grass fire that was not extinguished properly in the Berkeley Hills area. This fire quickly spread with the aid of wind gusts of up to 65 mph. In October 2003, the Cedar Fire in California burned 280,278 acres, killed 15 people, and burned 2820 structures.[13]

Consequences of Wildfires

Wildfires can have devastating effects, which include property damage and destruction to residential and business structures (leading to temporary housing and economic recovery issues), damage to infrastructure, mass casualties, and air quality issues. Occasionally, wildfires burn enough trees and other vegetation to cause erosion problems, which can lead to flooding and landslide hazards. Frequent wildfires can also upset or destroy the natural ecosystems that are in place. This can lead to displaced wildlife and long-term economic damage to industries such as timber, pulp, and tourism.

Preparation and Planning

Man-made wildfires can be prevented. Building awareness of wildfire dangers and wildfire prevention measures can help to possibly curb accidentally set wildfires. Although these types of fires can typically be prevented, naturally occurring wildfires cannot (such as those caused by lightning, which can strike in areas inaccessible to firefighting equipment). Along with public outreach to prevent wildfires, Emergency Managers must also look at ways that they can control or mitigate wildfires. There are a few methods for this:

> Controlled Burning—This involves the burning of vegetation in order to provide a shortage of fuel that can be burned should a wildfire ignite. These types of fires involve a set parcel of land, and should only be set by trained professionals who have been permitted to do so. Many jurisdictions require a permit in order to have a controlled burn. These are intended to be set under the right weather conditions so that unintentional spreading cannot occur. This can be very risky if not done under the correct circumstances.
>
> Fire Breaks—Clearing areas of vegetation can also reduce the amount of fuel that a wildfire has to burn. This can help to slow down a wildfire so that it can be controlled, but must be accessible to fire control equipment. Examples

of fire breaks include unimproved roads and fields. Natural fire breaks can include rivers, streams, and roads.

Public Outreach—*Encouraging* homeowners to develop a defensible space of 30 feet around their home, cleaning out gutters, using fire resistant shingles and siding, and planting fire resistant trees and shrubs are all ways that residents can protect their homes. Using residents as partners in your preparedness and mitigation efforts is good business practice for Emergency Management.

Earthquakes

Description and History

The Earth's crust is made up of many sections, called tectonic plates. These plates are consistently in motion due to the convective currents of the Earth, which is caused by the heating cycles in the planet's mantle (the liquid layer beneath the Earth's crust). At the deepest parts of the mantle, the temperatures get extremely hot due to the proximity to the Earth's core. As these parts of the mantle get hot, they rise through the Earth's mantle, where they cool and begin to sink toward the Earth's core again. This cycle causes convective currents. These currents are what cause the tectonic plates to move.

There are three different types of movements that occur with tectonic plates. The energy released as these plates move is what triggers earthquakes.

Divergent Plate Movement (or Divergent Fault Lines)—Any earthquake from this type of movement occurs when tectonic plates separate and drift apart. In most cases, magma will rise up and fill the void created by the separation. In some cases, a plate may slide downward to fill the gap. These types of plate movements typically occur in the ocean. One prime example of this type of plate movement is the Mid-Atlantic Ridge in the Atlantic Ocean. This divergent fault line reaches through Iceland, and is thought to eventually cause the country to split into at least two pieces.

Convergent Plate Movement (or Convergent Fault Line)—An earthquake occurring from this type of movement occurs when tectonic plates collide. During such a collision, a subduction zone can form, where the older plate begins to slide under a younger plate, forming a trench. These trenches can be many miles wide. The best example of convergent plate movement is on the Indonesian Island of Sumatra with the Great Sumatran fault. This fault last caused a large earthquake in December 2004 and caused devastation along its northern end.

Transforming Plate Movement (or Transform Fault Line)—An earthquake occurs at one of these fault lines when energy is released from two tectonic plates rubbing past one another. The San Andreas Fault in California is an

> example of this type of fault. At its deepest point, the San Andreas Fault is 10 miles below the surface of the Earth. This is the primary fault of a series of fault lines that stretch along the California Coast. This discounts the myth of California one day "falling off into the Pacific Ocean," because it is not a divergent plate fault (even then it allows for the question of this myth).

Once the movement of these plates takes place, energy is released causing seismic waves. Earthquakes are measured both in magnitude and intensity. For these purposes, there are two scales used. Magnitude reflects the amount of energy released at the hypocenter, or focus, of the earthquake. This measurement is determined by the amplitude of the earthquake energy wave (or seismic wave) and is measured on a seismograph. Magnitude is reflected on the Richter Scale and does not reflect damage. Intensity is measured based on observed effects on people, buildings, and natural features. Intensity may vary from place to place in relation to the observer's location to the epicenter—the point on the Earth directly above the focus, hypocenter. Intensity is measured on the Modified Mercalli Intensity Scale (see Table 5.3).

In the early morning hours of January 17, 1994, an earthquake struck Los Angeles, California. The magnitude for this earthquake was measured at 6.7, according to the United States Geological Survey. The quake was felt for 2000 square miles in Los Angeles, Orange, and Ventura counties. There were nearly 15,000 aftershocks following the main earthquake. The earthquake killed 57 people and injured nearly 12,000 people. The damage was extensive, damaging about 100,000 houses and businesses. Parking garages collapsed and some apartment buildings were reduced to rubble. The earthquake caused more than $40 billion in damage (Figure 5.6).[14]

Consequences of Earthquakes

Earthquakes can cause a myriad of secondary disasters and incidents. These include infrastructure damage or destruction, fires, hazardous material incidents, mass casualties, loss of housing and businesses, contaminated water supplies, lack of water for fire suppression (due to damaged infrastructure), and transportation accidents (including mass transit). Many of the damages associated with earthquakes correlate with the intensity of the earthquake. Smaller earthquakes may not cause any noticeable issues, whereas large earthquakes can cause total destruction.

Preparation and Planning

As mentioned with previous hazards, it is imperative that the Emergency Managers work to prepare their organization, as well as the public, well before an earthquake

Table 5.3 Modified Mercalli and Richter Scales

Modified Mercalli Scale		Magnitude Scale
I	Detected only by sensitive instruments	1.5
II	Felt by few persons at rest, especially on upper floors; delicately suspended objects may swing	2
III	Felt noticeably indoors, but not always recognized as earthquake; standing autos rock slightly, vibrations like passing truck	2.5
IV	Felt indoors by many, outdoors by few, at night some awaken; dishes, windows, doors disturbed; standing autos rock noticeably	3
V	Felt by most people; some breakage of dishes, windows, and plaster; disturbance of tall objects	3.5 4
VI	Felt by all, many frightened and run outdoors; falling plaster and chimneys, damage small	4.5
VII	Everybody runs outdoors; damage to buildings varies depending on quality of construction; noticed by drivers of autos	5
VIII	Panel walls thrown out of frames; walls, monuments, chimneys fall; sand and mud ejected; drivers of autos disturbed	5.5
IX	Buildings shifted off foundations, cracked, thrown out of plumb; ground cracked; underground pipes broken	6 6.5
X	Most masonry and frame structures destroyed; ground cracked, rails bent, landslides	7
XI	Few structures remain standing; bridges destroyed, fissures in ground, pipes broken, landslides, rails bent	7.5
XII	Damage total; waves seen on ground surface, lines of sight and level distorted, objects thrown up into air	8

Source: Modified from Steeples, D.W., Earthquakes: Kansas Geological Survey pamphlet, 1978.

Figure 5.6 Collapse of bridge following Northridge earthquake, January 1994. (California Emergency Management Agency.)

hits. Because earthquakes give very little advance warning, preparations should be made as soon as possible. These include encouraging residents and employees of businesses to build emergency kits that will sustain them for a minimum of 72 hours. These kits should include utility shut-off tools, nonperishable food, water, personal hygiene items, medications, cash, important papers, a battery-operated radio, supplies for pets, and supplies for infants and children (just for starters). Emergency Managers can also work with building officials to ensure that structures are built to be earthquake-resistant.

Winter Storms

Description and History

Winter storms affect regions across the United States each year. These storms cause mass amounts of snow, winds, coinciding blizzards, ice, avalanches, below freezing temperatures, and hypothermic conditions for those outdoors, those without heating abilities, children, and pets. Each year, dozens of people die due to exposure to winter weather conditions. This does not include the number of people who die from heater-related fires, fatalities from vehicle accidents, and deaths from carbon monoxide poisoning due to unmaintained or misused heating systems. All of these factors make winter storms a significant threat to the public. Severe weather storms can last for many days, weeks, and even months. These storms can also trap people in their homes or cars, with no utilities or assistance available to them.

Emergency Managers must know the definitions associated with winter storms in order to be able to prepare and respond to these events. These definitions include

> Blizzard—Winds of 35 miles per hour or more with snow and blowing snow, reducing visibility to less than ¼ mile for more than 3 hours.

Snow Squalls—Brief, intense snow showers accompanied by strong winds. Accumulation could be significant.
Snow Showers—Snow falling at various rates for brief periods of time. Some accumulation is possible.
Ice Jams—Long cold spells can cause freezing of lakes, rivers, and creeks. A rise in water level or a thaw can lead to the breaking up of the ice into chunks, which can act as dams when lodged up against man-made structures such as weirs or bridge pilings. This can lead to significant flooding.
Snow Melt—The sudden thawing of a heavy snow pack that often leads to flooding.
Wind Chill—How wind and cold temperatures feel on exposed skin. As the wind increases, heat is carried away from the body at a faster rate, causing the body temperature to drop.

Historical winter storms include what's become known as the "Storm of the Century" on March 15, 1993. The death toll from this winter storm is approximately 270, with 48 people lost at sea. Because of the widespread nature of this storm, the estimations of damages and casualties were difficult. Thousands of people were left isolated by record snowfalls, especially in the Georgia, North Carolina, and Virginia mountains. In general, all interstates from Atlanta northward had to be closed. Hundreds of roofs collapsed because of the weight of the wet snow. More than three million people were without power at one time. At least 18 homes on Long Island fell into the sea due to pounding surf eroding away the shore. Florida was struck by at least 15 tornadoes, and 44 deaths were attributed to the tornadoes and other severe weather. About 110 miles south of Cape Sable Island, Nova Scotia, a 177-meter ship sank in heavy seas, with all 33 of its crew lost at sea. Sixty-five-foot waves were reported in the area. The highest winds recorded for this storm included a 144 miles per hour reading at Mount Washington, New Hampshire, and 109 miles per hour in the Dry Tortugas. Snowfall accumulations included 56 inches on Mount LeConte, Tennessee, and 50 inches on Mount Mitchell, North Carolina (with 14-foot drifts). Some damage and cost estimates for this storm exceed $6 billion.[15]

Consequences of Winter Storms

As discussed in the preceding subsection, winter storms can have a myriad of effects on an area. These effects include flooding, storm surge, blocked roads, damaged homes and businesses (from roof collapses), blocked roads, homes isolated from assistance, power outages, agricultural losses, debris issues (primarily from downed trees and limbs), damaged infrastructure, structural fires, transportation accidents, and mass casualties due to exposure to the cold and carbon monoxide poisoning (Figure 5.7).

Figure 5.7 Downed tree and snow-covered automobile attest to the magnitude of snowfall in the Asheville, North Carolina, area. Photo taken March 14, 1993. (National Oceanic and Atmospheric Administration.)

Preparation and Planning

Emergency Managers must be knowledgeable of the different watches and warnings that can be issued for winter weather. Knowledge of these definitions will assist the Emergency Management official in determining appropriate action to take to protect their organization, as well as the public. Winter weather watches and warnings include

Winter Weather Advisory—Winter weather conditions are expected to cause significant inconveniences and may be hazardous.
Winter Storm Watch—A winter storm is possible in your area. Tune in to NOAA All Hazards Radio, commercial radio, or television for more information.
Winter Storm Warning—A winter storm is occurring or will soon occur in your area. Finish all precautionary measures immediately.
Blizzard Warning—Sustained winds or frequent gusts to 35 miles per hour or greater and considerable amounts of falling or blowing snow (reducing visibility to less than a quarter mile) are expected to prevail for a period of three hours or longer.
Severe Blizzard Warning—A severe blizzard warning means that very heavy snow is expected with winds exceeding 45 miles per hour and temperatures below 10°C. Visibility can be reduced to a few feet.
Frost/Freeze Warning—Below freezing temperatures are expected in the area.

Emergency Managers must work with their organizations and the public to prepare before a winter storm occurs. This includes encouraging constituents to build a winter storm survival kit for their home as well as their vehicle. The winter storm survival kit is like a typical family preparedness go-kit or disaster supplies kit. However, it should also include rock salt to melt ice on roadways, sand or cat litter to improve traction, snow shovels and/or any other snow removal equipment, and adequate clothing and blankets to keep warm. Emergency Managers can also work with the agency responsible for their roads and bridges to ensure that plows are ready, and that there is enough sand and salt to cover the roads as needed. As with most debris-causing hazards, it is also prudent that the Emergency Manager work with other stakeholders to develop a debris management plan for their jurisdiction. This includes debris collection, disposal, and reduction as warranted.

Drought

Description and History

A drought is a period of abnormally dry weather that persists long enough to produce a serious hydrologic imbalance. This can lead to a water supply shortage and crop damage. The severity of a drought depends on the amount of moisture deficiency, the duration, and the size of the affected area. There are three types of droughts:

> Meteorological drought—A period of time, generally ranging from months to years, during which time the actual moisture supply at a given location consistently falls short of the climatological moisture supply.
> Hydrological drought—Occurs when stream flows and reservoirs are low due to a lack of prolonged rainfall.
> Agricultural drought—Occurs when the amount of water needed for crops is more than that available in the soil.

In 1996, the state of Texas withstood a drought that lasted more than 10 months. This drought was both a meteorological drought (8 months) and a hydrological drought (10 months). This drought period saw significant drops in reservoir and aquifer levels over much of Texas. Agricultural impacts as a result of the drought were significant with estimates of total loss being in the range of $5 billion.[16]

Consequences of Droughts

Droughts can cause economic, environmental, and social effects. Economic effects typically come in the area of agricultural losses. Droughts typically lead to lower crop yields, which can lead to lower profits for farmers and higher market prices for agricultural products such as produce. Environmental impacts include erosion, dry vegetation increasing wildfire risks, insect infestation, plant diseases, deterioration of habitat and landscape, poor air quality, and degraded water quality of water that remains (due to stagnation). Social effects come from the lack of potable drinking water that is available. Social unrest can occur (and has occurred) due to perceived disparities in water distribution to particular segments of population in comparison with others.

Preparation and Planning

Emergency Managers can work with area horticulturists to encourage the planting of drought-resistant plants and trees. This will reduce the need for use of water for irrigation should a drought occur. The Emergency Manager can also work with officials to develop water conservation rules and regulations in the case a drought occurs. This can include limitation of water use for irrigation purposes, cutting off water to decorative ponds and fountains, and imposing fines for excessive water usage.

Extreme Heat/Heat Waves

Description and History

There is no universal definition for a heat wave because heat waves are relative based on the normal temperatures an area typically experiences. However, it can be said that heat waves are extended periods of extreme heat, which may or may not be accompanied by high humidity. Extreme events are only rivaled by cold weather events in the amount of deaths they cause on an annual basis. Historically, from 1979 to 2003, excessive heat exposure caused 8015 deaths in the United States. During this period, more people in this country died from extreme heat than from hurricanes, lightning, tornadoes, floods, and earthquakes combined. In 2001, 300 deaths were caused by excessive heat exposure.[17]

One historical extreme event occurred in Chicago, Illinois, between July 12–15, 1995. During this four-day period, the amount of deaths caused by factors related to the heat rose to more than 500. The closeness of the buildings in downtown Chicago, the widespread amount of asphalt pavement, the tendency for people to stay indoors due to fear of crime, and a lack of a lake breeze from Lake Michigan complicated the heat-related issues. During this time, heat indices rose to 125°F (it was thought to be higher in some pockets of Chicago).[18]

Consequences of Extreme Heat/Heat Waves

Extreme heat events can cause heat exhaustion, heat stroke, and dehydration in individuals who do not take precautions to protect themselves during these events. Each of these heat-related conditions can also lead to death if undetected or not treated. Children, elderly, people with breathing problems, and pets, are all particularly vulnerable to extreme heat events. Extreme heat events can also cause droughts (described above) and power outages due to the excessive demand on electricity for cooling.

Preparation and Planning

Although extreme heat can be common in certain areas, it is necessary for the Emergency Manager to become familiar with watches and warnings associated with extreme heat events. These include

> Excessive Heat Outlook—Issued when the potential exists for an excessive heat event in the next three to seven days. An outlook is used to indicate that a heat event may develop.
> Excessive Heat Watch—Issued when conditions are favorable for an excessive heat event in the next 12 to 48 hours. A watch is used when the risk of a heat wave has increased, but its occurrence and timing is still uncertain.
> Excessive Heat Warning—Issued when an excessive heat event is expected in the next 36 hours. A warning is issued when an excessive heat event is occurring, is imminent, or has a very high probability of occurrence.

Emergency Managers must work with the public to ensure that they have an emergency supplies kit that can assist them should an extreme heat event occur. This can include battery-operated fans for cooling, plenty of water, battery-operated radios, and plenty of nonperishable foods. The public must also be encouraged to stay indoors. If there is a need from a person to go outdoors, they should wear appropriate clothing and sunscreen to protect themselves, and stay hydrated. Because of the possibility of droughts, it may also be decided that water restrictions are necessary in order to conserve water.

Tsunamis

Description and History

A tsunami is a series of ocean waves triggered by a sudden shift in the ocean floor (earthquake), landslide, or volcanic eruption. Tsunami waves in the deep ocean can be only a few inches tall and can travel 400 miles per hour (and sometimes higher). Most ships in the ocean would not feel this type of wave because of the wavelength being potentially hundreds of miles long and the amplitude (wave height) being that of only a few feet. These types of waves are not usually visible from the air. However, as the wave moves ashore into shallower waters, these waves decrease in speed and their amplitude increases. The first wave of the series is typically not the largest. Damages from the tsunami waves vary from location to location, dependent on several factors. One area could be inundated with deadly waves, while another area not that far away could miss the effects of the tsunami.

On December 6, 2004, the Indian Ocean tsunami was triggered by a 9.3-magnitude earthquake that had its hypercenter (point on Earth's surface directly above the epicenter) located just north of Simeulue Island, off the western coast of

Figure 5.8 A mosque is left standing amid the rubble in Banda Aceh following the Indian Ocean tsunami in 2004. The tsunami waves reached the middle of the second floor. (United States Geological Survey/Photo by Guy Gelfenbaum.)

northern Sumatra. Many countries felt the effects of this earthquake and tsunami, including Australia, Bangladesh, India, Indonesia, Kenya (minor), Madagascar, Malaysia, the Maldives, Mauritias, Myanmar, Oman, Seychelles, Somalia, South Africa, Sri Lanka, Tanzania, Thailand, and Yemen. In total, 227,898 people were killed or were missing and presumed dead and about 1.7 million people were displaced by the earthquake and subsequent tsunami (Figure 5.8).[19]

Tsunamis can occur along the coastal areas of the United States, although the highest risk is along the Californian and northwestern parts of North America (including Alaska).

Consequences of Tsunamis

Tsunamis act much like the storm surge from a hurricane, except with the potential of significantly more force. There is an extensive threat of flooding, not just along the immediate coast, but for quite a way inland (depending on elevations above sea level versus height of the waves). The large amounts of rapidly moving water can produce violent currents that can sweep people, vehicles, structures, vegetation, and anything else away. As waters recede, there is an increased threat of disease due to stagnant, contaminated water and mold. Other effects that can occur as a result of tsunamis include structural damage to residences and businesses (leading to temporary housing and economic loss issues), infrastructure damage, and large amounts of debris. Some of the secondary effects of tsunamis include long-term power outages, mass evacuations of people and animals, mass casualties, infrastructure failure, agricultural damage or destruction, contamination of water supplies (through contaminated freshwater or saltwater inundation), and hazardous materials incidents. As with other hazards, the cascading effects of disasters depend on multiple variables, so this list is by no means comprehensive.

Preparation and Planning

As with other hazards, Emergency Managers should educate the public as to what precautions they should take should a tsunami warning be issued. Some may also see a receding of the ocean, exposing the coastal floor. This is a sign that a tsunami is coming, and immediate evacuation to higher ground should take place. If an earthquake is felt along the coast, people should be instructed to turn on a NOAA All-Hazards to see if a tsunami warning has been issued. Whatever the signal, the public should know to stay away from the coastline, get to high ground, and stay away from structures that could be torn down by the tsunami. Typically, there is not a lot of advance warning for a tsunami. Therefore, once a tsunami warning is issued, immediate action must be taken to save lives.

The NOAA has two tsunami warning centers that it operates as part of an international tsunami warning system. The Pacific Tsunami Warning Center (PTWC) is located in Ewa Beach, Hawaii, and issues warnings for all participating members of the international tsunami warning system in the in the Pacific. This includes other nations. The other tsunami warning center, the West Coast and Alaska Tsunami Warning Center (WCATWC), is located in Palmer, Alaska. This location issues warnings for all West Coast regions of Canada and the United States, except for Hawaii.

Volcanoes

Description and History

Volcanoes are impressive yet violent and can be very hazardous to life and property. Eruptions are explosive and can change the landscape for several miles around the volcano, as well as chase people from their homes (sometimes permanently). The droplets of sulfuric acid that are blown into the atmosphere can temporarily change the climate globally. Those not inhabiting the areas immediately surrounding the volcano may not lose everything to the explosion from the volcano, but their homes, modes of transportation, utilities, and businesses can be negatively affected by the byproducts of the volcano. These include

> Ash and tephra—Dust from pulverized rock and glass resulting from the volcanic eruption. Ash deposited on the ground following an eruption is called an ashfall deposit. Ash is the smallest form of tephra, which is debris emitted into the atmosphere from an erupted volcano. Ash deposits can potentially choke out ecosystems and, if thick enough, can collapse roofs on homes and businesses. These can also exacerbate breathing issues for those with respiratory problems and cause new respiratory problems for those who do not protect their respiratory system. Larger forms of tephra can cause injury and further damage several miles downwind from the volcano.

Lahars—Hot or cold mixture of water and rock fragments flowing down the slopes of a volcano. These vary in speed from a few feet per second to tens of meters per second. They can also vary in size. The size and consistency of a lahar can change as it moves down the slope of a volcano. At its origination point, it can include water from snow and ice melt, vegetation, and rocks. This collection of materials as it moves down the slope can increase the size and flow of the lahar to many times of what it was when it started. As the lahar moves further away from the volcano, it begins losing its heavier loads, and will slow down.

Volcanic gases—Numerous gases are emitted from a volcano's soil, magma, fumaroles (vents from which volcanic gas escapes into the atmosphere), and hydrothermic systems. All can induce serious health problems, which include respiratory issues, skin irritations, mucous membrane irritations, and death. Most gases are of such concentrations that they can be fatal to both humans and animals.

Lava Flows—Lava flows can be a result of an explosive or nonexplosive release from the volcano. Typically, lava flows are slow enough to enable people to get out of their way. However, some flows can move more than 20 miles per hour. Lava will incinerate, surround, or bury anything in its path, and there is no way of stopping it. Lava flows also bury agricultural land under several inches to several feet of black rock.

Pyrolclastic Flows—High-density mixtures of hot, dry rock fragments and hot gases that move at high speeds away from the vent where they originated. These flows will destroy almost everything in their path. They can move at speeds of more than 50 miles per hour.

One historic volcano that has erupted in the United States is Mount St. Helens, which last erupted on May 18, 1980 (Figure 5.9a and b). Scientists had been monitoring this volcano (and still do) after a series of hundreds of earthquakes occurred.

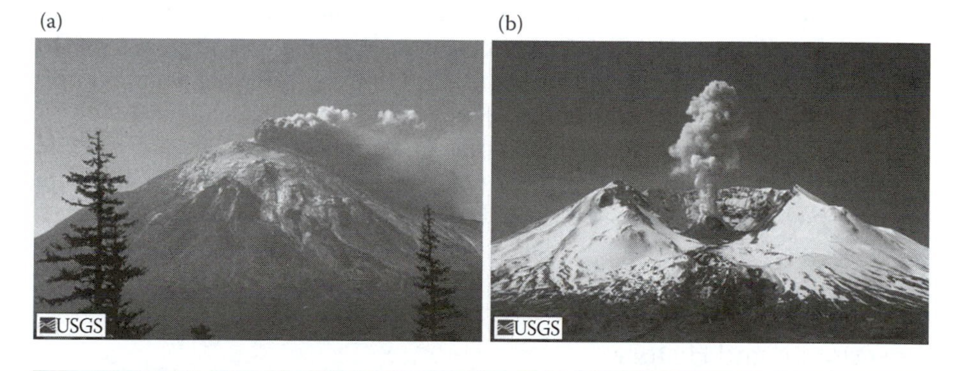

Figure 5.9 Mount St. Helens (a) one day before eruption in 1980 and (b) shortly after eruption. (United States Geological Survey.)

By May 17, 10,000 earthquakes had occurred on the volcano and the north flank had grown outward at least 450 feet to form a noticeable bulge. On May 18, following a magnitude 5.1 earthquake, the volcano's bulge and summit slid away in a huge landslide that depressurized the volcano's magma system, causing a massive eruption. Fifty-seven people perished as a result of this eruption.[20]

Consequences of Volcanoes

Volcanoes can be very devastating to the immediate areas surrounding the eruption site. Lava flows and pyroclastic flows will bury, knock down, and incinerate everything in their path. This can lead to the destruction of homes, businesses, utilities, modes of transportation, and agricultural lands. The gases emitted from a volcano can cause injury or be fatal to both humans and animals. The volcanic ash and other tephra emitted from the volcano can also lead to both injuries and fatalities. The lahars from volcanoes can cause dams to form in creeks, streams, and rivers, leading to flooding. There is a myriad of cascading effects from volcanoes ranging from poor air quality issues, to mass sheltering and feeding, and mass casualty (both human and animal). Massive amounts of volcanic ash, tephra, vegetative debris, and damages building and vehicles can lead to massive debris management efforts.

Preparation and Planning

If an Emergency Manager has volcanoes as one of the hazards in their community, it is important (as with all hazards) that the public is encouraged to develop individual and family disaster supplies kits. As part of the kit, an N95 respirator should be included. Also, citizens and/or members of the Emergency Manager's organization should have a plan to not only evacuate if told to do so by local officials, but also to shelter in place. Emergency Managers should work with stakeholders in their community to determine how the massive amounts of debris will be managed, as well as how citizens will be sheltered and fed both short and long term as required. In addition, it is a good practice to have a plan as to what roads will be cleared and opened (if possible) first. Typically, priority is given to roads that provide ingress and egress to critical facilities such as fire and police stations, hospitals, and utilities. Also, redundant communications plans (as with all hazards) should be put together before an eruption to ensure the ability to communicate with both internal and external stakeholders, including the public.

Epidemics and Pandemics

Description and History

An epidemic is defined as a disease outbreak that affects more than the expected number of cases of disease occurring in a community or region during a given period. A pandemic is defined as a sudden disease outbreak that becomes very widespread and affects a whole region, a continent, or the world. A pandemic can stem from an epidemic. It also must be noted that epidemics and pandemics *are not* based on the severity of the disease, but instead the geographic area that is affected by the disease outbreak. The World Health Organization (WHO) has developed a phased system for influenza to allow government and health officials to understand what steps need to be taken to protect people from further spread of influenza (regardless of type). The phases are developed based on findings in epidemiological surveillance. These phases are as follows:

Phase 1—No animal influenza virus circulating among animals have been reported to cause infection in humans.
Phase 2—An animal influenza virus circulating in domesticated or wild animals is known to have caused infection in humans and is therefore considered a specific potential pandemic threat.
Phase 3—An animal or human–animal influenza reassortant virus has caused sporadic cases or small clusters of disease in people, but has not resulted in human-to-human transmission sufficient to sustain community-level outbreaks.
Phase 4—Human-to-human transmission of an animal or human–animal influenza reassortant virus able to sustain community-level outbreaks has been verified.
Phase 5—The same identified virus has caused sustained community-level outbreaks in two or more countries in one WHO region.
Phase 6—In addition to the criteria defined in Phase 5, the same virus has caused sustained community-level outbreaks in at least one other country in another WHO region.
Post-Peak Period—Levels of pandemic influenza in most countries with adequate surveillance have dropped below peak levels.
Post-Pandemic Period—Levels of influenza activity have returned to the levels seen for seasonal influenza in most countries with adequate surveillance.

The above phases refer to influenza. However, other diseases can be part of an epidemic or pandemic, including but not be limited to cholera, typhus, smallpox, measles, tuberculosis, leprosy, malaria, plague, and yellow fever. Antibiotic-resistant diseases, sometimes known as "superbugs," can make an epidemic or pandemic difficult to control.

One pandemic that has received much attention is the H1N1 influenza pandemic of 2009. This pandemic reached more than 214 countries and overseas territories. In the United States alone, the Centers for Disease Control estimates that between 43 million and 88 million cases of the 2009 H1N1 occurred between April 2009 and March 13, 2010 (mid-range 60 million). The estimates for hospitalizations due to H1N1 were between 192,000 and 398,000 (mid-range about

270,000). The estimates on H1N1-related deaths in the United States range from 8720 and 18,050 (mid-range is about 12,270).[21] This is low compared to many epidemics and pandemics of the past. This can more than likely be attributed to advances in medical treatment, epidemiological surveillance, and public education on personal protection.

Consequences of Epidemics and Pandemics

Epidemics and pandemics can cause great economic and social problems with people not wanting to go to certain areas, such as grocery stores and other public gathering locations, for fear of contamination. Epidemics and pandemics can also tax hospital surge capacities by causing more demand for hospital beds than are actually available. Also, immunizations and medicines, along with medical supplies such as N95 masks, can go into short supply because of increased demand. Isolation of people may also cause psychological and other social issues due to lack of human interaction. Mass casualties can also be a result of an epidemic or pandemic, depending on the disease that has caused the outbreak.

Preparation and Planning

The Emergency Manager must develop a relationship with health officials within their community so that they can plan for, share information on, and respond to threatened or actual epidemics and pandemics. This information should be shared with those stakeholders who can potentially have gatherings of many people at one time, such as business owners, schools, health care facilities (including nursing homes and assisted living facilities), faith-based organizations, and convention centers. Also, planning for Continuity of Operations Planning should be considered. Many times, epidemics and pandemics can reduce the workforce of an organization (sometimes by more than 30%). Managers, supervisors, and directors should determine which functions are critical to their organizations so that they can cross train employees to maintain those functions should manpower be cut due to illnesses. This is good business practice for all organizations.

Space Weather

Description and History

Space weather is probably one of the more abstract hazards that Emergency Managers must think about and plan for. Although space weather is a rare occurrence, its effects can be very detrimental on a global basis. Space weather includes conditions in the magnetosphere, ionosphere, and thermosphere (as well as on the Sun) that can influence the performance of space and ground-based technologies. There are several types of space weather, but for the purposes of this section, the focus will be on space weather that causes geomagnetically induced currents, or GIC. These currents can affect the Earth's surface and cause damaging electrical currents that can flow through power grids, pipelines, and other technological networks. In space, these currents can disrupt and cause loss of satellites.

An example of a space weather event is the collapse of the Hydro-Quebec power network on March 12, 1989, due to geomagnetically induced currents. The failure started with a transformer failure, which led to a blackout for six million people over a period of nine hours. The geomagnetic storm causing this event was the result of a coronal mass ejection, ejected from the Sun on March 9, 1989. A coronal mass ejection is a burst of solar wind, or other light plasma, and magnetic fields. These are commonly associated with the Sun and solar activity (primarily solar flares).

Consequences of Space Weather

Space weather can cause both economic and sociological effects. A disruption in power utilities can be detrimental for businesses and for the health care industry. With the increased dependencies on technologies, a disruption or loss of these systems could lead to a delay or complete halt of certain businesses that do not have backup plans consisting of manual methods for doing business.

Planning and Preparation

Emergency Managers should ensure that their organizations, as well as the businesses in their communities, have Continuity of Operations Plans (COOP) that will allow them to sustain business following a disruption from any hazard. It is difficult to plan for space weather. However, if the Emergency Manager has developed, or is maintaining, a comprehensive all-hazards Emergency Management plan, then the planning for the effects of space weather events should be included. The Emergency Manager should also be educating the public on the development of a disaster supplies kit that will prepare them with the supplies to handle the effects of space weather.

Man-Made Hazards

Hazardous Materials Incidents

Description and History

Hazardous materials, commonly referred to as HazMat, are any solid, liquid, or gas that can cause long-lasting health effects, injury, and even death. Hazardous materials can also cause damage to buildings, homes, and other property. Many products containing hazardous chemicals are stored in the home on a regular basis. Hazardous chemicals are also used to purify our drinking water, increase crop production, and any number of day-to-day activities. However, they can become hazardous to humans and the environment if misused or accidentally released in an uncontrolled environment. Hazardous materials are transported via vehicle, rail, and plane along the nation's transportation routes every day. Therefore, there is no area in the country that is not susceptible to a hazardous material incident.

The United States Department of Transportation uses a placarding system to identify hazardous materials that are in containers as well as those in transport. The placards are diamond-shaped and are required to be places on both ends and both sides of trucks, railcars, and intermodal containers that carry hazardous materials. This helps regulators to ensure that proper shipping practices are being adhered to. These placards also help first responders understand the type of chemical or substance involved with an incident, and therefore allows them to take appropriate actions to protect themselves and the public from harm. In bulk and certain nonbulk shipments, a four-digit hazardous material identification number may be on the placard. These placards may be accompanied by an orange panel or white square sign that will further assist in identification of the hazardous property of the substance being transported (see Figure 5.10).

Hazardous materials accidents occur on a daily basis. Fortunately, most are not fatal, nor do they cause significant injury. However, hazardous materials certainly can cause large amounts of injuries and fatalities, and/or cause extensive damage to the environment.

The largest industrial accident to date that involved hazardous materials occurred in Bhopal, Madhya Pradesh, India, on the night of December 2 and 3, 1984, at the Union Carbide India Limited pesticide plant. This incident involved a release of more than 40 tons of methyl isocyanate gas, along with other chemicals over the city of Bhopol. This cloud of gas gradually descended into the city. The immediate death toll was about 3800, and the incident caused significant morbidity and premature death for many thousands more.[22] Livestock were killed or injured. Businesses were interrupted, and the environment was immensely affected.

Consequences of Hazardous Materials Incidents

The consequences of hazardous materials incidents can vary depending on what is released, the size of the release, cause of the release, and population around the release. Depending on the nature of the incident, it may be required that the public shelter-in-place or a mass evacuation may be necessary. In some cases, sheltering and mass feeding may be necessary for longer-term evacuations from an area. Mass decontamination may be necessary for both people and objects that may have been exposed to the hazardous material. Mass casualties are also a possibility with these types of incidents, which can also tax the surge capacity of health care facilities. Moreover, environmental issues can arise as a result of a hazardous material event (air, soil, and water). Therefore, it is necessary to have environmental remediation plans in place in case a hazardous material incident occurs.

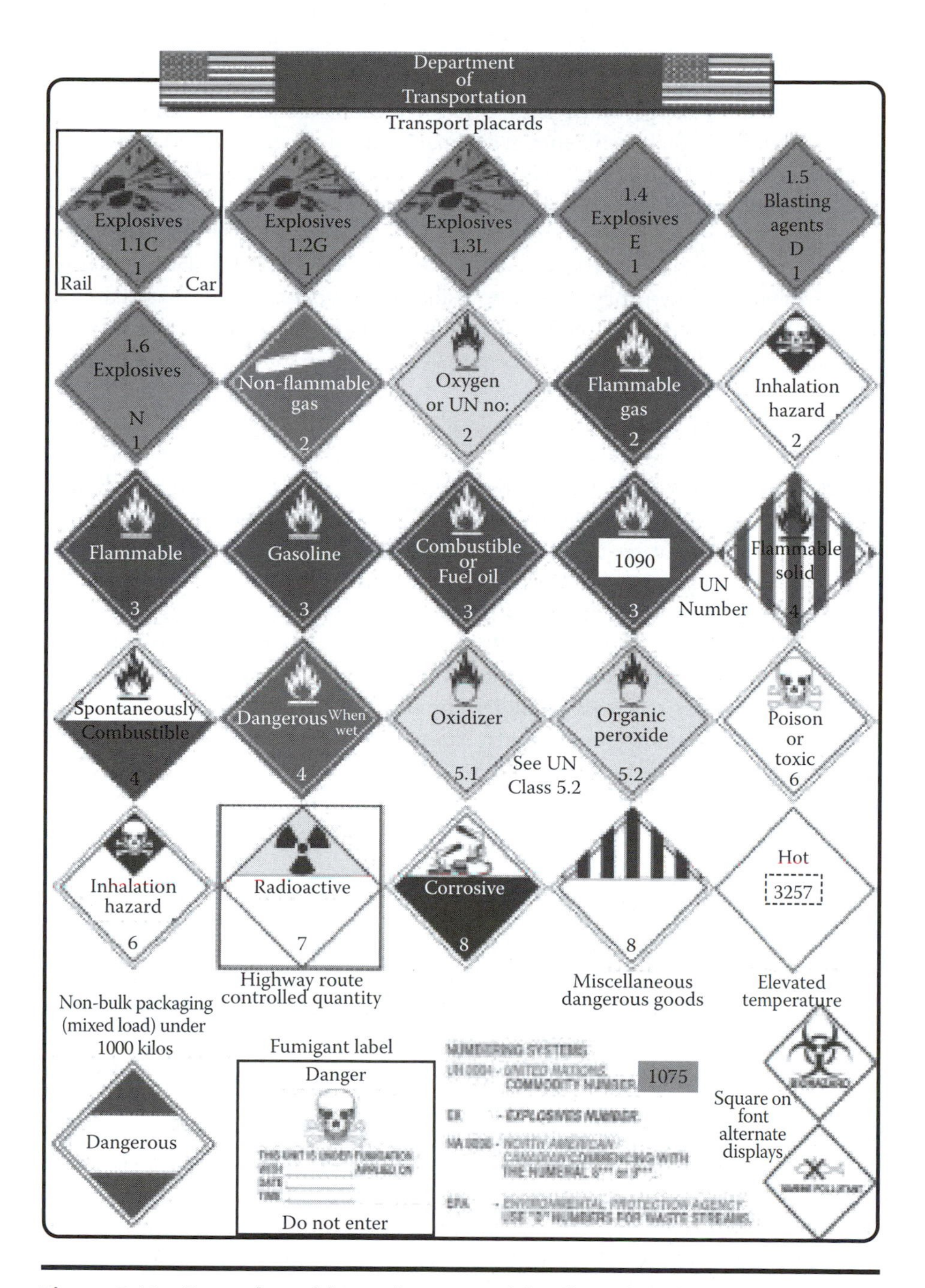

Figure 5.10 Examples of hazardous materials placards. (U.S. Department of Transportation.)

Planning and Preparation

The Emergency Manager must ensure that hazardous materials planning is part of their overall comprehensive emergency management program. There is no place that is not vulnerable to hazardous materials, whether it is a household chemical accident or an industrial chemical spill. Many communities have Local Emergency Planning Committees (LEPCs) that are formed in accordance with the Emergency Planning and Community Right to Know Act of 1986. This legislation requires LEPCs to prepare emergency plans for possible releases of hazardous materials, and for fixed facilities to be involved in the planning process. The Emergency Manager can contact their State Emergency Response Commission to identify active LEPCs in their area.

Emergency Managers must also engage the public in preparing for hazardous materials incidents in their community. This should include shelter-in-place actions, as well as actions needed for evacuations. Citizens and organizations should also build disaster supply kits that contain materials such as duct tape, plastic sheeting, and towels in case a shelter-in-place order is given and "buttoning up" of residences and businesses is required. Families should also develop a family disaster plan. The family disaster plan should include a communications plan, as well as an evacuation plan and a shelter-in-place plan. Emergency Managers must also be prepared to deal with sheltering and mass feeding issues should an evacuation be required. The Emergency Manager also should ensure that their organization has contracts for environmental remediation should a hazardous material incident affect soil or a waterway or storm drain.

Terrorism

Description and History

Terrorism is any act of force or violence (or threat of violence) used to coerce, intimidate, or seek ransom (i.e., cause terror). Acts of terrorism are also criminal activities. Oftentimes, politics, religious beliefs, and ideologies are the driving forces behind terrorism. Terrorism can take many forms, and can include threats of terrorism, kidnappings, assassinations, hijackings, cyber attacks (described in the next section), and bombings or other threatened attacks using chemical, biological, radiological, nuclear, or explosive weapons or devices. Terrorism can also be aimed toward economies, where particular actions are taken to cause economic disruption or depletion of funds. Targets of terrorist acts can include government officials and establishments, military personnel, people serving the interest of governments, and civilians representing some ideology that is different than that of the person(s) committing the terrorist act. Terrorism has occurred for centuries. However, it began receiving increased attention in the United States following the World Trade Center bombing in New York City in February 1993, the Oklahoma

City Bombing in April 1995 (domestic terrorism), and then the terrorist attack of September 11, 2001.

On September 11, 2001, 19 members of the militant Islamic group Al-Qaeda hijacked four American jetliners and flew two of them into the two towers of the World Trade Center in Manhattan, New York City, New York. The third plane was flown into the United States Department of Defense headquarters, the Pentagon, in Arlington, Virginia. The fourth plane crashed in a field near Shanksville, Pennsylvania, after passengers fought with the hijackers of the plane. In total, more than 2600 people died at the World Trade Center, 125 died at the Pentagon, and 256 died on the four planes. The death toll surpassed that of Pearl Harbor on December 7, 1941 (Figure 5.11).[23]

Consequences of Terrorism

Terrorism can cause cascading effects that can have global effects socially and economically. As the term alludes, terrorism is designed to evoke fear in its victims. This can lead to psychological as well as sociological effects, such as suspicion and exclusion of certain religious or ethnic groups, even though they may have no association with the terrorist or terrorist organization. The overall immediate effect of terrorism is largely based on the type of weapon or tactic used by the terrorist. An assassination or kidnapping can be very localized and be primarily a law enforcement investigative process. However, detonation of a biological weapon or a chemical weapon can cause mass fatalities and widespread health problems, as well as mass decontamination issues for areas that are exposed. Terrorist acts can also disrupt critical infrastructure such as communications, utilities (i.e., power, gas, water, and sewer), social services, public safety services, and other government services. Terrorism can also affect businesses by either physically disrupting them, or by disrupting the supply chain required for the business to function.

Figure 5.11 New York, September 21, 2001: These FEMA rescue workers are dwarfed by the pile of rubble at the site of the World Trade Center. (Photo courtesy of Michael Rieger, FEMA News Photo.)

Planning and Preparation

Emergency Managers typically are not responsible for counterterrorism actions or investigations. These roles are primarily law enforcement responsibilities. This is particularly the case because all terrorist acts are considered criminal acts. However, the Emergency Manager does have a role in preparing the public and for consequence management following an act of terrorism. Information that the Emergency Manager can use in preparing the public is typically covered under other hazards that the public should be prepared for. For example, in the case of a chemical weapon being used, the same preparatory measures should be taken as if preparing for a hazardous material event. For a biological weapon being used, the same preparatory actions should be taken as if preparing for an epidemic or pandemic. The only notable difference is the malicious and criminal intent behind the incident.

Preparing for a terrorism act is no different than preparing for other hazards. It is important that Emergency Managers provide information on developing disaster supplies kits, as well as information pertaining to developing a family disaster plan. The family disaster plan should include a communications plan as well as an evacuation plan and a shelter-in-place plan. Emergency Managers should ensure that plans for Continuity of Operations (COOP), mass casualties, health care facility surge plans, and other emergency plans are reviewed and ready for action should a terrorist incident occur. Because of the varying nature of terrorist activities, it is necessary that all plans be reviewed and maintained to manage the consequences of a terrorist act.

Cyber Threats and Attacks

Description and History

As more and more of the world relies on technologies, we raise our vulnerabilities to malicious attacks on our computer systems, including those that run our day-to-day businesses and processes for critical infrastructure. Threats to our computer systems can come from a variety of sources. These sources can include disgruntled employees, malicious intruders, and even hostile governments and terrorist groups. The following is a list of definitions used by the Department of Homeland Security in identifying cyber threats.[24]

> Hackers—Individuals who break into networks for the thrill of the challenge or for bragging rights in the hacker community. Although remote cracking once required a fair amount of skill or computer knowledge, hackers can now download attack scripts and protocols from the Internet and launch them against victim sites. Thus, while attack tools have become more sophisticated, they have also become easier to use. The worldwide population of hackers poses a relatively high threat of an isolated or brief disruption causing serious damage.

Phishers—Individuals, or small groups, who execute phishing schemes in an attempt to steal identities or information for monetary gain. Phishers may also use spam and spyware/malware to accomplish their objectives.
Spammers—Individuals or organizations who distribute unsolicited e-mail with hidden or false information in order to sell products, conduct phishing schemes, distribute spyware/malware, or attack organizations (i.e., denial of service).
Spyware/Malware Authors—Individuals or organizations with malicious intent carry out attacks against users by producing and distributing spyware and malware. Several destructive computer viruses and worms have harmed files and hard drives, including the Melissa Macro Virus, the Explore.Zip worm, the CIH (Chernobyl) Virus, Nimda, Code Red, Slammer, and Blaster.
Insiders—The disgruntled organization insider is a principal source of computer crime. Insiders may not need a great deal of knowledge about computer intrusions because their knowledge of a target system often allows them to gain unrestricted access to cause damage to the system or to steal system data. The insider threat also includes outsourcing vendors as well as employees who accidentally introduce malware into systems.
Foreign Intelligence Services—Foreign intelligence services use cyber tools as part of their information-gathering and espionage activities. In addition, several nations are aggressively working to develop information warfare doctrine, programs, and capabilities. Such capabilities enable a single entity to have a significant and serious impact by disrupting the supply, communications, and economic infrastructures that support military power—impacts that could affect the daily lives of U.S. citizens across the country.
Criminal Groups—Criminal groups seek to attack systems for monetary gain. Specifically, organized crime groups are using spam, phishing, and spyware/malware to commit identity theft and online fraud. International corporate spies and organized crime organizations also pose a threat to the United States through their ability to conduct industrial espionage and large-scale monetary theft and to hire or develop hacker talent.
Bot-Network Operators—Hackers who, instead of breaking into systems for the challenge or bragging rights, take over multiple systems in order to coordinate attacks and to distribute phishing schemes, spam, and malware attacks. The services of these networks are sometimes made available in underground markets (e.g., purchasing a denial-of-service attack, servers to relay spam, or phishing attacks).

A few more definitions help to explain the differences in the different types of malicious code that can be used to corrupt technological systems:

Virus—This type of code requires the victim to perform an action, such as open an e-mail, before it infects your computer.

Worm—This type of code propagates itself without user action or intervention. Worms typically start by exploiting a software security vulnerability. Once the victim's computer has been infected, the worm will attempt to find and infect other computers through address books, contact lists, or other connective means. The self-propagation of a worm distinguishes it from a virus.

Trojan Horse—This type of program claims to be one thing on the surface, but is totally different behind the scene. For example, it may be a program that says it will increase your computer speed, when in fact it may be sending confidential information to a remote intruder.

One case of a cyber attack occurred in early 1998, in which United States military systems were subject to an "electronic assault" noted as "Solar Sunrise." The intruders hid their tracks by routing their attack through computer systems in the United Arab Emirates. They accessed unclassified logistics, administration, and accounting systems that control the ability to manage and deploy military forces. The United States response to this incident requires a massive, cooperative effort by the Federal Bureau of Investigation, the United States Justice Department's Computer Crimes Section, the Air Force Office of Special Investigations, the National Aeronautics and Space Administration, the Defense Information Systems Agency, the National Security Agency, the Central Intelligence Agency, and various computer emergency response teams from the military services and government services.

In the end, it was found that two young hackers in California had carried out the attacks under the direction of a hacker in Israel, himself a teenager. They gained privileged access to computers using tools available from a university website and installed sniffer programs to collect user passwords. They created a backdoor to get back into the system, then used a patch available from another university website to fix the vulnerability and prevent others from repeating their exploit. Unlike most hackers, they did not explore the contents of their victim's computers.[25]

Consequences of Cyber Threats and Attacks

Typically, cyber threats and attacks are aimed at disruption of infrastructure, disruption of the economy, and exposure of sensitive strategic information. At the corporate levels, cyber attacks can create large liabilities and cause losses large enough to bankrupt most companies. Countries and organizations wishing harm upon the United States and its general ideologies, as well as organized crime, have introduced this type of attack as being profitable for the attacker. For this reason, American corporations and government entities need to be urgently informed, not just of the technical vulnerabilities that may exist, but also of the significant strategic and economic consequences of an attack on those vulnerabilities.

Planning and Preparation

The Emergency Manager must work with the information technology staff of their organization, or possibly a contractor, to perform a risk assessment on the technologies that their organization relies on to perform critical functions. The risk assessment will identify vulnerabilities to the technologies that should be addressed. This may include development of new information security policies, firewall development, antivirus and antispyware software installation, and developing a regularly scheduled backup of all of the organization's data stored on computers.

To prepare for a cyber threat or attack, the Emergency Manager must encourage their organization, as well as the public, to be prepared to do without typical services that they rely upon daily. This can include electricity, natural gas, telephone, gasoline pumps, ATM machines, cash registers, and Internet transactions. An organization's COOP should be developed to address these types of business disruptions. The Emergency Manager must also plan for the cascading effects of a cyber attack, which can include hazardous material releases, nuclear power plant incidents, and infrastructure failures (dams, water treatment plants, traffic signals, etc.).

Infrastructure Failures

Description and History

Physical infrastructure failure can occur for a variety of reasons including lack of maintenance, age of infrastructure, damage to infrastructure (accidental or intentional), and faulty construction or defective materials. Examples of infrastructure that can be involved include electrical power grids, bridges, roads, water treatment plants, dams, pipelines, and storage tanks (water or chemical).

One such infrastructure failure occurred in August 2007 in Minneapolis, Minnesota, when the I-35 W bridge (eight lanes) over the Mississippi River collapsed (Figure 5.12). One hundred and eleven vehicles were on the portion of the

Figure 5.12 Minneapolis, Minnesota, bridge collapse, August 1, 2007. (Minnesota Department of Transportation.)

bridge that collapsed, killing 13 people and injuring 145 people. The National Transportation Safety Board determined that the probable cause of the bridge collapse was inadequate load capacity, due to a design error of the gusset plates. These gusset plates failed due to a combination of (1) substantial increases in the weight of the bridge, which resulted from previous bridge modifications, and (2) the traffic and concentrated construction located on the bridge the day of collapse (four of the eight lanes of the bridge were under construction, and heavy machinery was present).[26]

One of the most historic examples of infrastructure failure came in the town of Johnstown, Pennsylvania, on May 31, 1889. On that date, the South Fork Dam gave way, flooding the town with 20 million tons of rushing water. The 36-foot wall of water containing debris, such as boulders, swept through the town, killing 2249 people and destroying hundreds of buildings. As a side note, this was the first disaster for the newly formed American Red Cross under the leadership of Clara Barton. Contributing factors to this disaster were (1) the town of Johnstown was built in a floodplain and (2) the South Fork Dam had been poorly maintained.

Consequences of Infrastructure Failure

The effects of infrastructure failure can vary depending on the location and type of infrastructure that has failed. For this reason, it is difficult to plan or prepare for a particular piece of infrastructure to fail. Also, the interdependencies of critical infrastructure (they rely on one another to function) can cause a number of secondary effects. Among the types of secondary effects that can occur are long-term power outages, contamination or complete disruption of water supplies, flooding, mass casualties, loss of communications, loss of critical technologies, disruption or destruction of pipelines, structural damage or destruction, hazardous material releases, and nuclear power plant accidents.

Planning and Preparation

There are many variables that come into play that make it difficult to plan and prepare for specific infrastructure failures. However, much of the planning that an Emergency Manager does for their community or organization, along with the preparedness information given to the public, should already include plans and steps to be ready in case of an infrastructure failure. These steps include developing a family disaster supplies kit that allows for at least 72 hours of self-sufficiency (one week may be more realistic). The development of individual and family disaster plans for evacuation, sheltering in place, and loss of communications can help the public to be ready in the case of an infrastructure failure. Emergency Managers should review Emergency Operations Plans and Standard Operating Procedures to ensure that they cover infrastructure failures and their potential consequences.

Nuclear Power Plant Emergencies

Description and History

Nuclear reactors or power plants develop heat from controlled nuclear fission. The heat, when produced, converts water into steam, which in turn powers generators to produce electricity. Nuclear power plants operate in most states in the country and produce about 20% of the nation's power. Nearly three million Americans live within 10 miles of an operating nuclear power plant.[27]

The construction and operation of nuclear power plants is regulated by the National Regulatory Commission. Although this regulation occurs, human error and other circumstances can lead to accidents. These accidents can lead to the release of radioactive materials that can be detrimental to the health and well-being of humans and animals in proximity to the nuclear power plant. This is typically characterized as a plume (cloudlike formation) of gases and particles that are radioactive. Radioactive materials are composed of atoms that have become unstable. An unstable atom gives off its excess energy until it becomes stable. The energy emitted is known as radiation. The hazard to people from these types of releases are related to radiation exposure to the body by inhaling, ingesting, or coming into contact with radioactive particles that have fallen to the ground. Radiation has a cumulative effect. The longer a person is exposed to radiation, the greater the effects on the body. A high exposure to radiation can cause serious illness or death.

One example of a nuclear power plant accident is that of the Chernobyl, Ukraine Nuclear Power Plant on April 26, 1986. During this accident, a sudden power surge caused a chain of events that destroyed Unit 4 at the nuclear power plant, and released massive amounts of radioactive material into the atmosphere. The Chernobyl accident caused many severe radiation effects almost immediately. Among the approximately 600 workers present on the site at the time of the accident, two died within hours of the reactor explosion and 134 received high radiation doses and suffered from acute radiation sickness. Of these, 28 workers died in the first four months after the accident. The Chernobyl accident also resulted in widespread contamination in areas of Belarus, the Russian Federation, and Ukraine inhabited by millions of residents. Apart from the increase in thyroid cancer after childhood exposure, no increase in overall cancer or noncancer diseases have been observed that can be attributed to the Chernobyl accident and exposure to radiation. However, it is estimated that approximately 4000 radiation-related cancer deaths may eventually be attributed to the Chernobyl accident over the lifetime of the 200,000 emergency workers, 116,000 evacuees, and 270,000 residents living in the most contaminated areas.[28]

Consequences of Nuclear Power Plant Emergencies

Accidents at nuclear power plants can potentially have long-lasting effects on human and animal health, as well as on the environment. Depending on the amount of radioactive materials released from the accident, areas may have to be evacuated for very long terms (many years) because of contamination. This may lead to long-term sheltering/housing and mass feeding issues. Businesses may be affected and may have to relocate operations in order to meet their customer demands. The local, regional, and even national economy can also be affected depending on what is located within the area(s) of contamination.

Planning and Preparation

Emergency Managers should understand as well as work with the public to understand that alerts that can be issued to areas around nuclear power plants. These alerts are as follows:

> Notification of an Unusual Event—A small problem has occurred at the plant. No radiation leak is expected. No action on your part will be necessary.
> Alert—A small problem has occurred, and small amounts of radiation could leak inside the plant. This will not affect you and no action is required.
> Site Area Emergency—Area sirens may be sounded. Listen to your radio or television for safety information.
> General Emergency—Radiation could leak outside the plant and off the plant site. The sirens will sound. Tune to your local radio or television station for reports. Be prepared to follow instructions promptly.

Public safety officials may have a different alerting system, so it is important to be aware of the public alerting system used in the community, or within the organization.

Emergency Managers must work with their constituents and organizations to build an understanding of protective actions that can be taken should a nuclear power plant accident occur. These include sheltering in place or evacuating the area, and taking action that will take into account

> Time—Most radioactive materials lose their reactivity quickly.
> Distance—The more distance between you and the source of the radiation, the better. This could involve evacuation or remaining indoors to minimize exposure.
> Shielding—The more heavy, dense material between you and the source of the radiation, the better.

As with other hazards, it is important that Emergency Managers also promote the development of a disaster supplies kit that can sustain the individuals of a household, including pets, for a minimum of 72 hours. It is also important that family disaster plans are developed so that family members know what actions to take should a hazard threaten or affect their family.

Summary

In this chapter, many different hazards have been identified and described. As mentioned before, this listing of hazards is by no means comprehensive. However, a couple of trends should be noted. First, it is important for Emergency Managers and their constituents and/or organizations to fully understand the hazards in their community, along with the potentially cascading effects of each hazard. This ensures the most comprehensive all-hazards approach that has proven to be the mark of a good Emergency Management program. Second, families and individuals should build a disaster supplies kit that will help them to sustain life for a minimum of 72 hours. A functional disaster supply kit contains several items that remain the same no matter what the hazard is. Other items can be added to this set of supplies for more specific hazards that the community may face. Families and individuals must also develop disaster plans for evacuation, sheltering in place, and communications. Without these items, families are more prone to becoming victims of a disaster rather than self-sustaining survivors. It must be realized that government services are limited and cannot possibly cover the needs of all people affected by a disaster. Therefore, it is important for people to understand the need for self-reliance immediately after a large emergency or disaster in their community or organization.

References

1. "Turn Around Don't Drown." United States National Oceanic and Atmospheric Administration, National Weather Service Southern Region Headquarters. National Weather Service, 12/12/2007. Web. 6 Nov 2010.
2. Larson, Lee. "The Great USA Flood of 1993." United States National Oceanic and Atmospheric Administration, National Weather Service, Northwest River Forecast Center. National Weather Service, 06/24/1996. Web. 6 Nov 2010.
3. Larsen, Matthew, Gerald Wieczorek, Scott Eaton, Benjamin Morgan, and Heriberto Torres-Sierra. "Natural Hazards on Alluvial Fans: The Venezuela Debris Flow and Flash Flood Disaster." United States Geological Survey, 10/23/2001. Web. 6 Nov 2010.
4. "Flash Floods and Floods ... The Awesome Power." United States National Oceanic and Atmospheric Administration, National Weather Service. National Weather Service, 07/01/1992. Web. 6 Nov 2010.
5. "Hurricane Preparedness." Hurricane History. National Hurricane Center, unknown. Web. 6 Nov 2010.
6. "When Thunder Roars Go Indoors Lightning Safety Awareness Toolkit." United States National Oceanic and Atmospheric Administration, National Weather Service Lightning Safety Outdoors. National Weather Service, n.d. Web. 7 Nov 2010.
7. "Hurricane Basics." United States National Oceanic and Atmospheric Administration, National Weather Service National Weather Service, 2010. Web. 14 Nov 2010.
8. Blake, Eric S., Edward N. Rappaport, and Christopher W. Landsea. Unites States. NOAA Technical Memorandum NWS TPC-5, The Deadliest, Costliest, and Most Intense United States Tropical Cyclones from 1851 to 2006 (and other Frequently

Requested Hurricane Facts). Miami: National Hurricane Center, 2007. Web. 14 Nov 2010.

9. Report on Hurricane Camille 14–22 August, 1969. Mobile: U.S. Army Corp of Engineers, 1970. Print.
10. Tornadoes, Nature's Most Violent Storms. National Weather Service, 1992. Print.
11. May 3, 1999 Oklahoma/Kansas Tornado Outbreak. National Oceanic and Atmospheric Administration, 2007. Web. 17 Nov 2010.
12. Ewell, P. Lamont. United States Oakland. Berkeley–Hills fire of 1991. U.S. Forest Service, 1995. Print.
13. City of San Diego, California. 2003—Cedar Fire. San Diego: Web. 17 Nov 2010.
14. Northridge Earthquake. Washington, DC: Federal Emergency Management Agency, 2010. Web. 2 Dec 2010.
15. Lott, Neal. United States. Technical Report 93-01—The Big One! A Review of the March 12–14, 1993 "Storm of the Century." National Climatic Data Center, 1993. Web. 17 Nov 2010.
16. State of Texas. Drought in Perspective 1996–1998. Austin: State of Texas, 1999. Web. 18 Nov 2010.
17. Extreme Heat: A Prevention Guide to Promote Your Personal Health and Safety. Atlanta: Centers for Disease Control, 2009. Web. 21 Nov 2010.
18. Labas, Kenneth. United States. Meteorological Diagnosis of the Chicago Killer Heat Event of July 13, 1995. Chicago: National Weather Service, 1995. Web. 21 Nov 2010.
19. Magnitude 9.1—Off the West Coast of Northern Sumatra. Denver: United States Geological Survey, 2009. Web. 21 Nov 2010.
20. Brantley, Steve, and Bobbie Myers. United States. Mount St. Helens—From the 1980 Eruption to 2000. Vancouver: United States Geological Survey, 2000. Web. 21 Nov 2010.
21. CDC Estimates of 2009 H1N1 Influenza Cases, Hospitalizations and Deaths in the United States, April 2009–March 13, 2010. Atlanta: Centers for Disease Control and Prevention, 2010. Web. 22 Nov 2010.
22. Broughton, Edward. 2005. "The Bhopal Disaster and Its Aftermath: A Review." *Environmental Health: A Global* 4.6: 1. Web. 24 Nov 2010.
23. Norton, W.W. United States. 9/11 Commission Report. New York, 2004. Print.
24. Department of Homeland Security's (DHS's) Role in Critical Infrastructure Protection (CIP) Cybersecurity GAO-05-434. Washington, DC: Government Accountability Office, 2005. Web. 28 Nov 2010.
25. Serabian, Jr., John A. "Cyber Threats and the U.S. Economy. Statement for the Record Before the Joint Economic Committee on Cyber Threats and the U.S. Economy," Central Intelligence Agency, Washington, DC, U.S.A. February 23, 2000. Speech.
26. Highway Accident Report Collapse of I-35W Highway Bridge Minneapolis, Minnesota, August 2007. Washington, DC: National Transportation Safety Board, 2008. Web. 29 Nov 2010.
27. Nuclear Power Plant Emergency. Washington, DC: Federal Emergency Management Agency, 2010. Web. 30 Nov 2010.
28. Backgrounder on Chernobyl Nuclear Power Plant Accident. Washington, DC: Nuclear Regulatory Commission, 2009. Web. 30 Nov 2010.

Chapter 6

The Role of the Public Health Official

Michael J. Fagel

Contents

The Emergency Management cycle is just as important to the Public Health Official (PHO) as it is to any other person or agency that has a role in dealing with incident response. It is only through planning and training, under the preparedness portion of the cycle, that PHOs will know how to adequately respond to a major incident that is clearly beyond the scope of a routine emergency. A PHO is not going to be involved in a routine emergency. These are almost always best handled by local fire, police, and EMS resources on a daily basis.

But for incidents beyond that, which are larger in scope, then the PHO is going to be involved in a variety of ways. Moreover, the PHO must also be able to answer a variety of questions immediately after an incident occurs to determine what kind of response is needed. The questions listed below are among those that the PHO needs to think about, and will be expounded upon later in this chapter.

What Is the Incident?

The answer to this question will of course determine all courses of action related to the questions below. What might be a significant and sufficient response for one type of incident (such as coordinating to assist first responders in getting to an explosion scene as quickly as possible to perform rescue/recovery operations) may be completely inappropriate in a suspected hazardous materials incident, for example, where the first goal would be to identify the problem and clear as many people from the area as possible.

When Can the PHO Help?

An Incident Commander under an Incident Command System (ICS) procedure is probably going to be in place already by the time the PHO gets word of a significant incident that will require his/her involvement and response. But when can the PHO help? Have training exercises and materials been developed that clearly indicate when the PHO will be a part of the emergency management effort? Are there resources that the PHO needs to call upon that will not be immediately available, such as supplies from the Strategic National Stockpile (SNS), that even with everything functioning properly, might see more than 12 hours pass before necessary medicines and supplies arrive to an affected area? How quickly can the PHO provide necessary resources to an incident commander if those resources need to come from outside his/her agency? For example, a medical or research doctor from outside the area who may have particular expertise with emerging threats from outbreaks such as H5N1, "Avian Influenza", or other biological/chemical agents that might be deployed in a terrorist attack (or unleashed as a result of a spill or unintentional incident).

Who Is Involved?

When state and federal resources are called in, who is the point of contact to make sure everything is delivered efficiently and effectively? Who does the PHO absolutely have to keep open lines of communication with in the wake of a major disaster or terrorist attack? Who coordinates the information gathering and dissemination in such a situation, making sure that the public is not overly panicked, not underestimating the gravity of the incident, and accurately informed of what their immediate, short-term, and long-term next steps should be?

Your Training Plan in Action

This section will discuss a sample incident that could occur in your jurisdiction and provide a walk-through of what actions must be taken, who must be contacted, and other important things to remember related to the PHO's role in an emergency response scenario. This will also help to summarize the information presented throughout the rest of this chapter.

What Is the Incident?

The type of incident at hand will always determine what type of response is needed. The time to decide how to respond to incidents that vary in type and severity is *not* after the incident occurs. This is where the preparedness stage of the Emergency Management cycle is critical.

As part of the training and preparedness process, the PHO must, through his/her agency and with the cooperation of other agencies and neighboring jurisdictions, lay out specific plans for how to respond to various types of incidents. Your department must understand the fundamental differences in responding during the aftermath of a tornado strike in your jurisdiction and a terrorist attack or other large-scale explosion that causes not only casualties, but disruption of services and supplies that the public would view as basic needs.

Although it may sound like painstaking work, the PHO must partner with those in his/her agency, surrounding jurisdictions, and other organizations to develop plans on how to handle each type of incident. For example, your agency may have the following types of plans (general examples): strong aftermath response, natural disaster reponse, nonnatural disaster response (nonterrorist), terrorist attack.

Storm Aftermath Response

This will cover how the PHO will be involved in Emergency Management responses to tornadoes, hurricanes, and even severe thunderstorms that may cause damage in critical areas such that a Public Health response is needed.

Natural Disaster Response

If you work in an area that lies on or near fault lines, you must have a plan in place to deal with emergency response in the wake of an earthquake. Or, if you work in a coastal area, plan for the possible response need in the wake of a tidal wave that could affect your jurisdiction—even if the earthquake or other event that causes the tidal wave occurs well away from your location, where you would not normally be providing assistance in an emergency response.

Nonnatural Disaster Response (Nonterrorist)

How will the Public Health response be coordinated for disasters not caused by nature, but that can and do happen on a regular basis? Examples of this would include chemical plant explosions, train derailments (is a hazardous materials incident involved?), plane crashes, or other similar incidents that could involved mass casualties. These will be incidents that will involve large-scale work from first responders, but significant follow-up work by Public Health officials to deal with what could be a large number of dead or injured, and the aftereffects of a hazardous materials release in the event of the chemical plant explosion or transportation incidents.

Terrorist Attack

The difficulty for the PHO, and for everyone in Emergency Management, is that the exact form of the terrorist attack will not be known, of course, until after it has already occurred. Hurricanes can be forecast with a great degree of accuracy. Although earthquakes remain difficult to predict, they have been well studied in terms of what kind of destruction they cause and what response is needed in order to help the most number of people in a timely, effective fashion. Even the minutes given by releasing a tornado warning just before one strikes can be beneficial in coordinating the necessary Emergency Response effort.

There is no such advantage with terrorist attacks; however, by planning for all the other major types of events listed here, and by working with local, state, and federal authorities to coordinate the Incident Command planning and the Public Health agency's role in such a response, the emergency response to a terrorist attack will be more efficient, effective, and will, in the end, save more lives.

This type of effort requires more than simple planning. It is always wise for the PHO to work in their agency to coordinate drills and exercises that will allow everyone who could be involved in a response the opportunity to practice for the real thing—keeping in mind that inevitably, the real thing will occur in some form. Training your agency for any kind of incident should definitely include training on how to utilize Personal Protective Equipment (PPE). In an emergency response, staff in your agency are either going to have know how to function using this equipment, or they are going to have to quickly be able to teach someone else how to do so. It may also fall on the Public Health agency to train staff from other agencies and even volunteer organizations in how to use PPE, and such training should be offered as broadly as possible.

Another thing to think about is that first responders and everyone else who will be offering some sort of assistance after an incident have probably developed their own plans for how to do so, based on what the incident is. The Public Health facet of the response cannot ignore the plans of everyone else. The PHO should train those staff in his/her agency (and volunteers, if necessary) in the entire emergency operations plan for your jurisdiction. It is important that everyone is working on the same page at all points during an emergency response to an incident. Success in this area prevents "turf wars" between agencies and jurisdictions over who is the right person or agency to carry out a particular part of a response. Another matter to think about is that the Public Health agency should be involved in, or create, a Continuity of Operations Plan, in the event of a disaster so significant that preplanned communication lines, incident command response, and coordination are interrupted.

For the most part, the Public Health Official/Agency is not going to be the lead in Incident Command. Instead, the PHO and his/her staff will be coordinating various aspects of the response that first responders, law enforcement, and EMS crews are not going to have the time nor ability to deal with—resources, supplies, coordinating a volunteer effort, infrastructure, public information, and possibly mass evacuation. Although Public Health is not going to be the lead, it should be involved at nearly all levels, and should coordinate with surrounding jurisdictions and necessary contacts at the local, state, and federal levels to be able to respond efficiently no matter what type of incident is at hand.

Beyond the preincident training, there are questions that need to be answered immediately after an incident occurs related to a possible Public Health response. For example, if an incident occurs for which it is known going in that a Public Health response is required, is the incident significant enough that some abilities of the Public Health agency have been cut off? For example, you may need to have a plan for alternate communications with key staff that will participate in the Public Health response should electricity and communications (particularly cell phone communications) be cut off in the wake of an incident. Has the incident caused outages or damages to Public Health facilities? Will area hospitals become areas where rescue is needed instead of where victims are taken, such as what occurred in New Orleans in the aftermath of Hurricane Katrina? If Public Health operations have been affected, what is the backup plan that can be put into action, and does such a plan require extra staff or volunteers of time and/or resources?

For Public Health staff training, a job aid is provided here that can be used by incident in the planning process (Figure 6.1). Note that this job aid includes just general questions to go through by incident—conditions will, of course, vary by the severity, location, and scope of an incident. But this job aid can help PHOs determine who should be involved in an incident response, what coordination needs to take place with outside agencies and jurisdictions, what safety measures need to be taken, and what other questions need to be asked immediately after an incident occurs so that the Public Health response can be efficient and effective. It is a starting point to effective planning that should also include instruction, exercises, and coordination with other likely responders.

Purpose: To describe the means, organization, and process by which the Public Health agency in a jurisdiction will coordinate its role in an emergency response for this particular type of incident.
Type of Incident: *(Brief description of the type of incident. This job aid can be used to draw out the necessary Public Health response for all types of incidents, based on the categories already mentioned.)*
Staff Needed (with contact information and particular areas of expertise): *(For each type of incident, list the Public Health staff in your agency that will be critical to the emergency response. Some staff will be needed no matter what the type of incident is, but they should still be listed here. Also list key people outside your agency that will be important contacts, such as the Emergency Operations coordinator; Police and Fire Chiefs; local, state, and Federal Government contacts; and key contacts from nearby jurisdictions.)*
Responsibilities for the staff listed above, based on the type of incident.
Supplies Needed: *(For each type of incident, list key supplies that either the Public Health response will need to provide, or that will need to be procured from either volunteer organizations (blood from the Red Cross), outside jurisdictions (vehicles, blankets, food, etc.), or the Federal Government (Strategic National Stockpile (SNS)).*
Emergency Operations Center Liaison: *(Name and contact information for someone in your agency will be within the EOC to help coordinate Public Health efforts related to the overall effort. Liaison may change based on the type of incident that occurred and the type of expertise necessary–for example, upon the discovery of a Avian Influenza case(s) in the United States, it would be wise to have an EOC liaison with research knowledge in that area, and/or knowledge of pandemics, etc.)*
Lessons Learned: *(Has your jurisdiction faced this kind of incident before? If so, what worked well, what didn't? What supplies and resources were found to be useful and whose expertise is required–even if that person is no longer in the area, the PHO may still need to call on him/her. Part of being prepared is learning from the past and this is a section that can provide a brief synopsis of the response to previous incidents–can be particularly useful in the case of train derailments, hurricanes, tornadoes, etc.)*
Purpose:
Type of Incident:
Staff Needed (with contact information and particular areas of expertise): Volunteer organizations to call on?
Responsibilities for the staff listed above, based on the type of incident.
Supplies Needed:
Emergency Operations Center Liaison:
Lessons Learned:

Figure 6.1 Job aid 1: Public Health response by incident.

When Can the PHO Help?

The type of response needed from a Public Health perspective varies by the amount of time that has elapsed after an incident occurs. For example, what is needed from the Public Health function is different in the first three hours after an incident than it is after the first 12 hours (Figure 6.2). In the first hours after an incident occurs, the most important function for the PHO and agency—based on the training that has been performed and the exercises that the Public Health staff has gone through—is to assess what the incident is (discussed above), then coordinate what the public health response is going to be.

The opening hours after an incident occurs are the time for the Public Health agency to coordinate with all necessary agencies and jurisdictions to determine the response. In the first hours, the Public Health agency should determine if their locality is affected (e.g., if the incident is beyond your jurisdiction's borders), and how many people in your area could be affected. If the incident is in another jurisdiction, is it a significant enough of a disaster that it will still affect an increased portion of the population that you serve? A PHO will also need to know who the other responders are to the incident, and once in the area, who the Incident Commander is—if the Incident Command System has been set up at that point. You will also need to know if an Emergency Operations Center (EOC) has been activated, and if so, or if it will be, providing information to the person in your agency who will work within the EOC to help coordinate the Public Health response to the disaster. Remember, Public Health will not lead the incident response except in rare cases, so this early coordination in the first three hours after an incident is critical to assuring a cooperative, efficient, and effective response plan.

As time passes from when the incident occurs, moving toward the period 3–6 hours afterward, there are additional responsibilities that the Public Health response must take on. This is done while continuing to carry out all the responsibilities with coordination and assessment that take place in the first few hours discussed above.

Figure 6.2 Time is of the essence in the immediate aftermath of an incident or crisis and different responses are required at different times postincident.

This next period of time serves as the first window to begin seriously updating the public about what has taken place, what rescue and stability plans are in place and being activated, if an evacuation is necessary or any other instructions the public needs to heed, and set up means for Public Health functions to receive donations and other volunteer work that may be needed in a disaster. For example, if a disaster renders a significant need for donated blood—this will need to be communicated to the public by a Public Information Officer (PIO), with directions on where to go (presuming the disaster has not made it unsafe to do so). This is almost automatic in any kind of significant disaster, given the regularly reported shortages of many blood types on hand for organizations such as the American Red Cross.

Another critical function of this postincident communication is providing direction for members of the public with special needs, or who are disabled, or otherwise unable to respond as needed after an incident occurs. There may be a specific Public Health need to order an evacuation of nursing homes and other large health care facilities, and this will also need to be made public. The PHO must also know the makeup of his/her community. Do you work in a largely diverse population area, where for many, English is not the first language? If so, you will need to have preplanned for communicating in English, Spanish, and possibly other languages (depending on the area) to your community regarding the incident, what has occurred and what next steps the public should take. This is especially critical in the event that a mass evacuation has to be ordered, or—in the case of a biological or chemical incident—that you must order people to stay in the homes and take certain precautions so as to not be adversely affected.

During this period, especially in the course of responding to a significant incident, there will be large, continuing coordination effort between agencies and jurisdictions. What may start as a local incident, depending on whether it is a terrorist attack or simply how many people are affected, may become an incident with state and federal response interest. There will need to be people available from the Public Health agency to work with these officials for coordination purposes. During this period, the local Public Health agency may also discover that because of the gravity of the incident, local supply resources and staffing needs are going to be insufficient for this level of response, and state and federal assistance is going to be needed.

This is the type of effort that will continue as the response reaches the 12- to 16-hour mark after occurrence. By this point, the local Public Health agency will be working in tandem with various federal agencies. Take the example of an epidemiological emergency, such as Avian Influenza. According to the World Health Organization as of August 2011, there have been over 500 cases of Avian Influenza A/H5N1 with nearly 350 human deaths reported so far world wide, but there are cases currently being reported in Asia on a more increased basis. It is not going to be likely that the local Public Health agency, no matter how well equipped and organized, is going to be able to handle such cases, or an outbreak, on its own. In such a case, the local Public Health agency is going to need to work with, at the very least, the CDC and the Department of Health and Human Services.

In addition, someone from the Public Health agency, probably the PHO, should be in contact with officials regarding the SNS. Very rarely will a locality have sufficient supplies of medicines and such needed in the case of Avian Influenza, or any significant epidemiological event, to handle the response, treatment, and possible vaccines on its own. The SNS was designed to be able to send needed supplies of medicines to localities within 12 hours of request. Beyond simply providing medicines, the SNS also houses supplies of antidotes, antitoxins, and medical/surgical items. It is specifically designed to bring necessary resources to localities that in most cases will not be expected to stock all such things—even though the public will need them in the case of specific types and degrees of disasters. Initial contact with SNS officials should be made within the first 6–12 hours after an incident has occurred or is discovered, to assure the most rapid delivery of supplies and medicines to those in your community that will need them.

As part of the Public Health response, there are also other concerns to look at as the time beyond an event changes from hours to days. Is there an environmental impact that the disaster has caused, and if so, what federal and state officials and agencies will need to be called in to assist in assessment, cleanup, and recovery? It is at this point in the recovery stage after an incident that the Public Health agency can also work in tandem with transportation, utility, and government agencies to assess other needs. Is a mass evacuation center needed and if so, how will affected people be transported to that facility? Have transportation routes been affected such that calling for a mass evacuation may cause more chaos and distress than that of the incident? Are there other federal agencies that have specialized teams with which the Public Health agency needs to be in contact?

Here is another job aid that features things for the PHO to think about in the aftermath of an incident, at varying time intervals. Although certain responsibilities are listed for certain time periods, keep in mind that as part of the response, those tasks that need to be taken care of early in a response will need to be continued as the response moves along, even as new responsibilities are added. This job aid should be used as part of training the entire Public Health agency staff, and shared with any volunteer organizations or other agencies that will be assisting as part of the Public Health response effort (Figure 6.3).

Who Is Involved?

As mentioned previously, the success of the Public Health agency's response to a disaster/incident will depend greatly on the cooperation of many people. And this involves more than just the other staff that make up the agency and work for the lead PHO. This part of the chapter will look at particular people that the PHO should look to for assistance and help in coordination, and note how these lines of communication should be opened well before any incident takes place, but rather during the Preparedness phase of the Emergency Management cycle. Trying to establish these contacts during the immediate aftermath of an incident

Purpose: To describe what, in general, actions should be taken in the aftermath of an incident to provide the most coordination with other local, state, and federal responders and to best serve the public's needs.
0–3 Hours after Incident: • Determine what localities are affected by the incident • Determine what parts of the possible Public Health response have also been affected and/or cut off by the incident itself • Coordinate with other local agencies and if necessary, Public Health agencies from other jurisdictions • Determine who has been assigned as the Incident Commander • Determine if an Emergency Operations Center is being opened • If so, who from the Public Health office will be the liaison to the EOC?
3–6 Hours after Incident: • Assign a Public Information Officer (PIO) to update the population on what has happened, the state of recovery, and any next steps they need to take (evacuation, vaccination, protecting their homes, etc.) • Begin coordination of volunteer effort for donated blood, food, water, and other supplies • Contact the Strategic National Stockpile if it's deemed necessary, based on the type of incident and medicinal supplies available in the locality • Begin serious coordination with other jurisdictions and state and federal agencies, as by this time their will be involvement from all levels of government depending on the severity and type of the incident
6–12 Hours after Incident and Beyond: • Have shelters up and running for the public • Have facilities available to handle mass casualties, if necessary • Begin consulting with environmental, transportation, utility and facilities experts to determine what long-term plans are going to be needed for recovery, in addition to the short-term plans for continuing to deal with the aftermath of the incident for those with immediate needs • Determine if any other specialized assistance from state or federal agencies are necessary based on the severity/type of the incident • Is a quarantine facility necessary in the case of Avian Influenza or other biological, chemical attack/incident

Figure 6.3 Public Health response: a timeline.

will only make the overall response less efficient, along with making the Public Health response less effective.

The types of people that the PHO needs to contact after an incident will always depend on the incident itself. A chemical plant or refinery explosion that spreads flames, smoke, and potentially hazardous fumes over a densely populated region in your jurisdiction is not going to require the expertise of an epidemiologist. But you can not discount the contribution that person could make in the event of the discovery of a Avian Influenza case in your jurisdiction, or something more standard, such as a disease outbreak of a malady that is well known, but that arrives on a large scale. As another example, the PHO will probably have no need to call on supplies from the SNS in the wake of a tornado, but there may be other disasters—ones that are not necessarily biological or chemical related, but simply happen on a grand scale—that could require that sort of contact.

On the next page is a list of possible contacts whom the PHO should have constantly open lines of communication with, so that when an incident occurs, the PHO, his/her staff, the Incident Commander, or other entities can more effectively and efficiently coordinate the response by calling in all the necessary expertise.

Note that, if you are in the Public Health agency in an urban area, or cover a large jurisdiction with a significant population, there may be even more agencies and contacts that you need to have to cover the communications lines for all possible incidents. The contacts listed in Figure 6.4 are suitable for any jurisdiction, but it is never unwise to add to it. The more prepared you are, the better.

Your Training Plan in Action

The final section of this chapter will carry you through the PHO's role in incident response based on a possible scenario. This scenario will deal with Avian Influenza, and the confirmation of cases in your jurisdiction. Your first thought may be to read on only casually—after all, as of December 2005, this malady had not been reported as striking a human in the United States, and the likelihood that it would seems very small.

But in emergency management, this is exactly where the training process begins instead of ends. The key to successful emergency management coordination is planning for everything, to the point of having a Plan X when all other plans have failed because of catastrophic conditions. Two cases of Avian Influenza, as you will see in this scenario, do not instantly set off the kind of response effort you would have were a jetliner to crash in a densely populated residential neighborhood, such as what occurred outside New York City several years ago. However, the sequence of events and contacts that the PHO must work on and with is similar and every bit as important.

One factor to consider in a Avian Influenza incident, or a similar type of disease outbreak, is that the typical first responders will not necessarily be in play. Police, fire, and EMS department personnel, who will be first to the scene in a transit accident, derailment, plane crash or explosion, will not be the first ones called in upon the discovery of a Avian Influenza case in your jurisdiction. Instead, the scenario would probably occur as described below.

- The PHO is likely to be first made aware of this situation once the first patient who has been infected is in the hospital. There are several concerns that the PHO must consider at this point.

Purpose: To generate as complete a list as possible of those who need to be called upon in the event of a disaster, either man-made or natural, that will lead to better coordination and a more efficient, effective emergency response. This list is designed to be useful for a Public Health Official who is new to the job, or who may be on his first PHO assignment in a smaller locality. Some of the entries may seem elementary, but all are necessary and the time to open the line of communications is before an incident occurs, not after it.

- Local police chief/sheriff/state police
  - Depending on the locality, these designations may differ, but open lines of communication with whoever is present.
- Local ambulance services and EMS personnel
  - Are these personnel in your area volunteers? If they are, they will be responding to an incident from many different directions, and the PHO will not have a central place to call to reach all of them. Get necessary mobile/text/Blackberry contact information as much as possible.
- All local and regional medical centers and emergency rooms
  - While all incidents are local, the incident that happens in your jurisdiction may require resources from outside your area. In addition, incidents that occur far from your area, that you may think won't affect you, may require evacuation of the wounded to facilities in your area and you must be ready for this eventuality.
- Nonemergency health care personnel
  - Any major incident is going to find the Public Health Official in need of nurses and other care givers, as well as doctors who may have specific areas of biological, chemical, physical or environmental health expertise. Track down these contacts in your area and make sure they are aware that you will look to them as resources in the incident response process. As a new PHO, you may question whether you would ever need the assistance of the pandemic flu expert in your area. The answer to this question is yes, you do. The time to find that out is not AFTER the discovery of multiple Avian Influenza cases in your jurisdiction, at which point every action and minute spent will be critical.
- State and federal agencies that will be critical in an emergency response
  - Examples include the Centers for Disease Control; officials who coordinate distribution of materials from the Strategic National Stockpile; the National Transportation Safety Board (major transportation incidents including highway, rail, and aviation incidents); the U.S. Department of Homeland Security/FEMA (local-regional and national contacts); the Department of Health and Human Services and their Emergency Response Teams; specialized emergency response teams at both the state and federal level; the Environmental Protection Agency, etc.
- Coroner's Office or medical examiners in your jurisdiction or nearby
- Local and state transportation departments
  - Consider the scenario if the PHO has to order an evacuation from an area after an incident occurs. This contact is necessary in order to notify the public of safe escape routes and ones that should be absolutely avoided.
- Local and regional animal control and veterinarians
  - The purpose of this is two-fold. The potential loss or displacement of pets is an issue in any incident. Post Katrina, the Pets Evacuation and Transportation Standards act sets new requirements for Pets in disasters, and will require integration of the PETS act into your planning process at ALL levels. The PHO can't be involved with this directly, but must know who to call in when the situation arises. On the other hand, some incidents will require the direct involvement of veterinarians, such as disease carried by birds or other animals that could cause a local epidemic.
- Local and regional utilities
  - The PHO will need to coordinate with these organizations to discuss possible power outages resulting from an incident, or if there are special instructions the public must follow in terms of drinking water, bathing water, etc., natural gas lines being turned off, etc.
- Volunteer organizations
  - The PHO should establish strong lines of communication with local, state, and federal volunteer organizations, such as the American Red Cross and the Salvation Army, as well as other community service organizations such as blood banks and shelters. This is often the public face of a response, in that requests for donated blood will be issued, and while the people affected by an incident will require assistance with supplies that these organizations can supply as part of the coordinated effort, others not directly affected by the incident will want to know where they can send donations and other assistance–or go and volunteer themselves. This extra resource of human help in an emergency response can never be discounted and it is important to know before an incident occurs how this is going to be coordinated from the Public Health perspective.
- Local, state, and federal government officials
  - Beyond dealing with any particular agency at these levels, the PHO must have emergency contacts that can provide a direct link to those that will need to make key management decisions in an incident. Is the incident significant enough that the federal government is going to officially designate your jurisdiction as a Disaster Area? If so, what does this mean for you and the Public Health response to the incident? These are issues that have to be discussed and coordinated during the preparedness stage.

Figure 6.4 Public Health response plan contacts.

- What medicines are available to treat the infected person in this particular case? You will discover this easily enough through dealing with the primary caregivers at the medical center where the patient is checked into. Medication to assist the patient (although not cure the disease) is available currently.
- Who else in that medical center may be infected?
- Surely, the personnel at the medical center did everything possible to keep this patient away from possibly infecting others. But did they all take the kind of precautions necessary in this situation? Could the disease have spread to either a caregiver or another patient? How many patients have checked *out* of that facility in the time that this patient has been there, thus raising the possibility (no matter how small it may be) that the disease could be carried into the general population in your area?
- Contacts: You are lucky here in the sense that if there has been no spread, you have your incident zone, and many of the people you need to have on site to handle this situation are already in place. However, it is at this point that the PHO would have to contact several individuals (see below).
- Local or regional doctors and/or knowledgeable research experts regarding Avian Influenza

  They will know how to treat it and how to contain it. You may find that this sort of assistance and expertise may only be available at the federal level, depending on your locality, and such contact needs to be made as quickly as possible so the proper authorities and such can be dispatched to your location.
- Neighboring Public Health officials

  It is entirely possible, of course, that if a Avian Influenza case has been discovered in your jurisdiction, that there may be others waiting in the jurisdictions around you. Word must be spread to their Public Health officials as quickly as possible to start any possible response there, and/or to help you with your work in your own jurisdiction.
- Your agency's PIO

  Someone is going to have to speak to the public about this incident. The public will need to know about possible symptoms, and where to go to get assistance if they are afraid of possibly having caught the disease. The public will also want to know what precautions to take if they do not have symptoms. Another thing to think about related to this is that in this age of instant information, news will leak about this and the media will converge on your locality, bombarding the Public Health Office with phone calls and requests for more detailed status reports and interviews. The PIO needs to be the person to handle all of this, as you will not have the time.
- Police and local and state government officials

  You will need to work with both of these entities to discuss a possible quarantine station if other cases exist, or to remove the current case from the medical center in order to avoid the risk of infecting other patients or caregivers. The public will need to know where this area is so as to avoid it, and the police will be needed to not only help set it up in some cases, but also to provide protection.

- EMS (local and regional)

  It is not unlikely that once word is released of this case, a sort of panic will fall over the population in your jurisdiction, and anyone who feels they are suffering from symptoms similar to those described for Avian Influenza may start calling for emergency assistance. The local and regional EMS units have to be put on alert for this possibility, both to quickly respond in the correct fashion, and also to have time to call in reinforcements and volunteers if necessary if the volume of requests for assistance proves to be tremendous.

- Centers for Disease Control

  Obviously, the CDC will want to know of these developments, and they will be the foremost experts in what steps the PHO and agency should take next in dealing with the patient, caregivers, and the public. They will also quickly dispatch experts into the locality and may well take the lead on this side of things. This will leave the PHO with more of the coordination and support response.

  In addition, experts from the CDC can also advise, based on the situation, if any kind of PPE is necessary and who should have it. You will need to know this so that you understand if PPE is readily available in your jurisdiction, or if you are going to need to call on health agencies in neighboring jurisdictions to provide resources.

- SNS Personnel

  While you have received a report that there are medicines and supplies available for the one case that you are aware of, will your locality be ready if you suddenly discover in the next few hours that there are four cases? Eight? Twenty? Officials from the SNS must be called on immediately. The SNS can have supplies, medicines, surgical and medical equipment, and other such needs delivered to your locality within 12 hours, with the help of state officials.

So, three hours have now passed. You have made all the contacts listed above, because the training you and your staff have gone through in the Public Health agency required opening lines of communications with all these contacts. Because they expected to hear from you in this situation, they are not getting their news from the TV or radio, but from those involved in the coordinated response effort that is operating efficiently and effectively to this point.

In the next three hours, the situation continues to unfold. The original patient has been successfully moved to a different facility and quarantined. The PIO has taken to the TV and radio airwaves with instructions for the population, both those who feel that they may have symptoms and want assistance and those who do not but want to protect themselves. A call for calm has been sounded, so as to not have a panicked population making bad decisions that could harm the emergency response effort. The SNS is sending supplies in case they are needed and a delivery method and location has been secured. Officials from the CDC are en route, and experts in how Avian Influenza develops and is transmitted are working with area doctors and staff in your agency to try and track down how this case ended up in your locality.

So what are the PHO's main responsibilities during this time?

- Maintain contact
  - Because of all the communications lines that are opened up, and all the various agents and facets of the response that you must deal with, it is important to be available to continue to receive these communications. You may or may not be near the quarantine site—such an emergency as this may not require direct, on-site involvement. But you will need to be constantly updated on the patient, the reaction, and all the steps that those involved in the response are taking. Constant communication with the Incident Commander—if one is designated for this case—is critical.
  - Reassure the public

    It is very important that the PIO is active in the public eye during this time. Although everyone is obviously worried about the possible spread of the disease—you also must coordinate an effort to soothe and educate the public. Something important to remember in this example is that much of the population in your jurisdiction will have no idea what they should or should not do in this situation in terms of staying safe. Panic is one of the leading risk factors to an effective emergency response, so information is going to be quick, accurate, and consistently available in multiple formats (and probably, multiple languages) so that everyone in your jurisdiction clearly understands the situation, how to stay safe, what to do if an evacuation is ordered, etc.
  - National Guard assistance?

    Depending on the response of the public to this situation, or if there is a lack of resources available in your jurisdiction, you may need to work with military authorities to receive National Guard assistance in maintaining order in your jurisdiction. Word spreading about the Avian Influenza case in your area may be as dangerous as the spread of the disease itself. The National Guard may be needed to make sure the response area is secured, while also possibly being called in to help bring in people who are suffering symptoms and want to be examined.
  - Finalize SNS delivery details

    At this point, it is too early to tell if you are going to have more than just an isolated case on your hands. Although primary caregivers on site have determined that what they need is currently available, that may not remain true should the situation change in the next 6–12 hours, and effectively coordinating the delivery of supplies and medicines with the

SNS will leave your jurisdiction ready to handle such an increase in cases in the coming hours and days should they occur.

- Another point to think about—although it may not necessarily be the case with Avian Influenza—is how will you coordinate the mass dispensing of vaccines if the SNS is able to deliver such an item that will keep a large portion of the population in your jurisdiction from being affected? This is where you need to continue to work with local police, government officials, and volunteer organizations to make sure such a large-scale effort to deliver a vaccine can be done in a calm, orderly fashion. Again, the PIO will be called on in this situation, as well, because information will have to be given to the public on where to go to receive such a treatment, as well as any special precautions that they should take.

One thing you will see in the 3- to 6-hour range, depending on the incident, is an increased federal presence of agencies and responders. But remember, the incident is still a local one. Only you have the first-hand information on what is happening, where it is happening, and who can potentially be affected. Any state or federal officials who join the response during this period will need this communicated to them before knowing what next steps they should take. Although you may not be the lead in such a response, you will most certainly play an important role and that role does not diminish, even when it appears that more of the response is being handled by officials and authorities from outside your jurisdiction.

As you enter the range beyond six hours, looking toward the end of the day and beyond, you now will need to consider future issues. You will need to advise local government officials on whether there needs to be a quarantine or condemning of the building where this case was first discovered. You may think that such an action would only happen in extreme cases, but extreme cases are exactly what you have to plan for in Emergency Management. All other facets of the response that have been discussed in the first hours after you received word of this case are progressing. To this point, you and your staff have followed their training very well.

It has been determined that area utilities, such as the water supply, are safe. There has been no need to call for an evacuation, and no major issues on the roads and with transit. SNS supplies and medicine reinforcements will arrive in the next couple of hours, and the public has been made aware of the process for receiving the attention they need, what time that will begin, where to go, and so on.

These times leading out to the 12-hour mark and beyond are also important for another reason. There will be shift changes taking effect and new people will be called into the response who perhaps are not up to speed on what has taken place to this point. You, or someone you designate, will need to cover that with new responders from a Public Health perspective. And at some point, you will also need to designate someone to fill in for you who can handle all of these responsibilities that you have carried out over the last day.

In this Avian Influenza scenario, one final task awaits for the PIO, as it will be him/her, or you, that will have to deliver the news either in the next couple of hours or in the coming days that the person diagnosed originally to set off this response has passed away. Again, this will be a time to prevent panic among the population in your jurisdiction, and to stress that although this is an extremely unfortunate event, everyone involved in the response is doing their job, and there is no further immediate threat to the public. Through your long-standing communications with a Avian Influenza expert in your jurisdiction, he has been able to report back to you that this was, in fact, an isolated case after all, and there is no threat to the population.

Upon this news, again, communication with the public is key. As the response winds down and any necessary cleanup and breakdown of equipment and such takes place, the Emergency Management cycle is not over. Now, the Preparedness stage begins all over again. Beyond practice drills and tabletop exercises, the PHO now has a live event to study, learn what went well, what went wrong, and what should be done, and how the emergency response should be altered should this situation occur again. In a sense, Emergency Management for the PHO never really ends, because the PHO is always either involved in a response, or training for the next one. Although it may seem overwhelming, this is in fact the key to being a successful contributor to the entire emergency response operation.

Chapter 7

Developing Public–Private Partnerships in the Twenty-First Century

S. Shane Stovall

Contents

The term "Public–Private Partnership" is often a confusing term that many in the Emergency Management community have grappled with for the past several years. However, it is important for Emergency Managers to have a consistent understanding of what Public–Private Partnerships are, and how critical they are to emergency and disaster prevention, mitigation, preparedness, response, and recovery efforts on a day-to-day basis. This chapter examines the types of Public–Private Partnerships, why they are needed, some of the obstacles that may be encountered when developing these partnerships, and how they can be developed. Public–Private Partnerships are sometimes known as PPPs or P3.

Types of Public–Private Partnerships

PPPs have existed in government and businesses for decades, including in Emergency Management and in its predecessor, Civil Defense. Here are the general definitions for some types of PPPs:

Joint Public–Private Partnership

This type of PPP occurs when a government entity(ies) and a private sector entity(ies) use their own personnel, funding, equipment, or other resources to collectively complete a project or enlist a service. Many times, such a partnership occurs via a contract between the two entities. An example of this could be a construction project that enhances ingress and egress of traffic to and from a particular location. From an Emergency Manager's standpoint, an emergency preparedness public outreach project may warrant a partnership that can be formed with private sector business owners, as well as with government agencies. Funding of the project could come from public and private sector entities. From the business standpoint, there could be some direct or indirect advertising, as well as an opportunity to provide community service. From the public sector standpoint, the project would be funded (in part or in full), allowing for a more comprehensive or complete public education effort.

Another example of this type of partnership, from an Emergency Management standpoint, is an information sharing partnership for weather information, disaster preparedness and recovery information, and postevent situation status.

Private to Government PPP

This type of PPP involves the private-sector interests making a capital investment with government entities to provide agreed-upon services. In most cases, the private sector pays the government for services. However, other arrangements that do not require funding can also take place.

An example of this type of partnership could be a private sector chemical company that contracts with a local fire department for hazardous materials planning and response services. Another example of this type of partnership is where a company may contract with a governmental entity for trash collection. From an Emergency Management perspective, an example of this type of partnership could include a health care company that is in need of a hazard vulnerability analysis for their local location's emergency plan. This could involve funding, or be done as a service as part of the Emergency Management agency's duties. Whatever the case, a partnership will need to be developed and maintained in order to ensure that the project is comprehensive and contains the details about the facility and the hazards to which it is vulnerable. Naturally, this partnership should be maintained as the health care facility may be considered a critical facility within the jurisdiction, and could provide an important role in disaster response and recovery.

Government to Private PPP

Another type of PPP occurs when the government contracts for services from the private sector. This can be an unfunded partnership, but is typically funded through general funds or grant funds. This type of partnership is probably the most commonly used PPP. However, many may not see it as such because the public sector is simply purchasing a product or service from the private sector. Although this may be the case, it is common sense for the public sector to want to partner with their vendor or private sector entity in order to ensure the success of the project. It is not wise for either party to have an agreement where no working partnership has been developed. Moreover, the partnership helps both parties understand goals, objectives, and expectations of each other on the partnering effort.

An example from an Emergency Manager's standpoint is a governmental jurisdiction that decides to contract private sector entities for debris management planning efforts. Although this will essentially result in a contract for services, it is wise to build a partnership between the public sector and private sector parties. This partnership will ensure that the project is successful and that relationships are built to ensure effective preparedness, response, and recovery efforts related to the project at hand.

As discussed above, there are several different types of PPPs that can be developed. They can be initiated from a private sector entity or from a public sector entity. They can be funded or nonfunded. Whichever the case, they are necessary for successful Emergency Management programs.

Need for PPPs

Following the events of September 11, 2001 and the events surrounding Hurricane Katrina in 2005, a need was identified for better-coordinated public–private sector planning and response to emergencies. Some excerpts identifying the needs are presented in the following subsections.

September 11, 2001

Private-Sector Preparedness

The mandate of the Department of Homeland Security does not end with government; the department is also responsible for working with the private sector to ensure preparedness. This is entirely appropriate, for the private sector controls 85% of the critical infrastructure in the nation. Indeed, unless a terrorist's target is a military or other secure government facility, the "first" first responders will almost certainly be civilians. Homeland security and national preparedness therefore often begins with the private sector.

Preparedness in the private sector and public sector for rescue, restart, and recovery of operations should include (1) a plan for evacuation, (2) adequate communications capabilities, and (3) a plan for continuity of operations. As we examined the emergency response to 9/11, witness after witness told us that despite 9/11, the private sector remains largely unprepared for a terrorist attack. We were also advised that the lack of a widely embraced private-sector preparedness standard was a principal contributing factor to this lack of preparedness.

We responded by asking the American National Standards Institute (ANSI) to develop a consensus on a "National Standard for Preparedness" for the private sector. ANSI convened safety, security, and business continuity experts from a wide range of industries and associations, as well as from federal, state, and local government stakeholders, to consider the need for standards for private sector emergency preparedness and business continuity.

The result of these sessions was ANSI's recommendation that the Commission endorse a voluntary National Preparedness Standard. Based on the existing American National Standard on Disaster/Emergency Management and Business Continuity Programs (NFPA 1600), the proposed National Preparedness Standard establishes a common set of criteria and terminology for preparedness, disaster management, emergency management, and business continuity programs. The experience of the private sector in the World Trade Center emergency demonstrated the need for these standards (Figure 7.1).

Recommendation: We endorse the ANSI's recommended standard for private preparedness. We were encouraged by Secretary Tom Ridge's praise of the standard, and urge the Department of Homeland Security to promote its adoption. We also encourage the insurance and credit-rating industries to look closely at a company's compliance with the ANSI standard in assessing its insurability and creditworthiness. We believe that compliance with the standard should define the standard of care owed by a company to its employees and the public for legal purposes. Private-sector preparedness is not a luxury; it is a cost of doing business in the post-9/11 world. It is ignored at a tremendous potential cost in lives, money, and national security.[1]

Figure 7.1 World Trade Center towers in New York City following terrorist attacks on September 11, 2001. (Photo courtesy of the National Park Service.)

Hurricane Katrina, August 2005

Lessons Learned

The Department of Homeland Security, in coordination with state and local governments and the private sector, should develop a modern, flexible, and transparent logistics system. This system should be based on established contracts for stockpiling commodities at the local level for emergencies and the provision of goods and services during emergencies. The federal government must develop the capacity to conduct large-scale logistical operations that supplement and, if necessary, replace state and local logistical systems by leveraging resources within both the public sector and the private sector.[2]

The federal response should better integrate the contributions of volunteers and nongovernmental organizations (NGOs) into the broader national effort. This integration would be best achieved at the state and local levels, before incidents occur. In particular, state and local governments must engage NGOs in the planning process, credential their personnel, and provide them with the necessary resource support for their involvement in a joint response.[3]

The Department of Homeland Security should develop a comprehensive program for the professional development and education of the nation's homeland security personnel, including federal, state and local employees as well as emergency management persons within the private sector, NGOs, as well as faith-based and community groups. This program should foster a "joint" federal interagency, state, local, and civilian team (Figure 7.2).[3]

Creating a Culture of Preparedness—Initiative

We must build on our initial successful efforts to partner with other homeland security stakeholders—private sector, NGOs, and faith-based groups. Each of these

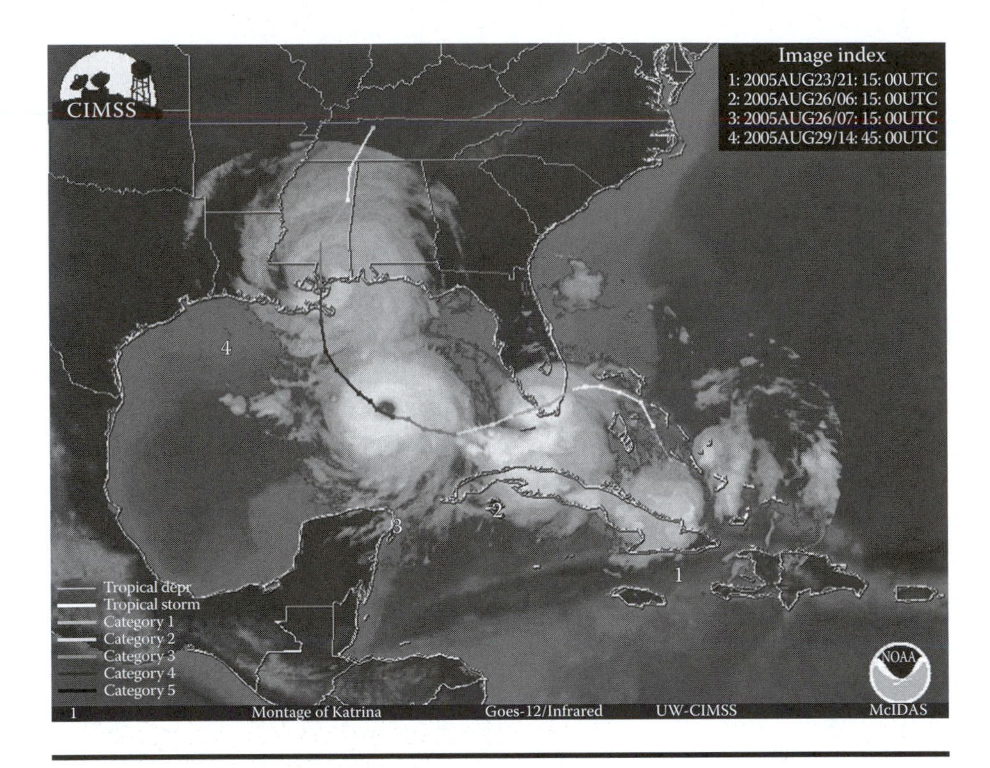

Figure 7.2 Hurricane Katrina track. (Photo courtesy of the NOAA.)

groups plays a critical role in preparedness. To the extent that we can incorporate them into the national effort, we will be reducing the burden on other response resources so that federal, state, and local responders can concentrate our energies on those with the greatest need.

Private sector companies own and operate 85% of the nation's critical infrastructure. Transportation, electricity, banking, telecommunications, food supply, and clean water are examples of services relying on infrastructure that have become basic aspects of our daily lives. Yet, these services are often only noticed when they are disrupted and when the American public expects speedy restoration. In fact, the nation relies on "critical infrastructure" to maintain its defense, continuity of government, economic prosperity, and quality of life. The services provided by these interconnected systems are so vital that their disruption will have a debilitating impact on national security, the economy, or public health and safety.

Companies are responsible for protecting their systems, which comprise the majority of critical infrastructure. Because of this, private sector preparation and response is vital to mitigating the national impact of disasters. Government actions in response to a disaster can help or hamper private sector efforts. However, governments cannot plan to adequately respond unless the private sector helps them understand what infrastructure truly is critical. Likewise, businesses cannot develop contingency plans without understanding how governments will respond. To maximize the nation's preparedness, federal, state, and local governments must join with the private sector to collaboratively develop plans to respond to major disasters. There are important initiatives in this area already underway by the Business Round Table and Business Executives for National Security (BENS) project. We must encourage and build upon these efforts. The private sector must be an explicit partner in and fully integrated across all levels of response—federal, state, and local.[4]

Filling Gaps

It is the responsibility of the Emergency Manager to develop continuing relationships between themselves and emergency and disaster management stakeholders in the public, private, and nonprofit sectors. Quite often, this is easier said than done with the increasing demands on the Emergency Manager's time. However, building these relationships is critical to a successful Emergency Management program.

Each Emergency Manager must examine his or her respective jurisdiction's Emergency Management Plan and determine what planning, training, exercise, personnel, and equipment gaps may exist. These gaps may exist due to lack of physical assets, personnel, or funding, but they also may exist due to lack of expertise. Whatever the case, it is important to identify these gaps before the emergency or disaster, and attempt to fill these gaps as soon as possible. PPPs can help Emergency Managers fill these gaps.

It can be assumed that there are PPPs to assist in filling virtually any operational gap that an Emergency Manager can identify. Many of these PPPs can be developed with entities within the Emergency Manager's communities and regions. Often, entities that can make up these partnerships (if not found within the Emergency Manager's community) can also be found at the numerous conferences and symposiums offered for Emergency Managers, Business Continuity Planners, and other Public Safety officials.

Shared Interest

Common interests for overall preparedness and an effective response and recovery frequently give impetus for PPPs. Frequently, these types of relationships can strengthen the ability of both parties to deal with a particular issue. One example of this is Pandemic Flu Planning. Pandemic Flu Planning cannot be done alone in the public sector or the private sector, for there are some interdependencies that exist. As we all know, businesses both large and small provide the economic backbone for most, if not all, of our communities. Private sector also plays a critical role in most of our critical infrastructure, such as utilities. If a Pandemic Flu incident were to occur, the U.S. Occupational Safety and Health Administration estimates that a pandemic could affect as much as 40% of the workforce.[5] This is due to absenteeism caused by the illness, taking care of children who may be stricken with the flu, and having to watch children who may be forced to stay home due to closed schools. Pandemic can also cause shifts in consumers, who may expend their money on items for infection control, rather than other items. A pandemic flu event can also cause more people to shop online, rather than visit retailers in the community. All of these can negatively affect the local economy. For these types of events, it is critical to the community that the Emergency Manager and the private sector entities (including health care providers, utilities, and other critical businesses) work together to develop plans and contingencies if such a situation were to arise. This is just one example, but shared interests in planning, training, and exercises can be found in all disasters. Therefore, it is critical to build PPPs now, rather than at the time of an emergency or disaster.

As can be seen, there is still a lot of opportunity for Emergency Managers to develop PPPs within their respective areas of responsibility. Since Hurricane Katrina, there have been significant efforts toward developing PPPs at the local, state, and federal levels. However, there is more that needs to be accomplished, and many more partnerships need to be forged. It must be recognized that development of future partnerships with our private sector stakeholders will strengthen both public and private sectors in all phases of Emergency Management.

Traditionally, PPPs in Emergency Management have been mostly limited to planning, training, or exercise contracts on a fee-for-service basis—based on specifications set forth by the jurisdiction. Some other examples of common PPPs in Emergency Management include debris management, temporary housing, and food, water, and ice supplies. However, Emergency Managers in both public and private sectors must broaden the scope of their PPP initiatives and begin looking to broaden these partnerships in the future in order to meet the challenges of future emergencies and disasters.

Challenges in Creating PPPs

As Emergency Managers begin developing additional PPPs, it is important to understand the challenges while examining potential partnerships (many of which can be overcome with some education).

Identifying Potential Stakeholders

Identifying a list of potential private sector (should also include volunteer) partners that could be stakeholders in emergency and disaster prevention, mitigation, preparedness, response, and recovery can be a daunting task. Table 7.1 illustrates some potential areas of expertise (not all-inclusive) under the Emergency Management umbrella that one may be able to find potential partnerships.

Table 7.1 Examples of Emergency Management Stakeholders and Potential Resources

Sector/Discipline	*Potential Resource*
• Utilities	Infrastructure, damage assessment
• Transportation Companies	Evacuation, transport
• Engineers	Infrastructure, staff augmentation for engineers within jurisdiction, damage assessment
• Building Inspectors	Damage assessment, staff augmentation for building inspectors within jurisdiction
• Communications	Equipment, damage assessment, staff augmentation for communications staff within jurisdiction
• Debris Monitoring and Management	Equipment, debris management planning
• Temporary Housing Manufacturers and Suppliers	Sheltering, temporary housing planning, portable offices
• Construction Companies	Equipment, materials, supplies, personnel for repairs
• Food, Water, Ice Retailers	Food, water, ice, other supplies
• Hardware Retailers	Repair supplies, protective supplies
• Health Care Facilities—hospitals, clinics, etc.	Nurses, doctors, medical supplies
• Temporary Staff Services	Skilled and unskilled personnel
• Corporations	Personnel, supplies, materials
• Private Security Companies	Security personnel
• Private Ambulance Services	Ambulances, medical supplies, personnel
• Generator Rental or Supply Outlets	Generators, power equipment

(*continued*)

Table 7.1 (*Continued*) Examples of Emergency Management Stakeholders and Potential Resources

Sector/Discipline	*Potential Resource*
• Retail and Commercial Rental Outlets	Supplies, equipment
• Restoration Companies	Repair materials, document recovery, facility restoration, cleanup
• Warehouse Space/ Temporary Storage	Mass storage

Developing Win–Win Partnership

Another challenge that Emergency Managers face when developing PPPs is developing a win–win scenario with the potential partner. One of the serious mistakes that can be made when developing a PPP is making a one-sided relationship. The Emergency Manager must identify joint benefits and educate the stakeholder on the advantages of the partnership. It is also important to stress how the partnership will benefit the community as a whole. Table 7.2 shows possible items for the Emergency Manager to keep in mind when developing a win–win partnership.

How to Start PPPs

Many Emergency Managers may be hesitant to develop PPPs. Reasons for this can include: (1) not understanding the need for PPPs; (2) not feeling that there is enough time to develop all of the relationships that are needed; or (3) not knowing who to approach within an organization to begin dialogue about a PPP. Regardless of the reason, it is important that the Emergency Manager look past these obstacles and forge ahead. In this section, the elements of the PPP will be discussed with the intent of giving the Emergency Manager (whether in public sector or private sector) several starting points for developing a PPP.

Establishing a Plan

Before contacting anyone about establishing the PPP, the Emergency Manager must ensure that they have a written plan that they can share with the other entity(ies) in the partnership. The written plan should include the following:

- What is the purpose of the partnership?
- What are its goals and objectives?

Table 7.2 Possible Areas for Win–Win Partnerships

Public Sector Benefit	*Advantages to Partnership*
• Sharing of common goals and expectations for completion of a joint project	Public and private sector entity can share goals, timelines, milestones, and expectations to achieve a successful result that will benefit both entities.
• Staff augmentation/added expertise	Partnership may bring additional personnel and skill sets to the relationship that may not otherwise be available.
• Innovative solutions	Partnership may bring new ideas to a project that may not have previously been considered by a single entity.
• Cost effectiveness	Pooling of resources may allow for less time to be spent on a project or effort, thus potentially saving costs.
• Enhanced level of preparedness	Planning, training, and drilling or exercising together will allow for a more effective joint response and recovery following a disaster.
• Exposure/indirect marketing	Although the partnership probably will not offer itself to direct marketing, the partnership may allow for some brand name recognition through sponsorships or other means.
• Community service	Companies typically want to give back to the community, and the partnership may allow them an opportunity to do so.
• Profit	If a public sector entity is contracting/paying for a service or product, then profit may serve as one of the benefits of the partnership.
• Relationship building	Inherently, building a partnership should build a relationship that will enhance future efforts dealing with disaster prevention, mitigation, preparedness, response, and recovery.

(*continued*)

Table 7.2 (*Continued*) Possible Areas for Win–Win Partnerships

Public Sector Benefit	*Advantages to Partnership*
• Expanded field of knowledge/areas of service	Partnerships will allow for the exchange of information and ideas, which may lead the way for new ways of doing business or service lines.
• Enhanced level of preparedness	Planning, training, and drilling or exercising together with public sector partners will allow for a familiarization of what to expect following a disaster. Additionally, such partnerships can enhance response and recovery following a disaster.

- What are the expectations of entities involved in the partnership?
- Are there costs or required funding for the project?
- What are the milestones, timelines, and/or deadlines?
- What are the benefits of the partnership (include benefits for each entity involved)?

There may be other information that can be included in the written plan. However, finding answers to these initial questions should give a clearer picture of the PPP to be developed (Figure 7.3).

Who to Involve

Once the written plan has been developed, it is important to determine the appropriate parties to develop the PPP. The type of PPP being developed will dictate the different types of people who will be brought into the relationship. If an Emergency Manager is developing the partnership with a private sector entity, it may be necessary to initially contact the following people to start a dialogue about a PPP:

- Owner
- Chief Executive Officer/President
- Chief Operating Officer
- Business Continuity Planner
- Risk Manager
- Security Director/Safety Director
- Discipline-specific personnel (i.e., Engineer, Facilities)

If the private sector entity, such as the Business Continuity Planner or Risk Manager, is looking to develop a partnership with a public sector entity, it may

Sample Partnership Plan

1. **What is the purpose of the partnership?**
   The purpose of this partnership is to develop a comprehensive emergency preparedness public outreach program.
2. **What are its goals and objectives?**
   a. Develop multilingual preparedness information for distribution by week 4 of project.
   b. Develop preparedness information for hearing and sight impaired by week 6 of project.
   c. Develop, print, and distribute community preparedness guide by week 12 of the project (count 70,000).
3. **What are the expectations of entities involved in the partnership?**
   The City will develop draft print material and will provide $15,000 toward printing. Multilingual and hearing and sight impaired information will be developed with City resources. YYY Company will assist with printing and provide remaining $10,000 toward project.
4. **Are there costs or required funding for the project?**
   Total $25,000 as mentioned in #3.
5. **What are the milestones, timelines, and/or deadlines?**
   See #2.
6. **What are the benefits of the partnership (include benefits for each entity involved)?**
   Community education materials printed and distributed. Company YYY logo placed on each copy with City logo.

Figure 7.3 Sample partnership development form.

be necessary to contact the following people to engage in a dialogue about a partnership:

- Emergency Management Director/Coordinator
- City/County Manager
- Fire Chief
- Police Chief
- Risk Manager
- Discipline-specific personnel (i.e., Planner, Road, and Bridge)

Maintenance of Partnership

Some partnerships may be finite in terms of what they set out to accomplish. However, it is important that all partnerships be fostered in order to maintain contact and communications for future efforts. One method of accomplishing this is for Emergency Managers to involve partners in drills and exercises, all-hazards planning, and training as appropriate. As previously discussed, developing and fostering these relationships before a disaster strikes allows for a more efficient and effective disaster response and recovery.

Resources for PPPs

The following is a list of primary agencies and organizations that directly support PPPs. Because website addresses change, along with the dynamic information about programs that each of these agencies and organizations support, this section will not go into details about current efforts. Further information can be found on the Internet for each of these agencies and organizations. There is no certain way to have a complete list of organizations and agencies because PPPs in Emergency Management are a growing effort on an international basis.

- Federal Emergency Management Agency (FEMA)—Private Sector Division

  FEMA established a Private Sector Division within the Office of External Affairs in October 2007. The Division's overarching goals include improving information sharing and coordination between FEMA and the private sector during disaster planning, response, and recovery efforts. The FEMA Private Sector Division cultivates public–private collaboration and networking in support of the various roles the private sector plays in emergency management, including: impacted organization, response resource, partner in preparedness, and component of the economy. The division also fosters internal collaboration and communication among FEMA programs that have an interest in private sector engagement.[6]
- Business Executives for National Security

  BENS is a nationwide, nonpartisan organization that is the primary channel through which senior business executives can help enhance the nation's security. BENS members use their business experience to help government leaders implement solutions to the most challenging national security problems. BENS has only one special interest: to help make America safe and secure.[7]
- National Incident Management Systems and Advanced Technologies (NIMSAT) Institute–University of Louisiana at Lafayette

  The mission of the NIMSAT Institute is "To enhance national resiliency to a full range of potential disasters by conducting research leading to innovative tools and applications that empower the homeland security and emergency management community through education, training, outreach, and operational support. The Institute seeks to improve the emergency preparedness, response, recovery, and mitigation activities for communities, supply chains and critical infrastructures that support our economy and our way of life."[8]
- International Association of Emergency Managers—Public–Private Partnership Caucus

  The Mission Statement for this Caucus states that it is designed to "Develop, identify, and promulgate best practices for creating effective partnerships among private, not-for-profit, and public sectors." Caucus objectives include

1. Establish information sharing forums regarding emergency and disaster planning, training, and exercises between the public, not-for-profit, and private sectors.
2. Identify relationships that can be developed between public, private, and not-for-profit entities that can enhance emergency and disaster prevention, mitigation, preparedness, response, and recovery.
3. Develop effective strategies for achieving common emergency and disaster goals and objectives in the public, private, and not-for-profit sectors using Best Practices.
4. Create an awareness of what public–private partnerships are, and how they can be leveraged in emergency and disaster management.[9]

- National Emergency Management Association—Private Sector Committee
  One of the missions for this Committee is to identify, promote, and foster public–private sector collaborations. This Committee focuses primarily on state-level efforts for its private sector members.

Conclusion

As Emergency Managers move their organization's or agency's programs into the future, PPPs are becoming increasingly critical for effective disaster prevention, mitigation, preparedness, response, and recovery. Emergency Managers, as a group of professionals, are becoming more cognizant of the rising demand placed on their programs, whether it is financial or operational. Emergency Managers are increasingly seeing patterns where many are asked to accomplish more with less—less people, less time, and less money. For these reasons, it is wise for the Emergency Manager to look for opportunities to develop PPPs to augment the programs in the public and private sectors. Integrating PPP into Emergency Management programs allows Emergency Managers to tap into technical, management, and financial resources in new ways to achieve their organization's goals and objectives.

References

1. United States. The 9/11 Commission. The 9/11 Commission Report: Final Report of the National Commission on Terrorist Attacks Upon the United States 1 vol. 2004 (pp. 397–398).
2. United States. Executive Office of the President. The Federal Response to Hurricane Katrina: Lessons Learned. 2006 (p. 44).
3. United States. Executive Office of the President. The Federal Response to Hurricane Katrina: Lessons Learned. 2006 (p. 49).

4. United States. Executive Office of the President. The Federal Response to Hurricane Katrina: Lessons Learned. 2006 (p. 81).
5. United States. Guidance on Preparing Workplaces for an Influenza Pandemic—OSHA 3327-02N 2007. Washington, DC. 2001. Web. 23 Jan 2011. http://www.osha.gov/Publications/influenza_pandemic.html#affect_workplaces.
6. United States. About FEMA Private Sector Division. Washington, DC: FEMA, 2010. Web. 24 Jan 2011. http://www.fema.gov/privatesector/about.shtm.
7. "Home Page." Business Executives for National Security. Business Executives for National Security, n.d. Web. 24 Jan 2011. http://www.bens.org/home.html.
8. "Home Page." National Incident Management Systems and Advanced Technologies (NIMSAT) Institute. National Incident Management Systems and Advanced Technologies (NIMSAT) Institute, 2011. Web. 24 Jan 2011. http://www.nimsat.org/.
9. "Caucus Mission." IAEM–USA Public–Private Partnership Caucus. International Association of Emergency Managers, December 2008. Web. 24 Jan 2011. http://www.iaem.com/committees/publicprivate/ppp.htm.

Chapter 8

Assessing Vulnerabilities

James Peerenboom and Ron Fisher

Contents

Introduction

The Homeland Security Act of 2002 provides the primary authority for the overall homeland security mission. This act charged the Department of Homeland Security (DHS) with primary responsibility for developing a comprehensive national plan to secure critical infrastructure and key resources (CIKR) and recommend "the measures necessary to protect the key resources and critical infrastructure of the United States." This comprehensive plan is the National Infrastructure Protection Plan (NIPP), first published by the DHS in June 2006. As defined in the 2009 NIPP, critical infrastructure are the systems and assets, whether physical or virtual, so vital that the incapacity or destruction of such may have a debilitating impact on the security, economy, public health or safety, environment, or any combination of these matters, across any federal, state, regional, territorial, or local jurisdiction. Key resources are publicly or privately controlled resources essential to the minimal operations of the economy and government. The NIPP provides the unifying structure for integrating a wide range of efforts for the protection of CIKR into a single national program.

Homeland Security Presidential Directive 7 (HSPD-7), Critical Infrastructure Identification, Prioritization, and Protection, was established as a national policy for federal departments and agencies to identify and prioritize United States CIKR and to protect them from terrorist attacks. The NIPP provided the follow-up plan to implement HSPD-7. The NIPP called out the need to conduct risk assessments to deter threats, mitigate vulnerabilities, and minimize consequences.

Vulnerability Assessment

Vulnerability assessment methodologies are generally intended to identify any weakness that can be exploited by an adversary to gain unauthorized access to or to disrupt an asset, facility, or system. Terrorism is often the primary focus; however, vulnerabilities can take an all-hazards approach. Vulnerabilities can result from, but are not limited to, the following:

- Asset, building, site, or system characteristics
- Equipment properties
- Personal behavior
- Operational and personnel practices
- Security weaknesses (physical and cyber)

Vulnerability assessment methodologies can be characterized in terms of four assessment elements—physical, cyber, operations security, and interdependencies. Each is briefly described next.

A *physical security assessment* typically evaluates the physical security systems in place or planned at a site, including access controls, barriers, locks and keys, badges and passes, intrusion detection devices and associated alarm reporting and display, closed-circuit television (CCTV) (assessment and surveillance), communications equipment (telephone, two-way radio, intercom, cellular), lighting (interior and exterior), power sources (line, battery, generator), inventory control, postings (signs), security system wiring, and protective force. These systems are generally reviewed for design, installation, operation, maintenance, and testing. It may also include an evaluation of sites housing critical equipment or information assets or networks dedicated to the operation of the physical systems.

A *cyber security assessment* evaluates the security features of the information network(s) associated with an organization's critical information systems. This could include an examination of network topology and connectivity, principal information assets, interface and communications protocols, function and linkage of major software and hardware components (especially those associated with information security such as intrusion detectors), and policies and procedures that govern security features of the network. It may also include internal and external scanning for vulnerabilities (penetration testing).

Operations security (OPSEC) is the systematic process of denying potential adversaries information about capabilities and intentions of the host organization. This is accomplished by identifying, controlling, and protecting generally nonsensitive activities concerning planning and execution of sensitive activities. An OPSEC assessment typically reviews the processes and practices used for denying adversary access to sensitive and nonsensitive information that might inappropriately aid or abet any individual's or organization's disproportionate influence over system operation. This should include a review of security training and awareness programs, a review of personnel policies and procedures, discussions with key staff, and tours of appropriate principal facilities. It should also include a review of information that may be available through public access (e.g., the Internet).

Infrastructure interdependencies refers to the physical and electronic (cyber) linkages within and among our nation's critical infrastructures (i.e., within and among the 13 critical infrastructure sectors and five key asset categories). An interdependencies assessment typically identifies the direct infrastructure linkages between and among both the internal infrastructures at a site as well as the linkages to external infrastructures outside the site. The process of identifying and analyzing these linkages requires a detailed understanding of how the components of each infrastructure and their associated functions or activities depend on, or are supported by, each of the other infrastructures. For example, a supervisory control and data acquisition (SCADA) system that operates a natural gas pipeline depends on the local electric power and telecommunications infrastructures to function. The failure of a separate, external infrastructure could prevent the SCADA system from operating, thus impacting natural gas deliveries to or within a system. Interdependencies can create subtle interactions and feedback mechanisms that often lead to unintended behaviors and consequences, including the potential disruption of critical infrastructures.

Critical Infrastructure and Key Resources

Attacks on CI-KR could significantly disrupt the functioning of government and business alike and produce cascading effects far beyond the targeted sector and physical location of the incident (see Figure 8.1). Direct terrorist attacks and natural, man-made, or technological hazards could produce catastrophic losses in terms of human casualties, property destruction, and economic effects, as

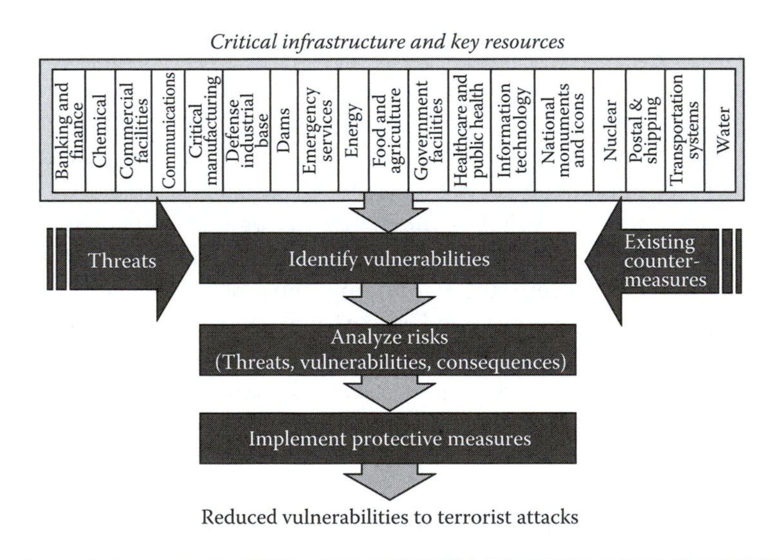

Figure 8.1 Protective security process for critical infrastructures and key resources.

well as profound damage to public morale and confidence. Finally, attacks using components of the nation's CIKR as weapons of mass destruction could have even more devastating physical and psychological consequences. CIKR sectors are described next.

Agriculture and Food

The Agriculture and Food Sector has the capacity to feed and clothe people well beyond the boundaries of the nation. The sector is almost entirely under private ownership and is composed of an estimated 2.1 million farms, approximately 880,500 firms, and more than one million facilities. This sector accounts for roughly one-fifth of the nation's economic activity and is overseen at the federal level by the U.S. Department of Agriculture and the Department of Health and Human Services, and Food and Drug Administration.

Banking and Finance

The Banking and Finance Sector, the backbone of the world economy, is a large and diverse sector primarily owned and operated by private entities. In 2007, the sector accounted for more than 8% of the U.S. gross domestic product.

Chemical

The Chemical Sector is an integral component of the U.S. economy, employing nearly one million people, and earning revenues of more than $637 billion per year. This sector can be divided into five main segments, based on the end product produced: basic chemicals, specialty chemicals, agricultural chemicals, pharmaceuticals, and consumer products.

Commercial

Facilities associated with the Commercial Facilities Sector operate on the principle of open public access, meaning that the general public can move freely throughout these facilities without the deterrent of highly visible security barriers. The majority of the facilities in this sector are privately owned and operated, with minimal interaction with the federal government and other regulatory entities.

Communications

The Communications Sector is an integral component of the U.S. economy as it underlies the operations of all businesses, public safety organizations, and government. Over 25 years, the sector has evolved from predominantly a provider of voice services into a diverse, competitive, and interconnected industry using terrestrial, satellite, and wireless transmission systems. The transmission of these services has become interconnected; satellite, wireless, and wireline providers depend on one another to carry and terminate their traffic, and companies routinely share facilities and technology to ensure interoperability. A majority of the Communications Sector is privately owned, requiring the DHS to work closely with the private sector and its industry associations to identify infrastructure, assess and prioritize risks, develop protective programs, and measure program effectiveness.

Critical Manufacturing

The Critical Manufacturing Sector is crucial to the economic prosperity and continuity of the United States. U.S. manufacturers design, produce, and distribute products that provide more than $1 of every $8 of the U.S. gross domestic product and employ more than 10% of the nation's workforce. A direct attack on or disruption of certain elements of the manufacturing industry could disrupt essential functions at the national level and across multiple other CIKR sectors.

Dams

The Dams Sector comprises the assets, systems, networks, and functions related to dam projects, navigation locks, levees, hurricane barriers, mine tailings impoundments, or other similar water retention and/or control facilities. The Dams Sector is a vital and beneficial part of the nation's infrastructure and continuously provides a wide range of economic, environmental, and social benefits, including hydroelectric power, river navigation, water supply, wildlife habitat, waste management, flood control, and recreation.

Defense Industrial Base

The Defense Industrial Base (DIB) Sector includes Department of Defense (DoD), government, and the private sector worldwide industrial complex with the capabilities of performing research and development, design, production, delivery, and maintenance of military weapons systems, subsystems, components, or parts to meet military requirements. The DIB Sector includes tens of thousands of companies and their subcontractors who perform under contract to DoD, and companies providing incidental materials and services to DoD, as well as government-owned/contractor-operated and government-owned/government-operated facilities. DIB companies include domestic and foreign entities, with production assets located in many countries. The DIB Sector provides products and services that are essential to mobilize, deploy, and sustain military operations.

Emergency Services

The Emergency Services Sector (ESS) is a system of response and recovery elements that forms the nation's first line of defense and prevention and reduction of consequences from any terrorist attack. It is a sector of trained and tested personnel, plans, redundant systems, agreements, and pacts that provide life safety and security services across the nation via the first-responder community comprising federal, state, local, tribal, and private partners. The ESS is representative of the following first-responder disciplines: emergency management, emergency medical services, fire, hazardous material, law enforcement, bomb squads, tactical operations/special weapons assault teams, and search and rescue.

Energy

The U.S. energy infrastructure fuels the economy of the twenty-first century. Without a stable energy supply, health and welfare are threatened and the U.S. economy cannot function. More than 80% of the country's energy infrastructure is owned by the private sector. The energy infrastructure is divided into three interrelated segments: electricity, petroleum, and natural gas. The sector's reliance on pipelines highlights the interdependency with the Transportation Sector, and the reliance on the Energy Sector for power means that virtually all sectors have dependencies on this sector.

Government Facilities

The Government Facilities Sector includes a wide variety of buildings, owned or leased by federal, state, territorial, local or tribal governments, located domestically and overseas. Many government facilities are open to the public for business activities, commercial transactions, or recreational activities. Others not open to the public contain highly sensitive information, materials, processes, and equipment. This includes general-use office buildings and special-use military installations, embassies, courthouses, national laboratories, and structures that may house critical equipment and systems, networks, and functions.

Healthcare and Public Health

The Healthcare and Public Health Sector constitutes approximately 15% of the gross national product with roughly 85% of the sector's assets privately owned and operated. Operating in all U.S. states, territories, and tribal areas, the Healthcare and Public Health Sector plays a significant role in response and recovery across all other sectors in the event of a natural or man-made disaster.

Information Technology

The Information Technology (IT) Sector is central to the nation's security, economy, and public health and safety. Businesses, governments, academia, and private citizens are increasingly dependent on IT Sector functions. These virtual and distributed functions produce and provide hardware, software, and IT systems and services, and—in collaboration with the Communications Sector—the Internet. The IT Sector functions are operated by a combination of entities—often owners and operators and their respective associations—that maintain and reconstitute the network, including the Internet. The Internet encompasses the global infrastructure of packet-based networks and databases that use a common set of protocols to communicate. The networks are connected by various transports, and the availability of these networks and services is the collective responsibility of the IT and Communications Sectors. The DHS is the sector-specific agency for the IT Sector.

National Monuments and Icons

The National Monuments and Icons (NM&I) Sector encompasses a diverse array of assets located throughout the United States and its territories. Although many of these assets are listed in either the National Register of Historic Places or the List of National Historic Landmarks, all share three common characteristics: they are a monument, physical structure, object, or geographic site; they are widely recognized to represent the nation's heritage, traditions, or values, or widely recognized to represent important national cultural, religious, historical, or political significance; and their primary purpose is to memorialize or represent some significant aspect of the nation's heritage, tradition, or values, and to serve as points of interest for visitors and educational activities. NM&I Sector assets are all physical structures, objects, or geographic sites. Included as part of each asset are the operational staff and visitors that may be impacted by an attack on the asset. There are minimal cyber and telecommunications issues associated with this sector because of the nature of the assets.

Nuclear Reactors, Materials, and Waste

Nuclear power accounts for approximately 20% of the nation's electrical use, provided by 104 commercial nuclear reactors licensed to operate in the United States. The Nuclear Reactors, Materials, and Waste (Nuclear) Sector includes: nuclear power plants; nonpower nuclear reactors used for research, testing, and training; nuclear materials used in medical, industrial, and academic settings; nuclear fuel fabrication facilities; decommissioning reactors; and the transportation, storage, and disposal of nuclear material and waste.

Postal and Shipping

The Postal and Shipping Sector is an integral component of the U.S. economy, employing more than 1.8 million people and earning direct revenues of more than $213 billion per year. The Postal and Shipping Sector moves more than 720 million messages, products, and financial transactions each day. Postal and shipping activity is differentiated from general cargo operations by its focus on small- and medium-size packages and by service from millions of senders to nearly 150 million destinations. The sector is highly concentrated, with a handful of providers holding roughly 94% of the market share.

Transportation Systems

The nation's transportation system quickly, safely, and securely moves people and goods through the country and overseas. The Transportation Systems Sector consists of six key subsectors, or modes: Aviation, Highway, Maritime Transportation, Mass Transit, Pipeline Systems, and Rail.

Water

HSPD-7 designates the Environmental Protection Agency (EPA) as the federal lead for the Water Sector's critical infrastructure protection activities. All activities are carried out in consultation with the DHS and the EPA's Water Sector partners.

Methodological Approaches to Vulnerability Assessment

There are different threat approaches as part of vulnerability assessment methodologies. These threat approaches can be summarized by two main approaches: asset based and scenario based. Asset based examines the impact on individual assets (e.g., loading dock, main lobby) if attacked, whereas the scenario-based approach considers multiple potential specific sequences of events (e.g., damage or destroy building).

Vulnerability assessment methodologies are categorized as using one (or more) of the following approaches: checklist, simple rating, risk matrix, or risk equation. In some cases, a methodology may be a hybrid that incorporates elements of multiple approaches, or a range of approach options may be available to the assessor. A description of each approach category is provided next.

Checklist

Checklist-based vulnerability assessments are the simplest methodological approach. They consist of a list of questions or criteria against which the assessor compares the characteristics of the facility or asset being evaluated. The checklists may be grouped according to the various assessment elements (e.g., physical, cyber, OPSEC, and interdependencies). The questions may be answered yes or no or may require a qualitative response. Generally, if the answer to a question is no or if the criterion is not met, a recommended action will be requested or required. An example of a checklist methodology is the Security Vulnerability Self-Assessment Guide for Small Drinking Water Systems. Under that methodology, the water system is evaluated by answering questions for each area of concern: general security, water source protection, treatment plant protection, hazardous materials, distribution plant protection, personnel controls, information/cyber controls, and public relations information controls. Table 8.1 provides an example of checklist items that are included in the self-assessment guide.

Based on the understanding that a methodology is a written documentation of a systematic process to be used by specialized teams or even normal employees to conduct an assessment of the risk or vulnerabilities of an asset or facility, certain other forms of tools or guidance documents were not included in this survey. Guidance documents may present a general discussion of the objectives or scope of a vulnerability, risk, or security assessment; may provide typical security measures or criteria for various types of facilities or assets; or may outline potential mitigation measures for various types of vulnerabilities or assets. They do not, however, provide a step-by-step systematic, documented process to be followed in order to

Table 8.1 Example Checklist Items from Security Vulnerability Self-Assessment Guide for Small Drinking Water Systems

Category	*Example Question*
General security	Are facilities fenced, including well houses and pump pits, and are gates locked where appropriate?
Water sources	Are well vents and caps screened and securely attached?
Treatment plant and suppliers	Can you isolate the storage tank from the rest of the system?
Distribution	Do you control the use of hydrants and valves?
Personnel	Are your personnel issued photo-identification cards?
Information storage/computers/controls/maps	Is computer access "password protected" or are maps, records, and other information stored in a secure location?
Public relations	Does your water system have a procedure to deal with public information requests and to restrict distribution of sensitive information?

conduct a vulnerability assessment and/or do not contain necessary specific evaluation techniques (e.g., checklists, ranking scales, matrix categorization, or quantifiable equations). There are also assessment tools, which can be used to support vulnerability assessments, but that are not, in and of themselves, assessment methodologies. These would include software platforms and formats that can store and manipulate data or predict impact severity.

Simple Rating

Many vulnerability assessment methodologies prioritize asset vulnerabilities for potential corrective action by defining a set of measurable criteria, rating each asset (and the associated vulnerability) on each criterion, and qualitatively or quantitatively combining the individual ratings. An example of a simple, broadly applicable rating approach is a target analysis process developed and practiced by special operations forces. This process, called CARVER analysis, has been adapted and used as part of numerous vulnerability assessment methodologies. CARVER is an acronym that stands for criticality, accessibility, recoverability, vulnerability, effect, and recognizability. Each factor in the acronym typically has an associated scale (e.g., a 5-point scale), and individual assets (i.e., potential targets) are numerically rated on each factor. A rank-order of critical assets is established on the basis of the overall CARVER score (determined by summing the points assigned to the individual factors). Other "rating and weighting" schemes also are used to provide a logical and consistent basis for prioritizing vulnerabilities for importance or potential corrective actions.

Risk Matrix

A risk matrix is often used to focus vulnerability assessment results and help categorize the assets, sites, and/or systems assessed into discrete levels of risk so that appropriate protection and mitigation measures can be applied. Figure 8.2 shows a typical risk matrix, which conveys the notion that risk is a function of event severity (i.e., the severity of consequences) and the likelihood of its occurrence. Likelihood is often determined by considering the attractiveness of the targeted assets, the degree of threat, and the degree of vulnerability.

As depicted in Figure 8.2, asset vulnerabilities that have the highest likelihood of being successfully exploited (i.e., frequent) and that result in the highest severity (i.e., catastrophic), have the highest priority for vulnerability reduction actions and protective measures to mitigate the risks. Similarly, asset vulnerabilities with the lowest likelihood of being exploited (i.e., unlikely) and that result in the lowest severity (i.e., negligible), have the lowest priority for mitigation. Many variations of this basic approach are used with different numbers of severity and likelihood levels, as well as definitions for those levels, to assist in focusing on the highest priority risks.

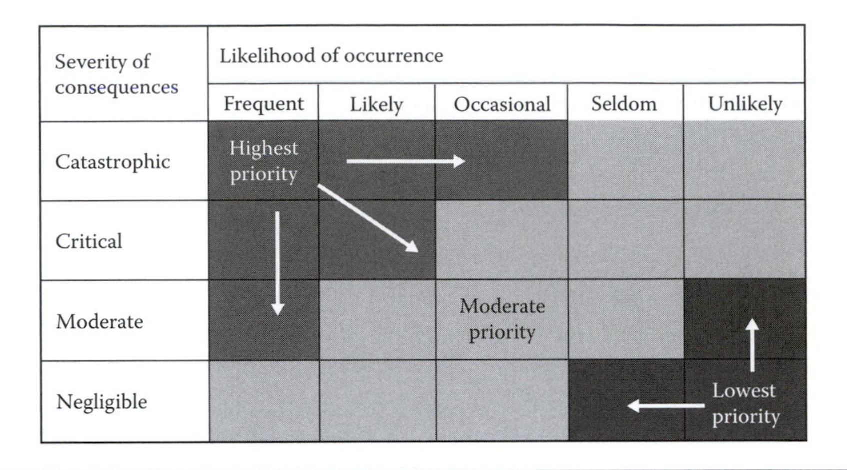

Figure 8.2 Illustrative risk matrix.

Risk Equation

Some methodologies seek a single measure that allows comparison of alternative countermeasures. Such a measure can be used to rank order or prioritize countermeasures. One approach is to calculate a risk number that is a function of probability of attack, system effectiveness, and consequence. A simple formula for risk defined in this manner is

$$\text{Risk} = R = P_A \times (1 - P_E) \times C$$

where

P_A = likelihood of occurrence (attack)
P_E = system effectiveness [therefore $(1 - P_E)$ = system ineffectiveness]
C = consequence value

In some approaches, consequence value is directly addressed by assigning, for example, a low, medium, or high value. Similarly, likelihood of occurrence and system effectiveness are combined by assigning, for example, a low, medium, or high value. Finally, to calculate a "risk value," one converts low, medium, and high assignments to 0.1, 0.5, and 0.9 values, respectively, and inserts the numbers into the risk equation to calculate a risk value. Or, numerical values are determined by constructing and running models that yield likelihoods and consequences.

Another variation calls for specification of several consequences and corresponding measures (e.g., economic loss in dollars, duration of loss in hours, number of customers impacted, fatalities, and illnesses). These must then be combined in some fashion to construct a single measure of consequence value. This can be done by considering all of the measures and simply assigning a single measure that represents the set of measures (e.g., "high" for economic loss, "medium" for duration of loss, and "low" for number of customers impacted may be rated as "medium" overall).

An even more thorough approach, advocated by some, is to carefully consider the ranges that may be obtained for each of the measures and assign "weights," which are used to construct a function that yields an overall measure of the desirability of each countermeasure (or portfolio of countermeasures). This function is sometimes called a "utility function." In this approach, the measure of risk is a "utility value" and high values are preferred over low values. Therefore, the portfolio of countermeasures that yields the largest expected utility is, by definition, the most desirable (best).

Required Expertise

To carry out a vulnerability assessment, a team of experts typically needs to visit the facility to ascertain the vulnerabilities of the critical assets and the anticipated results that would be caused by their physical destruction or impairment. Depending on the objectives and scope of the assessment, VA teams may include the following types of experts (these are representative of experts used by DHS during site assistance visits to review and/or conduct vulnerability assessments):

- Physical security experts who focus on the physical security of the facility, including access controls, barriers, locks and keys, badges and passes, intrusion detection devices and associated alarm reporting and display, CCTV assessment and surveillance, communications equipment, lighting, postings, security systems wiring, and protective force personnel.
- Explosive ordnance disposal (EOD) experts who examine and evaluate the vulnerability of critical assets to attacks that involve explosive devices of all kinds, including vulnerabilities to vehicle-delivered explosives and small charges.
- Special operations forces or assault planning experts who focus on terrorist strategies for the most likely method of attack, including physical security vulnerabilities (e.g., fencing or CCTV gaps) and outside surveillance/positioning vulnerabilities (e.g., areas of cover for clandestine operations and positions for using long-range weapons).
- Infrastructure systems experts who calculate the anticipated results of the loss of the asset as it pertains to the facility and the loss of the facility as it pertains to its specific infrastructure.
- Interdependencies experts who evaluate the dependence of the facility on outside infrastructures, such as electric power, water (potable and processed) and wastewater, natural gas, steam, petroleum products, telecommunications, transportation (e.g., roads, railroads, and marine links), and banking and finance.

- Operation security experts who evaluate human resources security procedures, facility engineering, facility operations, administrative support organizations, telecommunications and information technologies, publicly released information, and trash and waste handling.
- Intelligence operations experts who interact with local law enforcement and intelligence/security personnel to determine if there are potential terrorist or criminal elements in the region that may have an interest in the facility.

Outline of Risk Management Steps

This section presents an outline of the risk management process that has been applied by the Department of Energy, the DHS, and the private sector. Table 8.2 provides an overview of representative steps in a comprehensive, asset-based vulnerability assessment methodology. This includes countermeasure (actions taken to reduce or eliminate vulnerabilities) and risk management considerations. The methodologies included in this survey address, to a greater or lesser degree, some or all of these steps.

The following sections describe the steps of the risk management process in more detail. Where appropriate, the steps contain checklists of questions that could be used to guide the implementation of a risk management program.

Step 1: Identify Critical Assets and the Impacts of Their Loss

Estimates of the potential consequences, including economic implications, of not mitigating identified vulnerabilities or addressing security concerns are necessary to effectively apply risk management approaches to evaluate mitigation option and security recommendations. Outages because of security failures could degrade an energy facility's reputation and place the community served at risk to economic losses or even losses of property and life.

In addition, the modern energy facility's telecommunication and computer network has many external connections to public and private networks. Such connections are used to communicate with customers and offer new electronic services, such as online billing and payment. Cyber security should be a primary concern, especially for utilities that operate in this interconnected environment. An IT security architecture may need to be developed.

Possible critical assets include people, equipment, material, information, installations, and activities that have a positive value to an organization or facility. People include energy facility executives and managers, security personnel, contractors and vendors, and field personnel. Equipment includes vehicles and other transportation equipment, maintenance equipment, operational equipment, security equipment, and IT equipment (computers and servers). Material includes tools, spare parts, and specialized supplies. Information includes employee records; security plans; asset

Table 8.2 General Vulnerability Assessment Process

Step	*Description*	*Considerations*
1	Identify critical assets and the impacts of their loss	• Identify the critical functions of the facility. • Determine which assets perform or support the critical functions. • Evaluate the consequences or impacts to the critical functions of the facility from the disruption or loss of each of these critical assets.
2	Identify what protects and supports the critical assets	• Identify the components of the physical security system (e.g., perimeter barriers, building barriers, intrusion detection, access controls, and security forces) that protect each asset. • Identify the critical internal and external infrastructures (e.g., electric power, petroleum fuels, natural gas, telecommunications, transportation, water, emergency services, computer systems, air handling systems, fire systems, and SCADA systems) that support the critical operations of each asset (interdependencies). • Identify sensitive information about the facility and its operation that must be protected.
3	Identify and characterize the threat	• Gather threat information and identify threat categories and potential adversaries. • Identify the types of threat-related undesirable events or incidents that might be initiated by each threat or adversary. • Estimate the frequency or likelihood of each threat-related undesirable event or incident based on historical information. • Estimate the degree of threat to each critical asset for each threat-related undesirable event or incident.

(*continued*)

Table 8.2 (*Continued*) General Vulnerability Assessment Process

Step	*Description*	*Considerations*
4	Identify and analyze vulnerabilities	• Identify the existing measures intended to protect the critical assets and estimate their levels of effectiveness in reducing the vulnerabilities of each asset to each threat or adversary. (Step 2 provides a starting point for this activity.) • Estimate the degree of vulnerability of each critical asset for each threat-related undesirable event or incident and thus each threat or adversary.
5	Assess risk and determine priorities for asset protection	• Estimate the effect on each critical asset from each threat or adversary taking into account existing protective measures and their levels of effectiveness. • Determine the relative degree of risk to the facility in terms of the expected effect on each critical asset (a function of the consequences or impacts to the critical functions of the facility from the disruption or loss of the critical asset, as evaluated in Step 1) and the likelihood of a successful attack (a function of the threat or adversary, as evaluated in Step 3, and the degree of vulnerability of the asset, as evaluated in Step 4). • Prioritize the risks based on the relative degrees of risk and the likelihoods of successful attacks using an integrated assessment.
6	Identify mitigation options, costs, and trade-offs	• Identify potential mitigation options to further reduce the vulnerabilities and thus the risks. • Identify the capabilities and effectiveness of these mitigation options. • Identify the costs of the mitigation options. • Conduct cost–benefit and trade-off analyses for the various options. • Prioritize the alternatives for implementing the various options and prepare recommendations for decision makers.

lists; intellectual property; patents; engineering drawings and specifications; system capabilities and vulnerabilities; financial data; and operating, emergency, and contingency procedures. In addition to the operational installations that make up the energy infrastructure itself, installations include headquarters offices; field offices; training centers; contractor installations; and testing, research, and development laboratories. Activities include movement of personnel and property, training programs, communications and networking, negotiations, and technology research and development.

The energy facility, the local government, and energy industry associations have roles and responsibilities for identifying assets, effects of asset loss, vulnerabilities, threats, and risk mitigation options. Coordination among energy facilities; local, state, and federal agencies; and energy industry associations is crucial to this process.

Energy facilities need to identify the critical functions of the facility, and determine which physical and cyber assets perform or support the critical functions. The key assets identified should be related to the criticality of overall operations of the individual facilities. Potential assets include substations, transmission lines, pipelines, critical valve nests, power plants, pump stations, city gate stations, compressor stations, storage installations, interconnections, energy control centers, energy management systems, SCADA systems, remote monitoring and control units [remote terminal units (RTU)], communications systems linking RTUs and energy control centers, certain backup systems, and e-commerce capabilities. They should evaluate the consequences or impacts to the critical functions of the energy facility from the disruption or loss of each of these critical assets and prioritize the critical assets based on these.

Not all assets and activities warrant the same level of protection. The cost of reducing risk to an asset must be reasonable in relation to its overall value. The value, however, does not need to be expressed in dollars. A potential loss can be stated in terms of human lives or the impact on the local or state economy.

The first set of questions is designed to guide the process of identifying the critical functions of the energy facility and the assets that perform or support them, and evaluating the potential consequences of disruptions or loss of these critical assets.

Criticality Criteria (Functions and Assets)

- What critical mission activities take place at the energy facility or its remote sites?
- What critical or valuable equipment is present at the facility or its remote sites?
- Where are the critical assets located?
- Have people, installations, and operations been considered when assessing assets?
- Have cyber networks and system architectures (e.g., SCADA systems, business e-mail, and e-commerce) been documented fully?

Criticality Criteria (Impacts of Loss)

- What would the energy facility lose if an adversary obtained control of a specific asset?
- What affects would be expected if a specific asset were compromised?
- Is the asset still valuable to the energy facility once an adversary has it?
- What is the potential for immediate and significant local impacts due to the loss of the asset?
- What is the potential for loss of energy supply to civilian areas?
- What facility personnel, tenants, customers, and visitors could be affected by the loss of the asset?
- What would be the impact on people's lives and on national or local security due to the loss of the asset?
- What would be the financial impacts to the energy facility and the local community?

Criticality Criteria (Asset Value)

- Is there little or no redundant capacity or capability to mitigate the loss of the asset?
- What is the potential for cascading effects (e.g., to other interdependent infrastructures or industries) due to the loss of the asset?
- Do any special situations need to be considered regarding the loss of the asset, such as the status of hospitals, life support systems, or emergency services that depend on the energy infrastructure supported by the asset?
- What is the potential for catastrophic effects (weapons of mass destruction levels impact)?
- What did it cost to develop the asset?
- Would the energy facility need an extended period to make repairs to the asset?
- How does the need for protecting the asset compare with other assets also considered critical?

Once the assets critical to the operation of the energy facility have been identified and characterized, an impact assessment must be carried out to describe the consequences of losses if an undesirable event occurs. The degree of impact should be quantified by using a relative impact or criticality rating criteria and a consistent rating scale. (An example of a scale for rating criteria is presented in Step 5.) The assets are then ranked in terms of criticality.

Step 2: Identify What Protects and Supports the Critical Assets

The existing protection of critical assets provided by the physical security system and the dependence of the critical assets on both external and internal infrastructures must be known to evaluate the vulnerabilities of the assets to threats or adversaries. In addition, operating procedures and other sensitive information, which if available to adversaries might jeopardize critical assets, must be identified as their availability can also affect the vulnerabilities of assets.

Physical Security Systems

Physical security systems are used to protect energy facilities and their assets from unauthorized individuals and outside attacks. Such systems usually include perimeter barriers, building barriers, intrusion detection, access controls, and security forces.

Infrastructure Interdependencies

Today's energy facilities depend on many different infrastructures to support their critical functions and assets. These infrastructure interdependencies must be identified and the adequacy of security measures that are in place to protect and back up these infrastructures must be evaluated. Typically, these supporting infrastructures include

- Electric power supply and distribution
- Petroleum fuels supply and storage
- Natural gas supply
- Telecommunications
- Transportation (road, rail, air, and water)
- Water and wastewater
- Emergency services (fire, police, and emergency medical)
- Computers and servers
- Heating, ventilation, and air conditioning (HVAC) systems
- Fire suppression and firefighting systems
- SCADA systems

The electric power supply and distribution infrastructure can include the local electrical distribution utility, facility-operated electric generation equipment, backup generators fueled by natural gas or petroleum fuels, uninterruptible power supplies (UPSs), and the associated switching and distribution hardware. The petroleum fuels supply and storage infrastructure includes on-site storage as well as local suppliers, storage terminals, and the entire petroleum industry. The telecommunications infrastructure includes commercial telephone, fiber optic, and satellite networks and facility-owned radio, telephone, microwave, and fiber-optic pathways. Computers and servers, HVAC systems, fire suppression and firefighting systems, and SCADA systems tend to be operated by the energy facility and, in turn, depend on the other infrastructures such as telecommunication, electric power supply and distribution, petroleum fuels supply and storage, natural gas supply, water and wastewater, and emergency services.

Sensitive Information

Protecting operating procedures and other sensitive information, the release of which might jeopardize an energy facility and its assets, is the objective of OPSEC programs. OPSEC programs utilize tools such as employee background checks, trash handling procedures, telephone policies, and IT (computer) security to protect against both industrial espionage and deliberate disruption of critical assets and functions.

The second set of questions is designed to guide the process of identifying the existing components of the physical security system that protect the critical assets, the critical infrastructure systems that support the critical assets, and the operating procedures and sensitive information that must be protected to avoid jeopardizing the critical assets.

Energy Facility and Critical Asset Protection (Physical Assets)

- What department or person has overall responsibility for security or is that responsibility spread over many departments or people with shared responsibilities for security along with their other responsibilities?
- What perimeter barriers (e.g., fences, gates, vehicle barriers) protect the energy facility as a whole and the individual critical assets and what levels of protection do they provide?
- What building barriers (e.g., walls, roof/ceiling, windows, doors, locks) protect each critical asset and what levels of protection do they provide?
- What is the status of the intrusion detection that protects each critical asset (e.g., intrusion sensors, alarm deployment, alarm assessment, alarm maintenance) and what level of protection does it provide?
- What is the status of the access control used at each critical asset (e.g., personnel access, vehicle access, contraband detection, access point illumination) and what level of protection does it provide?
- What is the nature of the security force (both the protective force and appropriate local law enforcement agencies) that protects each critical asset (e.g., number, training, armament, communications) and what level of protection does it provide?
- What types of undesirable events (e.g., surreptitious forced entry, technical implant, theft of sensitive information or materials) are protected against?
- During which hours of the day and under what conditions are the various components of the physical security system effective?
- Over what areas do the various components of the physical security system provide protection?

- What is the history of the reported malfunctions of the various components of the physical security system?
- What is the correlation of the effectiveness of the various components of the physical security system to security incident reports that may indicate that the system was defeated?
- Have liaisons and working relationships been established with the local government and its departments, such as police, fire, emergency medical services, and public works?

Energy Facility and Critical Asset Protection (Infrastructure Interdependencies)

- Which infrastructures (both internal and external) are essential for a specific critical asset to be able to carry out its critical functions?
- What external utility or internal department and equipment is the normal provider of each essential infrastructure for each critical asset, and how is each infrastructure connected to each asset (e.g., the types and pathways of power lines, pipelines, and cables)?
- What alternatives (e.g., redundant systems, alternative suppliers, backup systems established work-around plans) are available if the regular providers of the essential infrastructures are disrupted, and how long can the alternatives support the critical functions of the assets?
- What is the potential for interdependency effects on external infrastructures (i.e., effects on the energy, telecommunications, transportation, water and wastewater, banking and finance, emergency services, and government services infrastructures)?

Energy Facility and Critical Asset Protection (Sensitive Information)

- What types of information about the energy facility, its assets, and its operations should be considered critical or sensitive information?
- What are the methods and means by which sensitive information might fall in the wrong hands, such as via disgruntled employees; access to the facility by the public; outside construction, repair, and maintenance contractors; press contacts; briefings and presentations; public testimony; Internet information; paper and material waste; telecommunications system taps; and cyber (computer) intrusions?
- What are the existing policies and procedures used to protect sensitive information, such as employee background checks, disciplinary procedures, security training, trash handling procedures, paper waste handling procedures, salvage material handling procedures, dumpster control, telephone policies, and IT (computer) security?

Once energy facilities have identified their existing physical protective measures, they should coordinate with their respective local governments and law enforcement agencies to ensure that the level of protection and response that they expect will be forthcoming. They should also coordinate with the critical external infrastructure providers to ensure that the robustness and redundancy that they depend on will continue to be provided. The objective of these coordination efforts is to ensure that roles for response and recovery from a disruption are understood by all so that quick and effective measures can be taken when problems occur.

In addition, local and state governments can assist energy facilities in infrastructure restoration activities. Potential support can come in many areas, such as maintaining critical spare parts, assisting with special equipment, working with the emergency telecommunications spectrum, securing easy access to the site of the disruption for repair crews and needed equipment, working out mutual assistance programs with other energy providers, and supplying temporary staffing.

The energy facility should also check with local and state governments to ensure that critical information about their facility, its assets, and its operations will not be released to the general public in any future additions to public Internet sites, press releases, or public hearings.

Step 3: Identify and Characterize the Threat

In order to put the information about the critical assets of the energy facility to use in a quantitative risk assessment, the potential threats and adversaries that may be expected must be identified and quantified. The set of questions provided in this section serves as guidance for evaluating the threat environment to which the energy facility could be exposed and establishing qualitative or quantitative threat ratings for each critical asset. The goals of the threat assessment are to understand, from the adversary's point of view, the adversary's capabilities and intent to collect critical information.

The federal agencies (e.g., Department of Energy [DOE] and the Federal Bureau of Investigation [FBI]), state governments, and energy industry associations each collect threat information. This information should be shared among these groups and with the local energy facilities in order to have the most comprehensive and updated threat information possible. In addition, threats to energy facilities could affect state and local assets. State and local governments have access to law enforcement and intelligence data. This information should be integrated and shared, together with any information that the energy industry associations and energy facilities collect.

This third set of questions is to be used to identify and evaluate the threat environment to which an energy facility may be exposed.

Intent and Capabilities of Adversaries

- What types of adversaries are expected?
- Who are the specific adversaries expected?
- What are the specific goals and objectives of each adversary?
- Which are the critical assets that each specific adversary is aware?

- Does each specific adversary know enough about the asset to plan an attack?
- What are the possible modes of attack (e.g., explosives or incendiary devices delivered by car, truck, boat, rail, mail, individuals, or standoff weapons; aircraft impacts; sabotage of equipment or operations; assaults by a lightly or heavily armed individual attacker or team of attackers; theft, alteration, or release of information, materials, or equipment; contamination by chemical agents, biological agents, or radioactive material; and cyber attacks) each adversary might use?
- Are there other, less risky means for a specific adversary to attain his or her goals?
- What is the probability that an adversary will choose one method of attack over another?
- What specific events might provoke a specific adversary to act?

Information concerning potential threats and adversaries can be gathered about potential threats and adversaries by

- Joining a threat analysis working group that includes local, county, state, and federal agencies, the military, and other industry partners.
- Obtaining access to the National Infrastructure Protection Center (NIPC), Analytical Services, Inc. (ANSER), FBI-sponsored InfraGuard, Carnegie Mellon University's CERT®, or other information system security warning notices.
- Initiating processes to obtain real-time information from the field (e.g., on-duty offices, civilian neighborhood watch programs, local businesses, other working groups in the area).
- Arranging for threat briefings by local, state, and federal agencies.
- Performing trend analyses of historical security events (both planned and actual).
- Creating possible threat scenarios based on input from the threat analysis working group and conducting related security exercises.

Step 4: Identify and Analyze Vulnerabilities

In addition to identifying the critical assets of the energy facility, the impact of their disruption, the present protection provided, and the potential threats against them, the vulnerability of those assets to the potential threats must be quantified, at least to some extent, to determine the overall risk to the assets.

There are various types of vulnerabilities, such as physical, technical/cyber, and operational. An energy facility, including perimeter barriers (fences, walls, gates, landscape, sewers, tunnels, parking areas, alarms), compound area surveillance (CCTV, motion detectors, lighting), building perimeters (walls, roofs, windows, doors, shipping docks, locks, shielded enclosures, access control, alarms), and building interiors (doors, locks, safes, vents, intrusion sensors, motion sensors), is subject to physical vulnerabilities. Electronic equipment, such as acoustic equipment, secure telephones, computers and computer networks, and radio-frequency equipment, are subject to technical or cyber vulnerabilities. The guard force, personnel procedures, and operational procedures are subject to operational vulnerabilities.

Various characteristics of assets, including any existing protection identified in Step 2, may affect their susceptibility to attacks and must be considered when identifying susceptibilities. Such asset characteristics include building design; equipment properties; personal behavior; locations of people, equipment, and buildings; and operational and personnel practices.

Both energy facilities and local governments should be concerned with identifying and analyzing vulnerabilities. Energy facilities should analyze the vulnerabilities of their physical and cyber systems. Local governments should coordinate management of the vulnerabilities of the energy infrastructure, including individual energy facilities, that support government and community operations and assets.

This fourth set of questions is to be used to evaluate the vulnerability of the critical energy infrastructure assets to the potential threats and to establish qualitative or quantitative vulnerability ratings for each asset.

Energy Facility and Critical Asset Vulnerabilities

- How susceptible is each critical asset to physical attack if readily available weapons (guns, normal ammunition, vehicle, simple explosives) were used?
- How susceptible is each critical asset to physical attack if difficult-to-acquire weapons (assault rifles, explosive ammunition, rocket launchers, biological or chemical agents, aircraft, sophisticated explosives) were used?
- How susceptible is each critical asset to physical attack from insiders?
- Are any of the critical assets unprotected? If so, describe them.
- Are any of the critical assets minimally protected? If so, describe them.
- How susceptible is each critical asset to cyber attack?

Step 5: Assess Risk and Determine Priorities for Asset Protection

Scales for the rating criteria identified in the first four steps (asset criticality in terms of the impact of loss or disruption, threat characteristics, and asset vulnerability) must be developed. The concept of criteria development is presented below in the form of a generic example. Those who conduct an actual assessment should define rating scales that are appropriate to the specific assessment.

Using the individual rating values assigned to each combination of asset criticality, threat, and vulnerability, a relative degree of risk or a risk rating can be established for each asset for one or more postulated adverse events or consequences that could result from an attack by the identified adversary. Often, a multiplicative approach involving the three rating criteria is used to obtain a risk rating:

Risk Rating = Impact Rating × Threat Rating × Vulnerability Rating

An additional scale must be developed to assign a qualitative overall risk level from the quantitative risk rating. The risk ratings or risk levels are used to prioritize the assets for the selection and implementation of security improvements to achieve an acceptable overall level of risk at an acceptable cost.

The following should be considered when developing and using rating criteria:

- Subject-matter expert opinions and perspectives should be documented. The team involved in the assessment should reflect a variety of different perspectives, and the team should work toward reaching a consensus regarding a set of priorities.
- Information should be presented in a usable format (e.g., table, matrix, or spreadsheet).
- Assumptions should be documented.

A generic example of possible scales for the rating criteria is presented next in the form of a set of tables to illustrate the concept. These or similar tables are used to establish qualitative or quantitative criticality, threat, and vulnerability ratings for each critical asset.

Asset Impact/Criticality Rating Criteria

Each critical asset identified in Step 1 of the risk management process is assigned an impact rating value that reflects the importance or criticality of a loss or disruption of that asset with regard to the continued operation of the energy facility or other organization being assessed. In Table 8.3, a quantitative criticality rating scale of 0% to 100% is used, which corresponds to qualitative criticality levels of critical, high, medium, and low.

Threat Rating Criteria

The individual potential threats against the assets of the energy facility or other organization being assessed identified in Step 3 are assigned a threat rating value that reflects the magnitude of the threat. In Table 8.4, a quantitative threat rating scale of 0% to 100% is used, which corresponds to qualitative threat levels of critical, high, medium, and low.

Vulnerability Rating Criteria

The vulnerabilities of the assets in terms of in-place measures to protect those assets identified in Step 2 are assigned a vulnerability rating value that reflects the extent

Table 8.3 Asset Impact/Criticality Rating Criteria

Criticality Level	*Description*	*Rating Scale (%)*
Critical	Indicates that compromise of the asset would have grave consequences leading to loss of life or serious injury to people and disruption of the operation of the energy facility. It is also possible to assign a monetary value or some other measure of criticality.	75–100
High	Indicates that compromise of the asset would have serious consequences that could impair continued operation of the energy facility.	50–75
Medium	Indicates that compromise of the asset would have moderate consequences that would impair operation of the energy facility for a limited period.	25–50
Low	Indicates little or no impact on human life or the continuation of the operation of the energy facility.	1–25

to which the asset is protected against each threat identified in Step 3. In Table 8.5, a quantitative vulnerability rating scale of 0% to 100% is used, which corresponds to qualitative vulnerability levels of critical, high, medium, and low.

Step 6: Identify Mitigation Options, Costs, and Trade-Offs

The ultimate goal of a risk management process is to select and implement security improvements to achieve an acceptable overall risk at an acceptable cost. Step 5 of the risk management process prioritizes the combinations of assets and threats by the risk ratings or risk levels. This, in turn, helps to identify where protective measures against risk are most needed.

In this step, potential measures to protect critical assets from recognized threats are identified, specific programs to ensure that appropriate protective measures are put in place are established, and appropriate agencies and mechanisms needed to put protective measures in place are identified. Protective measures that can address more than one threat or undesirable event should be given special attention.

A variety of approaches to developing protective measures exists. Protective measures can reduce the likelihood of a failure due to an attack by adding physical security. Protective measures can also be implemented to prevent or limit the

Table 8.4 Threat Rating Criteria

Threat Level	*Description*	*Rating Scale (%)*
Critical	Indicates that a definite threat exists against the asset and that the adversary has both the capability and intent to launch an attack, and that the subject or similar assets are targeted on a frequently recurring basis.	75–100
High	Indicates that a credible threat exists against the asset based on knowledge of the adversary's capability and intent to attack the asset and based on related incidents having taken place at similar assets or in similar situations.	50–75
Medium	Indicates that there is a possible threat to the asset based on the adversary's desire to compromise the asset and the possibility that the adversary could obtain the capability through a third party who has demonstrated the capability in related incidents.	25–50
Low	Indicates little or no credible evidence of capability or intent and no history of actual or planned threats against the asset.	1–25

consequences of a failure or to speed the recovery following a failure, regardless of the cause of that failure.

Best practices and lessons learned from DOE's Vulnerability Survey and Analysis Program provide some general actions, activities, and recommendations that can help identify appropriate potential mitigation measures. Some of these are listed next.

- The trend in IT until very recently has been to outsource more and more functions. Since the events of September 11, 2001, outsourcing is becoming less popular again. If possible, cyber security should remain as an enterprise function and should not become a contractor function.
- Logging and reporting should be enabled on IT network routers and firewalls to gain a better understanding of user access and interactions with remote systems.
- Sensitive and confidential documents should not be placed on websites. Appropriate document review, classification, and access controls should be implemented. This practice should apply to documents and other information found in newsgroups, media sites, and other linked sites.

Table 8.5 Vulnerability Rating Criteria

Vulnerability Level	*Description*	*Rating Scale (%)*
Critical	Indicates that there are no effective protective measures currently in place and adversaries would be capable of exploiting the critical asset.	75–100
High	Indicates that although there are some protective measures in place, there are still multiple weaknesses through which adversaries would be capable of exploiting the asset.	50–75
Medium	Indicates that there are effective protective measures in place; however, one weakness does exist that adversaries would be capable of exploiting.	25–50
Low	Indicates that multiple layers of effective protective measures exist and essentially no adversary would be capable of exploiting the asset.	1–25

- Security measures—such as traffic filtering, authorized controls, encryption and access controls, minimizing or disabling of unnecessary services and commands, minimizing banner information, and e-mail filtering and virus control—should be implemented.
- A formal process for accessing relevant threat information and for contacting the proper government and law enforcement agencies should be instituted (if it does not already exist), and reviewed and updated on a regular basis. The energy facility may need to work with the government to obtain security clearances for appropriate personnel.
- Appropriate security measures (e.g., access controls, barriers, badges, intrusion detection devices, alarm reporting and display, CCTV cameras, communication equipment, lighting, and security officers) should be implemented.
- Top management support is critical in ensuring a successful security program.
- Security training programs should be formalized.
- Procedures for escorting contractors and visitors into sensitive areas should be enhanced and enforced.
- Security should be incorporated in the company goals as well as in its corporate culture.
- The foundation for security is well-informed employees acting responsibly.

- A formal review process should be established for all information released to the public, particularly through the energy facility's website. A periodic review of "public" information should be performed to audit the effectiveness of information protection policies.
- The energy facility should be careful about disseminating sensitive information to the press or competitors. Only minimal information should be made available about personnel (especially executives).
- Security training and awareness should be provided to all employees on a regular basis.
- At a minimum, an annual audit of overall security should be conducted.

Some illustrations specific to energy facilities, including large utilities not specifically considered in this report, are listed next as an example of specific protective measures that can be implemented. The examples are grouped by type.

Measures to Prevent Damage

- Harden key installations and equipment—protect critical equipment with walls or below-grade installations, physically separate key pieces of equipment, and toughen the equipment itself to resist damage.
- Install surveillance systems (e.g., video cameras, motion detectors) around key installations that are monitored and coupled with rapid-response forces.
- Maintain security guards at key installations.
- Improve communication with law enforcement agencies, especially local law enforcement agencies and the local FBI office, to obtain threat information and coordinate responses to emergencies.

Measures to Limit Consequences

- Improve emergency plans and procedures for continued operation during undesirable events and ensure that operators are trained to implement these contingency plans.
- Modify the physical system—improve control centers and protective devices, increase redundancy of key equipment, and increase reserve margins.

Measures to Speed Recovery

- Conduct contingency planning for restoration of service, including identification of potential spare parts and resolution of legal uncertainties.
- Clarify the legal and institutional framework for sharing reserve equipment.
- Stockpile critical equipment (e.g., transformers, pumps, compressors, regulators) or any specialized materials (e.g., cables, pipe sections) needed to manufacture critical equipment or make repairs.

- Assure availability of adequate transportation for stockpiles of very heavy equipment by maintaining a database of rail and barge equipment and adapting Schnabel railcars to fit all needed types of large pieces of equipment, if necessary.
- Monitor domestic manufacturing capability to assure adequate repair and manufacture of key equipment in times of emergency.
- Investigate mutual aid agreements with vendors, industry associations, or large nearby energy companies.
- Establish backup arrangements with contractors for emergency services and other emergency support.

General Mitigation Measures to Reduce Vulnerability
- Emphasize inherently less vulnerable technologies and designs when practical, such as using standardized equipment.
- Move toward an inherently less vulnerable bulk energy system (e.g., smaller generators near loads, local storage) as new installations are planned and constructed.

As indicated earlier, local governments should coordinate energy facility activities related to risk management. The following questions can help guide local and state governments through the risk management process.

Local and State Government's Role in the Risk Management Process
- Has the local or state government identified any critical issues or vulnerabilities regarding its energy infrastructure?
- If the local or state government has identified critical issues, what are they and why are they critical?
- Has the local or state government developed plans to counter these vulnerabilities?
- Has the local or state government coordinated information with other local energy facilities, local law enforcement agencies, and others concerning these vulnerabilities?

Conclusion

Vulnerability assessment methodologies generally evaluate vulnerabilities by broadly considering the threat, existing protective measures, and consequences that result if an asset is attacked (asset-based approach). Alternatively, multiple potential sequences of attack events can be considered to evaluate the likelihood that the current protective measures at a facility will be able to successfully deter, detect, and/or delay an attack (scenario-based approach).

Protecting and ensuring the continuity of the CIKRs of the United States are essential to the nation's security, public health and safety, economic vitality, and way of life.

References

Critical Infrastructure and Key Resources. www.dhs.gov/files/programs/gc_1189168948944.shtm.

Energy Infrastructure Risk Management Checklists for Small and Medium Sized Energy Facilities, U.S. Department of Energy, Office of Energy Assurance, August 2002.

Survey of Vulnerability Assessment Methodologies, U.S. DHS Protective Security Division, October 2003.

The White House, Homeland Security Presidential Directive/HSPD-7: Critical Infrastructure Identification, Prioritization, and Protection, Washington, DC, December 17, 2003. www.fas.org/irp/offdocs/nspd/hspd-7.html.

Chapter 9

The Common-Sense Guide for the CEO

Michael J. Fagel

Contents

Introduction

As one result of rapidly increasing technology, the fact that a greater percentage of the population lives and works in areas that are at risk of disaster, the increased mobility of the worldwide population, and other factors, the risk of catastrophic disasters that result in large numbers of casualties and damage to property and the environment has never been greater, and the risk is expected to grow. Additionally, foreign terrorist groups that once seemed to pose only a remote threat to the U.S. homeland have demonstrated that they *can* strike us where we live—and strike with devastating consequences to our population, our infrastructure, and our way of life.

The increased risk that a catastrophic event may occur has placed new responsibilities on government at all levels. As the first line of defense against all types of incidents, you, as the elected leader in your community, must work actively to prevent, prepare for, respond to, recover from, and mitigate the effects of emergencies and disasters—and do so with a new sense of urgency. Emergency managers and responders need your support for a comprehensive emergency management program more than ever before. And the public is looking to you for reassurance that, should an incident occur, public officials and employees will be able to control the situation quickly and effectively.

Contrast these new expectations with the reality that most chief elected officials (CEOs)

- Are not typically involved in the emergency management process and are unaware of what is required to ensure that an effective Emergency Operations Plan (EOP) exists in their communities.
- Have little experience in managing emergencies.

- Do not have a clear perception of the key players' responsibilities in emergency preparedness or response.
- Do not understand the legal requirements and constraints placed on public officials by enabling legislation and legal governing framework.
- Do not play a significant role in public preparedness.

If the above-stated realities apply to your situation, this kit can help you fill the gaps in your understanding and become a leader in ensuring that your community is as prepared as possible.

The Public's Expectations and Your Responsibilities

A 2003 public opinion poll conducted for the Office of Citizen Corps surveyed 2000 adults about their concerns regarding emergencies. When asked about the types of emergencies that worried them most:

- Thirty-three percent stated that they were most concerned about terrorism.
- Twenty-six percent stated that they were most concerned about man-made accidents.
- Eighteen percent stated that they were most concerned about hurricanes and/or violent storms.

When asked who they would expect to rely on in these emergencies:

- Sixty-two percent stated that they would expect to rely on first responders.
- Thirty-four percent responded that they would expect to rely on government agencies.

Additionally, a high percentage of respondents indicated a desire to have a plan for emergencies. Seventy percent said that they would be more likely to develop plans if they had support from local government and community organizations.

What do these survey results mean for you as a CEO? Specifically, that means that today, with the influx of natural disasters that have struck the United States in 2011, citizens are demanding that local government provide adequate and timely leadership in times of disasters at ALL levels. ALL disaters ARE LOCAL.

What type of leadership are you providing to your community? Ask yourself:

> **Have I taken an active role in support of the emergency operations planning process?** Have I made it clear to the emergency management agency, first responders, and other personnel who are involved in emergency preparedness and response that their activities are important to me personally?

Have I supported the emergency management agency and response agencies by providing what they need to get the job done? Have I participated in "kick off" training and exercises to show my support?

Have I taken steps to ensure that the public is educated about the threats facing the community? Do citizens know what they need to do to protect themselves? Do they know what they can do to help prevent emergencies from occurring or to reduce the damage they cause? Do they know what they can do to help first responders?

Have I supported measures to mitigate the risk of damage from high-risk hazards? Are zoning requirements, building ordinances, and other measures in place to minimize the risk of physical and environmental damage from high-risk hazards?

If the answer to any of these questions is "No," you are not doing enough to protect your community from high-risk hazards.

Federal Expectations for Response to Domestic Incidents

The Department of Homeland Security (DHS), which has overall responsibility for coordinating the federal response to disasters and acts of terror within the United States, has recognized the need to improve domestic incident response at all levels of government. As such, DHS has developed and implemented the National Incident Management System (NIMS). Among the initiatives promulgated by NIMS are

- The requirement for governments at all levels to adopt the Incident Command System (ICS) as the model for managing all types of incidents, regardless of type, size, or complexity.
- The establishment of systems to ensure coordination among government entities and agencies.
- The establishment of standards for training and exercises for response personnel.
- The requirement that equipment be typed according to capability.
- The establishment of interoperability interoperability requirements, to be facilitated by the NIMS Integration Center, which will publish lists of equipment meeting established requirements.

To be compliant with NIMS initially, governments at all levels were required to adopt ICS not later than October 1, 2004, or risk being ineligible for preparedness funding. Timeframes for compliance with other NIMS requirements will be determined at a later date.

What Are Your Goals?

If, after reading the information presented above, you believe that you can and should do more to improve your community's preparedness for emergencies and disasters, you might begin by setting some goals for yourself. Some examples of goals that you might set are listed below.

- Understand the need to have a personal commitment to emergency management, to have the capability to judge its status, and develop the public policy to improve it.
- Understand how some emergencies differ from the management of normal operations and encourage the development of EOPs that all response agencies practice regularly and implement when responding to all major incidents.
- Ensure that ICS is adopted by all response agencies and implemented during exercises and in all responses.
- Understand the interaction among federal, state, and local governments during a disaster or severe emergency.
- Understand your responsibility to educate citizens about what they can do to prevent and prepare for emergencies that present a high risk to the community.
- Understand your responsibility to communicate to the community how different types of emergencies can affect their lives and property and how sound emergency management policy is a good investment in public safety.

Reducing Your Risk by Using This Kit

After reviewing and determining your personal emergency management goals, you need to follow up with several concrete actions. You should

- Review the contents of this kit.
- Conduct the self-assessment.
- Assign staff to prepare your survival kit.
- Review the survival kit.
- Take other actions, as necessary, including:
  - Direct senior staff to develop similar kits for their own use.
  - Chair monthly Emergency Management Council meetings to improve interagency working relationships for both day-to-day and disaster response.
- Participate—and direct your senior staff to participate—in disaster exercises at least once each year.
- Review and update the kit contents at least semiannually.
- Review the action steps included in Appendix C for more ideas to strengthen your community's comprehensive emergency management program.

Conducting Self-Assessment

This section presents an informal quiz that you can take as a confidential personal assessment of you and your community's emergency management risk. The questions reflect some of the survival capabilities that past experience has associated with effective CEO participation in emergency management.

Following the questions is a scoring section. The scoring section is designed to help you decide whether action is needed to reduce your risk.

1. Has your community updated its EOP within the past 12 months?

   ☐ Yes ☐ No
2. Have all response agencies adopted ICS as the single management system for all incidents?

   ☐ Yes ☐ No
3. Does your community have mutual aid agreements with surrounding communities, standby contracts for needed resources, and other mechanisms to ensure an adequate response to a major incident?

   ☐ Yes ☐ No
4. Has your community typed its response resources according to capability?

   ☐ Yes ☐ No
5. Does your community work with surrounding communities to improve interoperability among response resources, communications, and technology?

   ☐ Yes ☐ No
6. Have you personally reviewed your community's EOP within the past 12 months?

   ☐ Yes ☐ No
7. Have you and your senior staff participated in a disaster exercise within the past 12 months?

   ☐ Yes ☐ No
8. Has your community adopted a comprehensive program to educate the public on the risks faced by the community and how to prepare for them?

   ☐ Yes ☐ No
9. Has your community involved the media in planning, training, and exercising the EOP?

   ☐ Yes ☐ No
10. Does your community have adequate accounting and recordkeeping procedures to document requests for reimbursement from state and federal assistance programs?

    ☐ Yes ☐ No
11. Does your community ensure that the information needed to defend itself in a disaster-related lawsuit is maintained during an incident?

    ☐ Yes ☐ No

12. Have you committed time in the past year for face-to-face discussion with your emergency manager discussing ways to improve disaster management?
    ☐ Yes ☐ No
13. Have you supported zoning, building code, and other ordinances that would reduce your community's risk of damage from specific types of disasters?
    ☐ Yes ☐ No
14. Do you understand state and local emergency management law, particularly as it relates to the CEO's powers during an emergency?
    ☐ Yes ☐ No
15. Does your community's EOP include plans for continuity of operations in the event of a catastrophic disaster that renders critical facilities inoperable or incapacitates key personnel?
    ☐ Yes ☐ No

Scoring Your Self-Assessment

Obviously, the correct answer for all questions in the self-assessment is yes. To gauge how well you are doing, give yourself one point for each "yes" answer. Total your score. Then, grade your risk according to the scale below.

14–15 "yes" answers. Your community's risk is apparently well managed, but it can be managed better. Look over your "no" answers, and decide what you can do to reduce these areas of exposure.

12–13 "yes" answers. Your community is making good progress, but there are a number of actions you can take to reduce your risk. You may wish to focus your attention on the areas to which you answered "no." Based on the results of reviews in these areas, you can decide what additional steps you need to take.

9–11 "yes" answers. Your community may be at risk, but it is not too late. Scores in this range suggest that your community's emergency management responsibilities are being partially met, but there is room for (and need of) improvement. Start today by working with your emergency manager to improve your community's emergency management program.

Fewer than 9 "yes" answers. Your community is at risk! Prompt action is indicated. You need to take immediate action to improve your community's ability to respond effectively to a major emergency or disaster. Begin by conducting a complete review of the emergency management organization, and take an active role in the review. *Review the remainder of this kit*, and take the steps necessary to improve your community's capability.

Building Your Own Survival Kit and Your Community's EOP

Before building your own survival kit, it is important to understand the relationship between the kit and your community's EOP. Many progressive communities have developed a system of emergency response guides tailored to local needs. The basic EOP is the largest and most complete guide. It will contain the basic plan, functional annexes such as evacuation and mass care, and hazard-specific appendices such as flood and terrorism.

Because it is impossible for everyone to carry a copy of the complete EOP, some communities have developed a simplified guide to their EOPs. These guides may be called "Emergency Checklists," "Emergency Procedures," or another name but their purpose is the same—to provide a quick reference that spells out

- Which agency has primary responsibility for the different types of emergencies facing the community.
- The steps that are critical to take regardless of the emergency.

The guides are designed to provide department directors, response personnel, and other key personnel a short, easily readable, and readily available list of task assignments to be carried out as required for the categories of disasters faced by the community.

If your community has developed a similar guide, you should include it in your personal survival kit which, in turn, would be stored in your car or elsewhere. The CEO Checklist is designed for you to carry in your wallet or purse so that it will be available to you immediately in an emergency.

What to Put into Your Survival Kit

Your survival kit should include the tools and equipment that you, as an elected official, need with you during the first hours following a disaster, when critical decisions often must be made. Use the list on the next page as a guide to assembling your survival kit.

Kit Contents

Information:

- ☐ A pamphlet-sized mini emergency plan or other checklists
- ☐ A CEO Checklist
- ☐ Maps of the community (e.g., by street, topography, satellite photo)

Supplies:

- ☐ A tape recorder
- ☐ A notebook or log book
- ☐ Spare pencils or pens

Identification:

- ☐ Your government-issued credential with photograph

Contact Lists:

- ☐ The call-down list for all key personnel
- ☐ Contact information for key state counterparts or liaisons

Clothing:

- ☐ Seasonable outerwear

Optional Items:

- ☐ A portable two-way radio with extra batteries (not to be used if a bomb is suspected)
- ☐ A flashlight with lithium batteries and a spare bulb
- ☐ A cellular phone
- ☐ Protective clothing (e.g., a hard hat, goggles, gloves, boots)
- ☐ A change of clothing
- ☐ Personal comfort items

Suggestions for Developing and Using Your Kit

Use the guidelines below when putting your survival kit together:

- Keep it simple and small. Try to keep the kit compact—small enough to carry comfortably with you in your car.
- Direct all key subordinates to develop their own kits.
- Test the kits during preparedness exercises.
- Send enabling memos to all departments directing them to support plan development, revision, training, and exercises.
- Attend important training and exercises personally.
- Call for multiagency critiques after each significant incident.
- Use the checklist and other materials in the kit as a guide to reviewing your emergency management knowledge and the status of your community's preparedness.
- Review and direct the modification of the EOP and survival kits, as necessary—but not less often than once each year.
- Consider testing the response of your senior staff through unannounced drills.
- Establish a sequence for reporting critical information to you, and use this sequence as a guide for designing your own checklist.

The "Take-It-with-You" CEO Checklist

No commercial pilot would think of taking even a routine flight without going over the preflight checklist carefully. If pilots with hundreds of hours of training and practice need them for safety reasons, shouldn't you, as a CEO, have a checklist available when you are "flying solo" during a major emergency?

You can use the CEO Checklist included in this kit as it is—but it will be better and more effective if you use it as a prototype for addressing your community's specific needs. The Checklist folds to credit card size, so you can take it anywhere with you. Essentially, the Checklist is a set of reminders for the following:

- Information you should gather
- Questions you should ask
- Immediate steps you should take
- Key points for you to remember as you begin to manage your community's response

The Checklist is designed to take you through the first hours following a disaster. Make your version of it a synthesis of your role in your community's EOP.

Remember that neither this nor any other tool will substitute for direct involvement in your community's emergency planning process. Rather, the Checklist should be a reflection of your involvement in planning, development, training, and exercising.

A sample of the Checklist is included in this kit for your use. The Checklist is organized into six panels:

- Background Information
- Immediate Actions
- Personal Actions
- Legal Requirements
- Political Requirements
- Public Information

Each panel lists key reminders in brief phrases. To help you understand the phrases, each panel is explained in more detail on the pages that follow. If the meaning of any phrase is not readily apparent to you, refer to the appropriate section in this kit for a reminder as to its meaning.

If you see the value of the Checklist and are convinced that it or one similar to it would be an important tool for you in meeting your emergency management responsibilities, you may want to skip to the final item in this section. That item is a draft memo that outlines points you might like to raise with your staff when you forward this material to them for action.

BOX 9.1 THE CEO'S DISASTER SURVIVAL KIT

Background Information

Notified by:
Time:
Type of
Emergency:

Location

- Reporting point, open routes and means; communication channels

Sizeup

- Magnitude
- Impacted area
- Best/Worst case

Damage

- Injuries/Deaths
- Property damage
- Other impacts

Resources

- Incident command status
- Resources committed (int/ext)
- Resources required
- EOC status and location
- Other authorities notified

Background Information

The Checklist's first panel gives you a quick list of background information you will need when the incident is first reported to you. You will need this information to assess the incident and to determine your immediate actions.

This information is organized in a logical sequence that fits most situations. Because there is always the chance that you could be cut off from the official alerting you, the Checklist has been designed to get the most critical information first. You can help yourself while helping the reporting official if you make it clear that you want the official to report the facts in the order presented on the Checklist.

A description of each item on the Checklist is included on the following pages.

Notified by—For legal reasons, it is important to document who first notified you about the incident. Obviously, it is important to validate the authenticity of the call by password or other means.
Time—Be sure to note exactly when you were first notified of the incident. Use the space on the Checklist to note the time of notification. Liability judgments may well rest on how you respond based on information you were given in this first call.
Type of Emergency—Understand the type of emergency scenario you are dealing with and the actual or, if unknown, potential scale of it.
Location—Get the best description of the location and extent of the incident as possible.
CEO Reporting Point—Clarify whether to go to the Emergency Operations Center (EOC), and alternate EOC, or another reporting point.
Open Routes—Determine the best available travel route and type of transportation required.
Available Communication—Confirm the availability of primary and backup communication channels and frequencies available to you. Remaining in communication at all times is critical to effective leadership during an emergency.
Damage—The scope of damage is often unreliable in the initial reports. Despite this reality, try to document what you can. The information may prove crucial in how soon a state or federal disaster declaration is made. Be sure to gather information about

- Injuries and deaths
- Property damage
- Other impacts (e.g., damage to critical facilities or the infrastructure)

Resources—You need to gather information about resources to get an immediate sense of the resources committed, the resources required, and the possibility of the need for outside resources. Often, mutual aid agreements require the CEO or a designee to formally request the assistance.

Even if no immediate action is required from you to summon additional resources, it is important to know what resource requirements exist or might be needed later. This information allows the CEO to give immediate direction to authorize the use of additional resources.

Gather the information below as a minimum:

- **Incident command status.** Under the requirements of NIMS, your community is required to adopt ICS as its incident management system. One of the key aspects of successful emergency management is knowing who is in overall command at the incident and at the EOC.

- **Internal/external resources committed.** At this point, you will find out what local agencies are involved in the response and their degree of commitment (i.e., partial or full). You will also want to know if outside agencies have responded and, if so, whether they are engaged in operations or serving as backup resources.
- **Internal/external resources required.** Ascertain if there already is, or is about to be, a shortage of internal or external resources. Such a shortage may require you to take some immediate action, such as calling the CEO of a nearby community to request assistance.
- **EOC status and location.** The EOC may require your authorization to be activated—or an alternate EOC site may need to be designated. By checking the status of the EOC, you will be certain that everyone is clear about where they should be.
- **Other authorities notified.** Gathering this information will help you determine whether all appropriate officials have been notified of the incident. By determining what officials have been notified, you can determine whether other personnel should be notified immediately.

BOX 9.2 THE CEO'S DISASTER SURVIVAL KIT

Immediate Actions

- Begin personal log.
- Establish contact with Office of Emergency Management.
- Direct staff to assess and report on problems, resources, shortfalls, policy needs and options.
- Chair assessment meeting.
- Issue emergency declarations, as needed.
- Set reporting procedures.
- Remind staff to keep complete logs of actions and financial records.
- Begin liaison with other officials.

Immediate Actions

The Checklist's second panel will be your guide to actions you must take immediately to save lives and protect property.

A description of the Immediate Actions is included below and on the next page.

Begin personal log—Keep a log of all key information, factors weighed, and decisions reached from the time you are notified that an incident has occurred. The log can be written or recorded using a minicassette recorder.

The log should include all information and orders given under background information and should be used continuously throughout the incident. This log will document the amount of information you had when making decisions and will protect you if liability issues are raised later. The log will also be essential when preparing your after-action report.

Establish contact with the Emergency Management Agency—As the situation permits, and if you have not already done so, contact the emergency manager to ascertain the Emergency Management Agency's status.

Direct staff to assess and report on problems, resource shortfalls, policy needs, and options—Direct the emergency management staff to compile an initial assessment as it is reported from the incident site. The assessment should take the form of a situation report. It should give the status of dispatched resources, coordination efforts, and appropriate measures of problems, shortfalls, and options for the initial response.

Chair assessment meeting—If officials are conforming to local protocols—and if those protocols are sound, this initial meeting should begin to answer some of the questions include

- Who is in charge? Of what? Where?
- Has there been a proper vesting of authority?
- Is continuity of government assured?
- What is the status of intergovernmental coordination?
- What options are open to deal with resource shortfalls?
- What financial issues are surfacing?
- What parameters should be established in contacting outside public officials?
- Is there a need to place other personnel on alert?
- How much media interest is there in the incident?

Based on the information provided at this meeting, you may decide to place additional resources on standby or to activate them.

Issue emergency declarations, as needed—Decide if an emergency declaration should be issued or remain in force. Issue emergency declarations as appropriate, and be sure official documentation is initiated and continued throughout the emergency. This documentation may be critical if liability issues arise and will be necessary if state and federal assistance is sought.

Establish reporting procedures—It is important to establish a regular schedule for bringing your senior staff together to hear from those planning the next tactical response steps. You and your key advisors will need this information to establish strategic priorities and assign resources.

Remind staff to keep complete logs and financial records—Each key official should maintain a log that records actions taken, information received, and any deviation from policy, together with the rational for that decision.

Begin liaison with other officials—Consider this one of your primary responsibilities. Maintaining contact will foster cooperation.

BOX 9.3 THE CEO'S DISASTER SURVIVAL KIT

Personal Actions

- Provide family with contact information.
- Take personal survival kit.
- Other things to remember:

 ____

 ____

 ____

Remember! Your role is not operational.

Personal Actions

In the midst of thinking about the danger threatening your community, it may be difficult to recognize the need to take the time to consider your own and your family's needs. Experience has shown that public officials function better if they have made adequate provisions for their families' safety and to ensure that they have the personal items that they will need during the incident. The Checklist's third panel describes these actions.

A description of the personal actions you should take is listed below.

Provide family with contact information—Alert your family to the incident, and assure that they know how to respond if they are in any danger. Be sure that they know where you will be and how to contact you. Taking the time to leave word with your family about your whereabouts and how you plan to maintain contact will be reassuring to them and to you.

Take your personal survival kit—If you have planned ahead, your personal survival kit will be assembled and in your car or in another place where you can reach it easily. Be sure to include any personal items that you may need should the incident continue for an extended period.

Other things to remember—Jot down other personal reminders in the space provided.

BOX 9.4 THE CEO'S DISASTER SURVIVAL KIT

Legal Issues

- Contact legal advisors. Review legal responsibilities and authorities:
  - Making emergency declarations
  - Line of succession

- Mutual aid agreements
- Other legal issues:

  ____

  ____

- Monitor equity of service based on needs and risks. Maintain balance between public welfare and citizens' rights.
- Have status of contracts reviewed.

Legal Issues

The fourth panel of the Checklist includes legal issues that you will want to check on early in the incident. Checking potential legal issues early may very well save you and the community time, effort, and money later.

A description of the legal issues you should address is listed below.

Contact legal advisors—Set up communications links with the community's legal advisor(s). The earlier, the better. Review legal delegations and legally binding authorities for:

- Declaring emergencies
- Delegating authorities
- Securing intergovernmental assistance through mutual aid
- Protecting the population within the appropriate legal safeguards

Other issues may arise based on the incident type, location, cause, and severity.

Monitor equity of service based on needs and risks—Defend against charges of discrimination by establishing and following criteria for treating all sectors of the community equitably. This means keeping the public informed of what is being done to restore the community's essential services and monitoring service restoration to see that all neighborhoods receive equal treatment.

Have status of contracts reviewed—Have your community's legal advisors review any current contract with suppliers of emergency goods or services as necessary.

Political Issues

Regardless of how politically savvy you are, some issues will invariably arise. Staying aware of what issues are likely will save you political capital and allow you to focus your energy on the incident at hand.

BOX 9.5 THE CEO'S DISASTER SURVIVAL KIT

Political Issues

- Recognize accountability. Check provisions for public officials:
  - Space at the EOC
  - Periodic updates
  - Staff updates on politically sensitive issues, such as life and property losses and service interruptions
- Establish and evaluate policy decisions frequently.
- Confer with other selected officials when problems arise.
- Use elected officials to request assistance from public and private organizations.

Panel five of the Checklist includes some of the political issues you should consider. A description of these issues is provided below and on the next page.

Recognize accountability—Ultimately, the public will hold elected officials responsible for perceived response and recovery failures. Therefore, the response strategy must be carefully analyzed, without interfering with operational tactics. Recognize that accountability is a continuing issue and keep your senior staff and those at higher governmental levels informed of the ongoing situation.

Check provisions for public officials—Ensure that public officials have space at the EOC. Also be sure that they receive periodic updates and that senior staff are kept up to date on politically sensitive issues.

Establish and evaluate policy decisions frequently—Identify and consider the political aspects of declaring an emergency and other policy decisions.

Confer with other elected officials when problems arise—Get advice in anticipation of or as problems arise. A good rule of thumb here is that a good time to seek counsel from city council members or their equivalent is when things start to fall apart. By having these officials assigned in one place, these consultations can be undertaken promptly.

Consider also contacting any peer or advisor who has handled a similar incident for their advice and guidance.

Use elected officials to request assistance from public and private organizations—Often, a key to cutting through red tape and obtaining a quick response from other public and private resources is to contact the authority that controls the resource directly. This is a particularly appropriate role for public officials if normal channels are not responsive enough. Political connections can often expedite a special request.

Public Information

The impact of the media on response operations is sometimes referred to as the "disaster after the disaster." Disasters often have no greater challenge than that of attempting to use the media to inform the public accurately and in a timely manner. To succeed, the CEO must be prepared to deal with both the local and national media.

> **BOX 9.6 THE CEO'S DISASTER SURVIVAL KIT**
>
> **Public Information**
>
> - Designate a single PIO.
> - Ensure that the PIO establishes a Joint Information Center (JIC).
> - Approve all media releases.
> - Establish media update and access policies.
> - Enlist the help of experts.

The sixth panel of the Checklist presents key points for dealing with the media. A description of each of these points is covered below and on the next page.

Designate a single Public Information Officer (PIO)—Appoint one PIO to avoid conflicts in official statements that could result in confusion, panic, or misdirected public outcry about the handling of the incident. Appointing a single PIO may mean reaffirming your day-to-day spokesperson or it may mean appointing someone with experience in the type of incident that is occurring. Whomever you appoint, be sure that this individual coordinates with the PIOs from response agencies and other communities that may be involved in the incident—and with you—before releasing information to the public.

Ensure that the PIO establishes a Joint Information Center—A JIC is a single location for disseminating incident-related information to the media. Operating through a JIC helps to ensure a coordinated message and reduces "freelancing" by journalists who are looking for a story.

Approve all media releases—One of the worst situations that can occur during a response is for you to be blindsided by questions about media releases that you did not know about. Establish an approval process for media releases so that they are routed through you and coordinated with other agencies as they are developed. Take particular care with evacuation and return announcements.

Establish a media update and access policy—Consider the type of incident and the hazards at the scene when determining media update and access policy. For some incidents, it may be necessary to designate a pool of a few reporters and camera crews to allow reporting from the scene rather than granting blanket access. In any event, update the media regularly, especially early in the response when things are moving quickly. You can gradually reduce the frequency of media updates as the incident site is brought under control and especially during recovery.

Enlist the help of experts—It is not possible for you to be an expert in every type of incident response. Enlist the help of experts and let them answer technical questions about the incident. Experts may be available from agencies within the local or state government, from local colleges and universities, or from private industry.

A Note about the Checklist

Before using this or a modified checklist, review local procedures and your responsibilities from the basic EOP. You may find one of the Checklist's greatest benefits is that it facilitates your thinking about your and your community's readiness to handle a major incident.

Finally, to make this kit easy for you to use, a sample action memo is included here (Figure 9.1) for you to consider in taking your next steps to reduce your risk.

MEMORANDUM

Date:

To: **Emergency Program Manager**

From:

Re: **CEO Disaster Survival Kit**

Attached, please find the CEO's Disaster Survival Kit. Having reviewed the material, I am very interested in your opinion concerning the feasibility of tailoring this kit to meet our community-specific needs. The kit appears to fill some voids in my preparation. In addition to a CEO survival kit for myself, I think it may be appropriate to develop checklists for the following agency directors:

- Emergency Manager
- Director of Public Works
- Emergency Services Director
- Police Chief
- Fire Chief

Such a kit cannot help but facilitate our response to any community disaster by delineating agreed upon essential actions required in the first 30 to 60 minutes of the disaster and assigning responsibility for those actions.

This kit will not only provide a safer, more structured response by our response agencies, but at a time when liability suits threatening community officials and departments are so prevalent, using this kit could provide us with some improved armor.

Let's see if we can develop these materials before we conduct our next disaster exercise. I am looking forward to seeing what you can do to improve on ideas in this kit and would appreciate your initial response by ____________.

Thank you for your time and attention to this important matter.

Figure 9.1 Sample action memo.

Appendix A: IEMS and the CEO

Background: Nature of the Problem

We live with a wide range of potential hazards—floods, hurricanes, tornadoes, earthquakes, fires, chemical and toxic materials spills, civil disruption, public service strikes, major accidents, hostage situations, and terrorist attacks—that could affect any community no matter how large or small. Seventy percent of our nation's population resides in cities, providing high-density targets for disasters and, thus, greatly increasing the likelihood that these disasters will cause extensive loss of life and damage to property. One of the major problems in dealing with these hazards is the duplication of emergency response efforts at all levels of government. Currently, these emergencies are most often addressed by type and independently by the individual agencies that might respond. Responding to emergencies using this strategy has proven inefficient and ineffective, usually resulting in unnecessary and unacceptable loss of lives and damage to property and the environment.

Your Role in IEMS

When a disaster strikes, the CEO immediately becomes the focal point for leadership, and the community's success in responding to an emergency situation will depend on the effectiveness of the local emergency management system. The foundation of such a system should be an integrated emergency management plan and the capability of implementing the plan successfully.

One of the major roles of the CEO is to provide leadership in the development, training, and exercising of an effective integrated emergency management system (IEMS).

Often, CEOs take office with little or no experience in managing emergencies and without knowing what is involved in a comprehensive IEMS. The CEO may have little understanding of the laws that relate to emergency management or of liability issues revolving around action, or inaction, during an emergency. Community policies and systems that affect community risks and the potential response requirements may be unclear.

It is imperative to understand one's responsibilities to communicate to the public about the impact of emergencies on their lives and property and the need for emergency policy.

FEMA's Role in IEMS

The Federal Emergency Management Agency (FEMA) developed the IEMS to assist CEOs in establishing an effective system. IEMS has proven effective in communities throughout the country.

What Is IEMS?

IEMS is a comprehensive system that integrates and coordinates vital agencies and resources into a program of disaster prevention, preparedness, response, recovery, and mitigation. The establishment of an IEMS requires a systematic process to

- Identify risks and potential vulnerabilities
- Inventory community resources
- Outline roles and responsibilities of key agencies
- Ensure strict coordination and communication among agencies, businesses, and nongovernmental organizations (NGOs)

What Is the Cost of an IEMS?

A comprehensive IEMS does not require extraordinary allocations of community funds. Because it is a management process, the program strives to use existing resources in a more efficient and effective manner rather than seek additional resources and generate new programs.

Phases of IEMS

When fully developed, IEMS is all hazard and all phase. IEMS is all hazard in that it addresses natural, man-made, and technological emergencies. IEMS is all phase in that it addresses prevention, preparedness, response, recovery, and mitigation. Each of the phases of IEMS are listed below:

- *Prevention* includes all activities initiated to prevent a disaster from occurring. Prevention includes measures such as ensuring the safe handling, transport and storage of hazardous materials; safe storage of flammable materials; and enforcing building codes directed toward rendering structures safe from identified hazards.
- *Preparedness* readies governments to respond to emergencies. A response plan cannot be developed during a disaster and, if an emergency cannot be avoided, government must be prepared to cope with it. Planning, training, and exercises are essential elements of preparedness, as are proper and adequate supplies, equipment, facilities, and personnel.
- *Response* to a disaster depends on effective implementation of emergency operations that provide for an immediate, coordinated effort involving decisive actions that will eliminate or reduce the severity of the incident or will prevent it from intensifying. Response operations may include warning, evacuation, fire suppression, search and rescue, apprehension, treatment, and, in extreme cases, withdrawal for safety reasons.

- *Recovery* involves returning the community to its predisaster state. Recovery applies to individuals and organizations and includes physical, mental, and financial aspects such as repairing, replacing, or rebuilding property; regaining health and state of mind; and reopening damaged businesses.
- *Mitigation* includes all activities designed to reduce the recurring damage from known hazards. Mitigation and prevention are the best forms of emergency management but are often the most neglected.

IEMS: A System to Provide for Coordinated, Effective Response

IEMS is based on the recognition that there are common elements that form the foundation for responding to any emergency and increasing capabilities in these areas to improve the ability to deal with any type of emergency. The most important of these elements is the central management of all emergency activities under the ICS and Multiagency Coordination Systems.

Who is in charge during an emergency? *You*, as the CEO of your community, have the responsibility to provide for the protection of the lives and the property of your community's residents. Are *you* prepared to meet this responsibility? To carry out your responsibilities to the citizens of your community, you must ensure that local agencies are prepared to respond to any type of emergency.

Appendix B: Camera-Ready Copy of the CEO Checklist

Your printer can print these six panels back to front. Fold the Checklist (Figure 9.2) as shown on the next page, and laminate it for longer use.

Appendix C: Action Steps to Reduce Your Community's Risk and Implement IEMS

Getting Organized

1. Participate in monthly emergency management meetings.
2. Schedule presentations for elected and senior management officials.
3. Appoint representatives of other organizations to create an emergency management council or task force.
4. Hold a local IEMS workshop.
5. Assign an individual to be responsible for emergency management.
6. Provide line item budgets for emergency management activities.

<table>
<tr>
<td>
CEO's disaster survival kit

Immediate actions

▪ Begin personal log
▪ Establish contact with office of emergency management
▪ Direct staff to assess and report on problems, resources, shortfalls, policy needs and options
▪ Chair assessment meeting
▪ Issue emergency declarations, as needed
▪ Set reporting procedures
▪ Remind staff to keep complete logs of actions and financial records
▪ Begin liaison with other officials
</td>
<td>
CEO's disaster survival kit

Immediate actions

▪ Begin personal log
▪ Establish contact with office of emergency management
▪ Direct staff to assess and report on problems, resources, shortfalls, policy needs and options
▪ Chair assessment meeting
▪ Issue emergency declarations, as needed
▪ Set reporting procedures
▪ Remind staff to keep complete logs of actions and financial records
▪ Begin liaison with other officials
</td>
<td>
CEO's disaster survival kit

Personal actions

▪ Provide family with contact information
▪ Take personal survival kit
▪ Other things to remember:
</td>
</tr>
<tr>
<td>
CEO's disaster survival kit

Legal issues

▪ Contact legal advisors. Review legal responsibilities and authorities:
  ▪ Making emergency declarations
  ▪ Line of succession
  ▪ Mutual aid agreements
  ▪ Other legal issues:
▪ Monitor equity of service based on needs and risks. Maintain balance between public welfare and citizens' rights.
▪ Have status of contracts reviewed.
</td>
<td>
CEO's disaster survival kit

Political issues

▪ Recognize accountability. Check provisions for public officials:
  ▪ Space at the EOC
  ▪ Periodic updates
  ▪ Staff updates on politically sensitive issues, such as life and property losses and service interruptions
▪ Establish and evaluate policy decisions frequently.
▪ Confer with other selected officials when problems arise.
▪ Use elected officials to request assistance from public and private
</td>
<td>
CEO's disaster survival kit

Public information

▪ Designate a single PIO.
▪ Ensure that the PIO establishes a joint information center (JIC).
▪ Approve all media releases.
▪ Establish media update and access policies.
▪ Enlist the help of experts.
</td>
</tr>
</table>

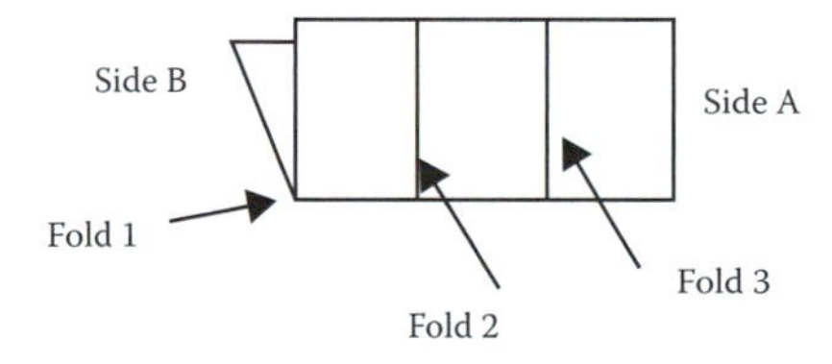

Trim away page except for the six panels. Fold (fold 1) and glue side B to the back of side A. Fold side A panels (folds 2 and 3) as an accordion fold.

Figure 9.2 CEO checklist.

Planning

1. Conduct or participate in community-wide planning.
2. Conduct or participate in multijurisdictional planning.
3. Conduct facility planning.
4. Conduct a hazard analysis and capability assessment.

5. Become involved in planning.
6. Involve the media, NGOs, and private entities in planning.
7. Evaluate and upgrade the EOP.
8. Designate a lead agency for each type of disaster.
9. Implement ICS for all responses.

Training and Exercises

1. Conduct or participate in community-wide training.
2. Conduct or participate in multijurisdictional training.
3. Conduct facility training.
4. Involve the media, NGOs, and private entities in training.
5. Conduct tabletop exercises every six months.
6. Conduct all-agency, multijurisdictional exercises at least annually.
7. Conduct ICS training.

Preparedness

1. Develop warning systems for the entire population.
2. Improve mutual aid agreements.
3. Give a customized Checklist to all senior officials.
4. Review legal responsibilities and authorities.
5. Develop procedures, media releases, and other emergency authorities before an emergency occurs.

Capability Development

1. Improve prevention and mitigation measures for all types of disasters.
2. Develop hazard-specific appendices to your plan for high-risk hazards.
3. Develop and maintain disaster resource lists.
4. Review liability and risk exposure.

Public Information

1. Conduct presentations on IEMS for senior officials.
2. Hold one-on-one meetings with senior officials to discuss emergency needs and procedures.
3. Conduct public and employee education programs on evacuation, sheltering in place, and disaster preparedness and response measures.
4. Improve relations with the media, NGOs, and the private sector.
5. Conduct seasonal public education campaigns on specific hazards.

Acknowledgments

This form was largely developed in the 1980s by the Federal Emergency Management Agency and the International Association of Fire Chiefs under contract to the U.S. Government. The author gratefully acknowledges their role in providing the initial format (that has been out of print since 1982).

MAKE this document YOUR own by using it as a TEMPLATE for your Community.

Chapter 10

Planning and Exercise

Kim Morgan and Michael J. Fagel

Contents

Introduction

I often hear "I wish there was an exercise in a box I could go purchase somewhere," or "I have a great idea for my exercise but don't know how to do it," or "My organization is required to have exercises but I don't have the time or money" and many other requests.

The choices are few; for example, hire a contractor to design and deliver an exercise for you, participate in someone else's exercise, or stop what you are doing for a week and take an expensive exercise course. Another option is to study the many resources and guides available for exercise design and implementation that, honestly, are often difficult to follow and apply, if not complicated and scary. Most people do not have the time, training, or funding. Even exercise experts are sometimes driven to find more simplified tools and templates that they can modify for their own use.

This publication addresses discussion-based exercises including seminars, workshops, tabletops, and games. The sections that are presented include discussion-based exercises and operations-based exercises. This is followed by appendices with templates that can be adapted for your use.

The material is designed to provide, in a very concise and direct manner, the steps and processes that can be taken to carry out a discussion-based exercise. It allows the user to be creative, use their own ideas, and adapt the information to suit their needs and organizational structure. It is designed for public and private organizations including corporations, municipalities, public health organizations, and hospitals, in addition to the many roles that may be involved including students, emergency managers, firefighters, planners, utility engineers, and municipal managers. Regardless of the need or organization, this material is meant to be applicable and modified to meet your specific needs. This material will allow you to make an assessment of your broader emergency management, crisis management, continuity of operations capabilities, and capacity to develop a programmatic approach to enhance and improve the strength of your operations.

This material will guide you through a step-by-step process to achieve the following:

- Steps to evaluate and enhance your organizations' crisis management, emergency management, and continuity of operations

- Steps in the design and implementation of an exercise
- The process and tools used for evaluation
- How to track the exercise results to maximize the efforts

This material does *not* address operations-based exercises, Homeland Security Exercise, and Evaluation guidance, and does *not* address how to develop and implement a multiyear exercise program.

Each type of exercise is presented in detail along with a guide that will refer you to the correct materials and templates in the appendices. These materials can then be adapted to your organization's specific needs.

Why Have an Exercise?

The world is always changing, and the risks and hazards that confront our organizations and communities also evolve. Private and public organizations including jurisdictions, agencies, departments, and corporations are expected or required to mitigate risks, prepare, respond, and recover in a timely manner. In order to do so, organizations must have a comprehensive and integrated capability and capacity in place. The best way to validate the capability and capacity to meet any worst-case scenario is through exercises.

BOX 10.1 REGULATORY REASONS TO EXERCISE

Governments, agencies, corporations, and certifying entities mandate preparedness training and exercising. The following is a very brief list:

- State and local governments are required to comply with the DHS HSEEP multiyear exercise and evaluation program.
- Nuclear power plants must exercise their plans yearly with a full-scale exercise every two years evaluated by the Nuclear Regulatory Commission (NRC).
- Superfund Amendment Reauthorization Act of 1986 (SARA) Title III requires that certain facilities that produce, use, or store hazardous materials conduct yearly exercises and evaluation of their hazardous materials response recovery plan.
- Airports, hospitals, and other healthcare facilities must conduct a full-scale exercise once every two years to maintain their certification or license to operate.
- Many employers are required by the Occupational Safety and Health Administration (OSHA) to develop an emergency action plan. OSHA recommends that such plans be exercised at least annually.

- Some critical infrastructures and key resources (CIKR), such as communications, energy, and cyber facilities, are required to plan and exercise to strengthen their ability to mitigate, withstand a disaster and recover quickly.

An exercise provides a focused simulated scenario where participants have the opportunity to function as they would in a real-world incident. Exercises serve as a valuable tool in the identification of enhancements, improvements to performance, and capabilities. This foresight provides the opportunity to train, practice, validate, and enhance activities within a realistic environment. In addition to meeting some regulatory requirements, exercises serve to

- Clarify roles and responsibilities
- Improve organizational and interagency coordination and communications
- Find resource gaps
- Develop and improve individual performance
- Identify opportunities for improvement

The focus of an exercise should always be on practice, as well as identifying and eliminating problems *before* an actual disaster. The benefit of an exercise comes not only from the practice or identification of something that needs to be improved, but in actually making a change or an improvement. The recommendations developed from the exercise must be acted on to make the effort worthwhile. Identified corrective actions are a critical part of exercise design and follow-up.

Multiyear Exercise Program

Exercise program development is not addressed here but deserves a strong emphasis. A single exercise will not cover all functions and eventualities. It will also not involve all of the potential people and organizations that might be involved in a real-world incident. If you are going through the effort to design and deliver an exercise, take a few extra steps and establish a more beneficial and lasting exercise program—one based on a long-term, carefully constructed plan. By creating a comprehensive program, your exercises can build on one another to maximize the organization's goals and capabilities, provide training to a wider group of participants, and vastly improve your organizations' ability to manage a disaster.

How This Material Is Organized

Exercise design, implementation, evaluation, and improvement can appear complex and overwhelming with a wide range of steps, tasks, and operations. The material is meant to aid in navigating to the exercise you need and the steps you will need to take to develop and implement your own exercise. These sections are arranged so that you only need to address the tasks that are appropriate for the type of exercise that you intend to develop and implement. For each type of exercise, a list of the materials and products are provided in text boxes for quick reference.

How FEMA and DHS Organize the Sequence of Main Tasks

The Federal Emergency Management Agency (FEMA) and the Department of Homeland Security (DHS) offer various guides in the development and implementation of exercises. For the most part, they are useful in providing a high-level overview of the general steps that they suggest you to follow. In the following graphic illustrating FEM IS-139 Exercise Design (Figure 10.1), the main tasks are shown in chronological order and provide a picture of the entire sequence. It does not provide the details of functions and tasks that would be expected of various types of exercises.

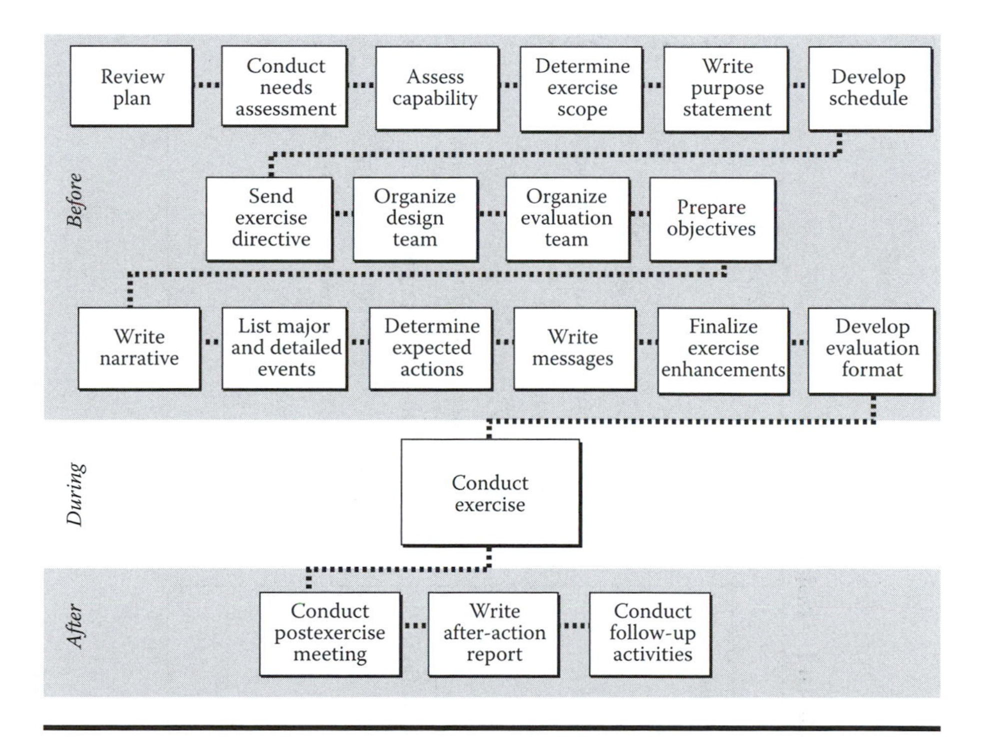

Figure 10.1 Sequence of tasks for a successful exercise.

Table 10.1 Task Categories

	Preexercise Phase	*Exercise Phase*	*Postexercise Phase*
Design	• Review plan • Assess capability • Address costs and liabilities • Gain support/issue exercise directive • Organize design team • Draw up a schedule • Design exercise (eight design steps)	• Prepare facility • Assemble props and other enhancements • Brief participants • Conduct exercise	
Evaluation	• Select evaluation team leader • Develop evaluation methodology • Select and organize evaluation team • Train evaluators	• Observe assigned objectives • Document actions	• Assess achievement of objectives • Participate in postexercise meetings • Prepare evaluation report • Participate in follow-up activities

Categories of Tasks

Another way to view the various steps is provided in a matrix of task categories (Table 10.1).

Major Task Accomplishments

One of the simplest ways to envision the exercise process is by major accomplishments. As shown in Figure 10.2, the process is factored into five major accomplishments that make up the design cycle:

1. Establishing the base
2. Exercise development
3. Exercise conduct
4. Exercise critique and evaluation
5. Exercise follow-up

Each accomplishment is the outgrowth of a set of specific tasks and subtasks (similar to those listed in the earlier models). The process is circular, with the results of one exercise providing input for the next.

Types of Exercises

Exercises may range widely in cost, size, scope, complexity, purpose, and approach. Each exercise should build on the scale and lessons learned from any previous exercises.

There are a total of seven different types of exercises that are either discussion-based or operation-based. A director or presenter leads the discussion-based

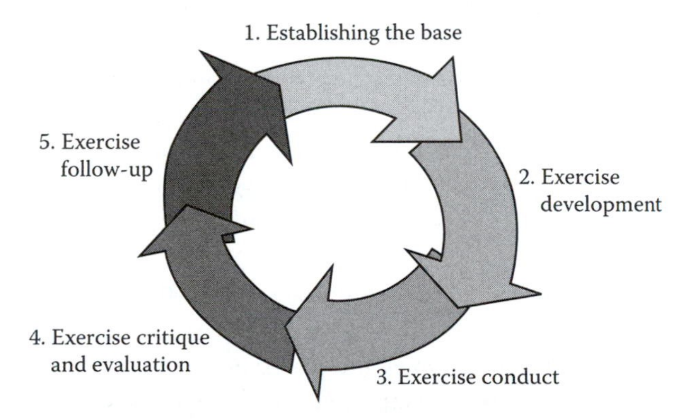

Figure 10.2 The five major accomplishments of the design cycle.

exercises. Here, exploration of assumptions and plans of participants are reviewed. Because discussion-based exercises are not tied to an actual response timeline, participants have time to address items as a group, such as plans and procedures. Items could include familiarization with, or revision of, plans and procedures.

These exercises serve to

- Familiarize participants with current plans, policies, agreements, and procedures (training)
- Develop new plans, policies, agreements, and procedures
- Identify opportunities for enhancements, integration, or improvements to existing plans, policies, and procedures
- Do not involve dispatch to the field of manpower, equipment or other resources

Discussion-based exercises are available in four different formats and are the subject of this material:

- **Seminar:** An informal discussion to familiarize participants to new or updated plans, policies, or procedures (e.g., Winter Weather Advisory Procedures or Building Evacuation Procedures).
- **Workshop:** Is very similar to a seminar but the participants are gathered to provide their input into a new draft plan, policy, or procedures (e.g., IT Workshop to define disaster procedures or a Multi County Workshop to develop a comprehensive plan and exercise program).
- **Tabletop Exercise (TTX):** Involves key decision-making personnel from different departments, roles, agencies, and organizations discussing plans, policy, and procedures within the context of one or more scenarios (e.g., County Flood TTX, Corporate Disaster Relocation TTX, Water Contamination TTX, Pandemic Outbreak TTX). Activities in a tabletop do not involve the activation of outside procedures or field activities such as setting up of triage tents, calls to outside agencies, or sending directions to staff outside the room such as field staff.
- **Games:** Involve two or more teams, typically competing, that are conducting a simulation or a practice of specific operations and must follow specific rules, data, and procedures to mirror real-world events (e.g., information security defend and breach, financial growth, Emergency Operations Center (EOC) data/GIS output, laboratory testing and results).

Operations-based exercises offer much of the same as discussion-based exercises:

- Involve deployment of resources and personnel
- Are typically more complex than discussion-based types
- Require execution of plans, policies, agreements, and procedures

- Clarify roles and responsibilities
- Improve individual and team performances
- Include drills and both functional and full-scale exercises

There are three types of operations-based exercises:

- **Drill:** A coordinated, supervised activity that tests a single, specific, operation or function within a single entity (e.g., evacuation drill, drill to set up a hospital decontamination tent, drill to test how quickly air or water samples can be collected, drill to identify how quickly backup power systems or IT backup records can be launched, or how quickly can a school go into lockdown). This is the only type of exercise where the word "test" is used.
- **Functional Exercise:** Examines and/or validates the coordination, command, and control between various multiagency coordination centers (e.g., emergency operation center to the joint field office, region to region, laboratory to laboratory, hospital to hospital). Field staff are not utilized in this type of exercise, for example, first responders or emergency officials responding in real time.
- **Full-Scale Exercises:** A full-scale exercise is a multiagency, multijurisdictional, multidiscipline exercise involving functional (e.g., joint field office, emergency operation centers) and real-time response activities (activation of equipment and resources, first response to mock victims, hospital triage of mock victims, laboratory testing of spiked samples, mock IT intrusion attack and response).

Exercise Documentation for Discussion-Based Exercises

Exercise documents are developed by the Exercise Planning Team and serve as tools to guide and capture information during the exercise process. Some documents that you may need to develop are obvious; they include sign in sheets and invitation letters or emails. The following documents are specific to the type of exercise that you elect to develop.

Situation Manual

A Situation Manual (SITMAN) is the handbook for discussion-based exercises and is provided to *all* participants. It provides background information on the scope, schedule, and objectives for the exercise. The SITMAN will also contain the scenario and typically includes the questions or issues that the presenter or Exercise Director will be asking. Ideally, the participants will receive this in advance of the exercise so that they can read the material and become familiar and comfortable with the topics and questions.

TIP BOX 10.2

If all exercise types and all documents that go with them seem complicated, do not worry. Once you start the process, you can use our checklist to ensure that you select the correct documents for the type of exercise you pick. Examples in the Appendix can be used as a base for your exercise.

After Action Report and Implementation Plan

The After Action Report (AAR) serves to capture the findings and input from the exercise. Based on the findings, the implementation plan (IP), which is sometimes included in the AAR, provides a prioritization of issues with assignments and timelines for reaching the goal in order to correct and improve operations and plans based on the exercise outcomes.

Participant Roles

Participants in discussion-based exercises play different roles depending on their level of participation in the exercise.

- **Players** are personnel who have an active role in responding to the simulated emergency and perform their regular roles and responsibilities during the exercise. Players initiate actions that will respond to and mitigate the simulated emergency.
- **Facilitator or Exercise Director** is a member of the controllers who manages the overall delivery and flow of the exercise. This role is found in all exercises since this is the person who guides and delivers the materials.
- **Evaluators** are chosen to evaluate and provide feedback on a designated functional area of the exercise. They are chosen based on their expertise in the functional area(s) they have been assigned to review during the exercise, as well as their familiarity with local emergency response procedures. Evaluators assess and document the participants' performance against established emergency plans and exercise evaluation criteria. They are typically chosen from among planning committee members or the agencies/organizations participating in the exercise.
  For most discussion-based exercises, it is very common for the players to self-evaluate and conduct their own evaluation and determine opportunities for improvement rather than having outside evaluators do this effort.
- **Observers** visit or view selected segments of the exercise. Observers do not play in the exercise, and do not perform any control or evaluation functions.
- **Exercise support staff** includes individuals who are assigned administrative and logistical support tasks during the exercise (i.e., registration, catering).

Defining Exercise Project Management

Exercise project management is the next step after program management. In this step, project managers are responsible for the design, development, and execution of a specific exercise, followed by evaluation and improvement planning. Good project management involves

- Developing a project management timeline
- Establishing project milestones
- Identifying the exercise planning team
- Scheduling planning conferences

These tasks are the foundation of every exercise—without them, other tasks and stages of the exercise planning cycle could not happen.

Exercise Timeline

In program management, the Multiyear Exercise Schedule provides a long-term calendar for multiple exercises. Exercise project managers build timelines to include

- A schedule of key conferences and milestones
- A Master Task List
- Planning team task assignments

Generally, timelines for discussion-based exercises are shorter and have fewer tasks than timelines for operations-based exercises.

Design and Development

Think of design as the framework of an exercise, and development as the building of that exercise.

Exercise design includes

- Assessing exercise needs
- Defining the scope of the exercise
- Writing a statement of purpose
- Defining exercise objectives
- Creating a scenario for the exercise

Exercise development includes

- Creating exercise documentation
- Arranging logistics, actors, and safety

- Coordinating participants and media
- Other supporting planning tasks (e.g., training controllers, evaluators, and exercise staff)

Needs Assessment

Exercises are designed to motivate participants to think and act as they would in a real incident. You will need to select the appropriate exercise design that best meets your organization's needs. If you have developed a comprehensive exercise program, you will have already evaluated the needs for your organization. In addition to developing a work plan and organizing a design team in some cases, getting ready for the exercise involves several steps (see below).

1. **Review of the current plans**
   Is there a plan or does it need to be developed? What does it tell us about ideal performance—i.e., how are we supposed to implement policies and procedures in the event of an emergency?
2. **Conduct a needs assessment**
   What are our risks and vulnerabilities, and where do we need to focus our training efforts? Use the checklist in Appendix A to assist you in gathering all necessary information, developing your needs assessment, and getting ready.
   Your needs assessment will help you identify the following:
   - What functions and activities need practice in the exercise, and which ones have priority?
   - Based on the priority, who will participate?
   - What does the exercise need to do and what are you or your organization capable of doing?
   - What are the most likely hazards and the priority levels of those hazards?

   For example, if your organization recently updated one of their plans, then that plan should be validated. Therefore, your needs assessment should reflect the desire to design and conduct an exercise that would then test that plan.
3. **Assess the organization's capability to conduct an exercise**
   What resources can we draw from to design and implement an exercise?
   As you begin the process of designing your exercise, consider the response activities that your organization would take, what resources would be needed, and whether your organization is ready for an exercise. Then you can select the exercise that suits your needs. If you are already sure what type of exercise you need, skip to the section addressing that type of exercise.

   Do you have the resources? Consider the management support, skills, personnel, time facilities, equipment, and funding that may be required. The questions below, although very general, can help you assess your organization's level of capability. For example, you may find that before you consider planning an exercise, you will need to develop support and train people.

- Managerial Support

  What is the level of support and demand for your exercise? Do you have managerial support and is the need for a exercise recognized?
- Experience and Organization Maturity

  When was the last exercise? If your organization has never had one, consider starting with workshop style meetings or tabletops. If your organization has had one before, did it go well and can you build on it, or is the organization preparedness mature enough to move on to a more complex exercise?

  What level of exercise experience is available to pull this together? If you have never designed an exercise, start with a less complicated exercise design that would have the same impact than a multiagency full-scale exercise.

  What physical facilities, equipment, communications systems, and other hard resources would be used in a real incident? Will they be available for an exercise? If you want to run an exercise of your EOC or your corporate headquarters but they are not available, then you will need to consider another design and scope for your exercise.
- Time and Effort

  How much time do you have and when do you expect to deliver the exercise? Some exercises can be pulled together in a matter of minutes and have no impact on regular operations. Consider drills, casual workshops, and very simplified tabletops. Some exercises can take a year or longer to implement. It all depends on your design and exercise type.

4. **Define the exercise scope**

   What are the limits of the exercise efforts? Scope refers to the extent and boundaries of the exercise. Typically, it defines the kind of exercise participants (e.g., levels of government, private sectors, corporate levels) rather than the actual number of people.

   Other parameters of scope may include
   - Geographic size (local, national, regional)
   - Number of participants
   - Responder functions
   - Hazard type
5. **Consider the exercise type**

   What type of exercise best meets your training needs within the available resources?
6. **Address the costs and liabilities**

   What will the exercise cost in terms of funding, human resources, and organizational liability?

   You should consider the liabilities and costs early in the exercise design.

Liabilities

For all exercises, consider the risk and exposure of personal injury and equipment damage. For all drills and full-scale exercises, you must evaluate the risks and liabilities, the insurance coverage, and safety considerations.

Costs

All exercises entail expenses. Make sure you do not reach too far in your exercise design! Keep the scope manageable, not too large or too complex, and select the participants who are best suited and would actively be involved according to the exercise program, type, budget, and objectives.

Consider the following items to see if these are items you may need and whether there is a cost limitation on the type of exercise you are considering.

- Your time
- Other people's time
- Equipment and materials
- Contract services
- Miscellaneous items (e.g., snack, lunch, coffee, pencils)
- Fuel to run equipment and transport volunteers

Some examples to consider:

- Staff with emergency management roles and responsibilities should have those actions reflected in their job description. Therefore, an exercise is a function of their department and the role of that person.
- Every employee who has an emergency management role of responsibility should have time set aside to participate in training, planning, and exercising.
- Costs for routine participation in exercises should be recognized by agency or organization officials.
- If the exercise takes place in the evening or over a weekend, will overtime pay be covered, or will the staff schedules need to be adjusted to avoid overtime pay?
- If the exercise is to meet a regulatory requirement such as a certification, who, or which department, will cover the costs?

If you do not have clear support for any of the above items, this needs to be addressed when you speak to top management and leadership and gather their support.

7. **Develop a statement of purpose**

   What do you expect to gain from the exercise?

8. **Gain support and announce the exercise**

   Determine the level of support from those in authority, and then use that support to garner support among participants.

Sample Needs Assessment Worksheet

Below is a job aid to help you identify and clarify your organization's exercise needs and capabilities, in addition to focusing the exercise design efforts.

1. Hazards
   List the various hazards in your community or organization. What risks are you most likely to face? You can use the following checklist as a starting point. If your community or organization already has a hazard analysis, utilize this resource.
   - ☐ Airplane crash
   - ☐ Dam failure
   - ☐ Drought
   - ☐ Epidemic or biological attack
   - ☐ Earthquake
   - ☐ Fire/Firestorm
   - ☐ Flood
   - ☐ Hazardous material spill/release
   - ☐ Hostage/Shooting
   - ☐ Hurricane
   - ☐ Landslide/Mudslide
   - ☐ Mass fatality incident
   - ☐ Radiological release
   - ☐ Sustained power failure
   - ☐ Terrorism
   - ☐ Tornado
   - ☐ Train derailment
   - ☐ Tsunami
   - ☐ Volcanic eruption
   - ☐ Wildfire
   - ☐ Winter storm
   - ☐ Workplace violence
   - ☐ Other ____________
   - ☐ Other ____________
   - ☐ Other ____________
   - ☐ Other ____________
2. Secondary Hazards
   What secondary effects from those hazards are likely to affect your organization?
   - ☐ Communication system breakdown
   - ☐ Overwhelmed medical/mortuary
   - ☐ Power outages
   - ☐ Transportation blockages
   - ☐ Epidemic outbreaks
   - ☐ Other
   - ☐ Other ____________
   - ☐ Mass evacuations/displaced population
   - ☐ Business interruptions
   - ☐ Supply chain interruption
   - ☐ Sustained power failure
   - ☐ Other ____________
   - ☐ Other ____________
   - ☐ Other ____________
3. Hazard Priority
   What are the highest priority hazards? Consider such factors as:
   - Frequency of occurrence
   - Relative likelihood of occurrence
   - Magnitude and intensity
   - Location (affecting critical areas or infrastructure)
   - Spatial extent
   - Speed of onset and availability of warning

- Potential severity of consequences to people, critical facilities, community functions, and property
- Potential cascading events (e.g., damage to chemical processing plant, dam failure)

  #1 Priority hazard:
  #2 Priority hazard:
  #3 Priority hazard:

4. Area
   What geographic area(s) or facility location(s) is(are) most vulnerable to the high priority hazards?
5. Plans and Procedures
   What plans and procedures will guide your organization's response to an emergency? This includes standard operating procedures, emergency response plans, contingency plans, operational plans, or any plan that may be developed in the future based on your present exercise.
6. Functions
   What emergency management functions are most in need of review? What functions have not been exercised recently? Where have difficulties occurred in the past?

   - ☐ Alert Notification (Emergency Response)
   - ☐ Warning (Public)
   - ☐ Communications
   - ☐ Coordination and Control
   - ☐ Emergency Public Information (EPI)
   - ☐ Damage Assessment
   - ☐ Health and Medical
   - ☐ Individual/Family Assistance
   - ☐ Public Safety
   - ☐ Public Works/Engineering
   - ☐ Transportation
   - ☐ Resource Management
   - ☐ Continuity of Government Operations
   - ☐ Other ____________
   - ☐ Other ____________
   - ☐ Other ____________
7. Participants
   What agencies, departments, operational units, or personnel need to participate in an exercise?
   - Have any entities updated their plans and procedures?
   - Have any changed policies or staff?
   - Who is designated for emergency management responsibility in your plans and procedures?
   - With whom does your organization need to coordinate in an emergency?
   - What do your regulatory requirements call for?
   - What personnel can you reasonably expect to devote to developing an exercise?
8. Program Areas
   Mark the status of your emergency program in these and other areas to identify those most in need of exercising.

Identification of Program Areas for Consideration in the Exercise

	New	*Updated*	*Exercised*	*Used in Emergency*	*N/A*
Emergency Plan					
Plan Annex(es)					
Standard Operating Procedures					
Resource List					
Maps, Displays					
Reporting Requirements					
Notification Procedures					
Mutual Aid Pacts					
Policy-Making Officials					
Coordinating Personnel					
Operations Staff					
Volunteer Organizations					
EOC/Command Center					
Communication Facility					
Warning Systems					
Utility Emergency Preparedness					
Industrial Emergency Preparedness					
Damage Assessment Techniques					
Other:					

9. Past Exercises
   Identify what was learned from previous exercises, if any, and how issues need to be addressed:
   - Who were the previous participants and who needs to be involved in the new exercise? (Are there new people or portions of your organization that need to be involved this time around?)
   - To what extent were the exercise objectives achieved? (Should you consider doing the same objectives again?)
   - What lessons were learned?
   - What problems were revealed and what was needed to fix them?
   - What improvements were made following past exercises and do they need to be included in the exercise?
   - Have there been structural changes in your plans, procedures policies, organizational structure, people and resources? Consider all of these to include in your exercise objectives.

Select and Conduct Your Discussion-Based Exercise

In this step, you will select the exercise that is most suited to your needs and requirements. The different types of exercises are described in detail and each has a box listing the process, requirements, materials, development time, and other factors that may influence your choice.

Seminars

The seminar is an overview or introduction to provide information on roles and responsibilities, plans, procedures, and/or equipment. Because there is rarely any simulation or scenario-based analysis, and the flow of information is from one direction only (the presenter), it is typically not considered a complete exercise.

Level of Difficulty: easy
Level of Effort: minimal
Development Time Frame: short
Delivery Time Frame: 1–2 hours
Suggested Process
- Review plans, policies procedures.
- Develop presentation materials.
- Establish meeting time, date, location, and attendees.
- Present the seminar.
- Collect input on how to improve the presentation.
- Make changes to the presentation materials.

Materials

- Presentation

Goals

- Orient the participants to new or existing plans, policies, or procedures.
- Research or assess interagency capabilities or interjurisdictional operations.
- Construct a common framework of understanding.

Roles

- Facilitator
- Players

Conduct Characteristics

- Casual atmosphere
- Minimal time constraints
- Lecture-based

Format

- Lecture
- Discussion
- Slide or video presentation
- Computer demonstration
- Panel discussion or guest lecturers

Products

- Human resources training record updates
- Regulatory requirement certification

Examples

- Presentation to discuss a topic or problem
- Introduction of new changes, policies or plan
- Orientation of new employees to policies, plans, and procedures
- Introducing a cycle of exercises or preparing participants for success in more complex exercises

Leadership

A facilitator or instructor presents information at seminars. The facilitator should have some leadership skills, but little other training is required.

Games

A game is a simulation of operations using rules, data, and procedures designed to depict an actual or assumed real-life situation. Games are similar to drills in that the participants run through a series of predetermined and planned steps. Games differ since the participants are typically on teams competing against each other and are required to analyze and implement data and assumptions.

Level of Difficulty: easy to moderate
Level of Effort: minimal to moderate
Development Time Frame: 1 day–3 months

Delivery Time Frame: 1 hour–3 days

Suggested Process

- Review plans, policies procedures.
- Develop game and presentation materials.
- Establish meeting time, date, location, and attendees.
- Determine equipment and communication needs.
- Establish gaming teams.
- Conduct games and measure outcomes.
- Collect and present findings.

Materials

- Presentation
- Gaming materials including databases, equipment, and other resources
- Activity lists, injects, and scenarios

Goals

- Explore the processes and consequences of decision-making.
- Practice procedures and identify alternate methods to improve results and speed of operations.
- Test existing and potential strategies.

Conduct Characteristics

- Requires an experienced facilitator familiar with the subject matter
- May require assistants to support multiple locations
- Focuses on the ability of one group to correctly conduct operations or procedures more accurately or rapidly than another group
- Time stress and sometimes the uncertainty of not knowing how your group is performing in relation to other participating groups

Products

- AAR or exercise summary that provides findings including evaluation and changes to current systems and operations
- Identification of new methods and procedures that may improve accuracy, capability, and/or speed

Examples

- Laboratory Operations with two teams to establish rapid testing methods of unknown materials: identify new testing methods, verify procedures, and identify the specimen before the other team using a wide variety of resources.
- Data compilation: have teams identify trends in stock markets, bank records, or other databases, and use the information to disable or disrupt the other team's capabilities or finances.
- Strategic Operations: Teams compete to identify weaknesses in other teams' operations and strategies. The goal is to identify weaknesses in operations and plans, and identify methods to prevent real-world attacks. An example would be two teams that are established to protect the databases and security of an operation while two other teams attempt to breach security.

- War gaming: Teams are established to compete against each other in war gaming scenarios to identify probable outcomes and mitigation strategies to prevent or reduce the observed outcomes.

Workshops

A workshop is a formal discussion-based exercise led by a facilitator or presenter, used to build or achieve a product, and often used for the development of new procedures or plans. Typically, a facilitator will guide participants through the design and development of a document or policy allowing for group collaboration and buy-in.

Level of Difficulty: moderate
Level of Effort: moderate to difficult
Development Time Frame:
Delivery Time Frame: 1–2 hours
Suggested Process:
- Review plans, policies, and procedures.
- Develop presentation materials.
- Establish meeting time, date, location, and attendees.
- Present the workshop.
- Collect input including information, issues, topics, action items, and materials developed during the workshop.
- Compile and draft the materials developed in the workshop such as draft plans, procedures, contact information, or rosters.

Materials
- Worksheets or templates for plans
- Presentations to guide the participants through the workshop
- Specific assignments, issues or talking points that need to be provided by breakout groups

Roles
- Facilitator
- Players

Goals
In a workshop, participants:
- Develop new ideas, processes, or procedures
- Develop a written product as a group in coordinated activities
- Obtain consensus
- Collect or share information

Products
- Develop and refine drafts of SOPs, policies, and plans
- Sharing of contact information, databases and other materials
- Build relationships between participants

Conduct Characteristics
- Presented and guided by a facilitator
- Involves participant discussion led by the facilitator
- Often uses break-out sessions and smaller groups to explore parts of an issue

Format
- Lecture
- Discussion
- Slide or video presentation

Examples
- Presentation to discuss a topic or problem
- Discussion to obtain input for policies, procedures, and plans
- Build consensus and agreement
- Practice, prepare, and review for more complex exercises

Leadership

A facilitator presents information and guides any discussion during workshops. The facilitator should have leadership skills, an understanding of the relationships between the players, and have the capability to guide the workshop players to the goal.

Examples of Workshop Exercises

1. **Example Workshop Exercise**

   Georges Hospital has an emergency operations plan that is very old, needs to be updated, and does not address new departments and processes at the hospital. The design team, consisting of the emergency manager and the head of security, has determined that the goals are to confirm the material in the plan and define what the roles and responsibilities for each hospital department would be during an emergency. The design team identifies specific parts of the plan that need to be discussed during the workshop and drafts a list of roles and responsibilities for each of the departments. During the workshop, the hospital emergency manager takes the role of facilitator and guides the group through the steps in the plan to confirm whether they are valid or need changes. The facilitator then presents the drafts of roles and responsibilities (the steps that the role would have to take during an emergency or disaster) to confirm that his best guess is correct and whether those departments wish to add or change the list.

   The design team determines that two very brief scenarios of a pandemic outbreak and hazardous materials release (5–6 bullets each) will serve as the background for the discussion.

   The design team discusses how they are going to present the information to the participants. One choice is to develop an informal handout with the scenario and list of questions or develop a more formal SITMAN that contains the background for the exercise, the scenarios, questions, objectives, and improvement plan that would be addressed. Another option is to present the information through a slide show presentation and have the participants take notes. Another option that the design team discusses is to have the participants view the overall scenarios on a slide show and then work out the issues assigned to them.

The design team determines that it is beneficial for the participants to be aware of the materials that will be covered so that valuable workshop time will not be wasted. They decide to provide the material in a three-page handout. Examples of some of the questions that will be presented in the workshop are listed below.

Questions addressing a pandemic outbreak:

- At what point will it be determined that there is an outbreak problem and who will make this decision?
- Who will be notified inside and outside of the hospital?
- What are the roles and responsibilities before, during and after it has been determined there is an outbreak? What are the roles and responsibilities of the new departments?
- What steps will your department be responsible for at each phase of mitigation, preparedness, response, and recovery?
- What outside agencies and organizations need to involved and coordinated with?
- Who needs to be alerted and notified, what are the regulatory and planning requirements for notification, who will make the notification, and when does the notification need to occur?
- Who is responsible for public information, what message will need to be scripted for release to the media, who does this need to be coordinated with inside and outside the hospital, and how often and when will this information need to be updated?

Questions addressing a material release:

- How, and when, will the hospital be notified that there is a hazardous material release that involves contaminated victims?
- If the hospital has advance notice, what can be done to prepare for the arrival of victims?
- How will the hospital handle victims that self-present and drive themselves to the hospital?
- What are the roles and responsibilities before, during, and after it has been determined that there is a hazardous materials release with multiple victims? What are the roles and responsibilities of the new departments? Is this different from a pandemic outbreak?
- What steps will your department be responsible for at each phase of mitigation, preparedness, response, and recovery during this scenario?

- At what point will the hospital resources become overwhelmed?
- How will public information be handled?
- How will decontaminated be handled of the victims, ambulances, and hospital equipment?

Potential Outcome of This Workshop

- This forum allows for all participants in the room to understand and learn the responsibilities and tasks of the other hospital departments.
- Provides the opportunity for corrections and updates to be made to the old plan.
- Provides the opportunity for all of the departments to provide input to their roles and responsibilities as well as provide comment on other department responsibilities.
- Prepares and trains the participants for more advanced exercises.

2. **Example Workshop Exercise**

The City of Denmar has existing plans and procedures that are exercised and reviewed on a yearly basis. In six months, the city will be participating in a TTX with other nearby cities. In one year, the city will be participating in a large full-scale regional and statewide exercise. Additionally, the city recently annexed land and purchased water and sewer utility plants and services. The design team consists of the emergency manager, city mayor, and a representative from the police, fire, public health, water, and wastewater utilities.

The design team determines that the objectives will be to ensure updates of existing plans, address the addition of the water and wastewater utilities, practice topics that may be addressed in the upcoming exercises, and begin a discussion of media and public information issues that may be used as a public information and media plan annex to the existing emergency operations plan. The design team finds that the amount of material and topics is extensive and unproductive to present to a single group. They design the workshop to start as a presentation to all participants and then have the participants break into work groups in the same room to discuss among themselves the issues that have been assigned to them.

The design team uses the list of invites to determine who will be assigned to which breakout group. They also define the breakout groups to be Preparedness, Response, Recovery and Media, and Public Information. The design team develops a list of questions and topics that each breakout group will need to address during the workshop. During the last two hours of the workshop, each group will be provided with time to present their findings to the entire room.

Action items will be compiled and assigned to the various groups, as needed. The participants determine that their action items and list of things to be completed include updating a communications and contact list, add the new water and wastewater facilities to the existing city emergency operations plan, update the plan list, and create a draft for a new public information plan.

The design team determines that they will use a brief earthquake scenario to serve as a background for the discussion because upcoming exercises will be based on earthquake scenarios. They also determine that additional questions will include other scenarios in order to make sure the plan information is as flexible as possible and not disaster specific.

The design team decides that the information will not be provided to the players before the exercise. They decide that the exercise conduct will involve having the introduction and scenario provided to the entire room at the same time. Once the breakout groups are ready, the list of breakout group questions will be provided to each table. The breakout groups are asked to document the findings, suggestions, and improvements that they identify for presentation at the end of the exercise. The following lists are examples of some of the questions that might be considered during this type of workshop:

Questions for the preparedness breakout group

- What can be done to prepare for the scenario that is presented?
- What if the scenario were for a large storm, hazardous material release, or outbreak?
- What has been done in the past and what can be improved?
- What plans and training are in place?

Questions for the response breakout group

- At what point will it be determined that there is a problem, and who will make this decision?
- Who will be notified inside and outside of your organization?
- What are the roles and responsibilities during response? Are the response roles and responsibilities accurate?
- Is the response portion of the city plan consistent with the plans of the other organizations including fire, police, and public health?
- What changes need to be made in the existing plan?
- What outside agencies and organizations need to involved and coordinated with during response activities?
- At what point during the scenario does your department need to escalate its operations and call in additional support from other agencies?
- Who needs to be alerted and notified, what are the regulatory and planning requirements for notification, who will make the notification, and when does the notification need to occur?

Questions for the recovery breakout group

- At what point do recovery operations start?
- Who needs to be involved in recovery, and who should be on the recovery team?

- What are the roles and responsibilities during recovery? Are the response roles and responsibilities accurate?
- What are the specific recovery problems that need to be addressed for this type of scenario (debris, damaged infrastructure, transportation problems, communications problems, and interruption)? What if the scenario is a flood or major storm?
- Is the recovery portion of the city plan consistent with the plans of the other organizations including fire, police, and public health?
- What changes need to be made in the existing plan? Does the plan need to be changed to ensure that there is integration and consistency between the city plan and other plans such as police, fire, hospital, or outside plans such as counties and state?
- What outside agencies and organizations need to be involved and coordinated with during recovery activities?
- Do you have enough resources, where can more resources be located, how will they be tracked, how will this be paid for? Form the list of resources that are identified, which ones are the most important?
- At what point during the scenario does your department need to escalate its operations and call in additional support from other agencies?
- Who needs to be alerted and notified, what are the regulatory and planning requirements for notification, who will make the notification, and when does the notification need to occur?

Questions for the media and public information breakout group

- Given the scenario and the various organizations that would be involved, who is responsible for public information?
- How will public information be coordinated?
- At what point would a media and public information group be established in the EOC, who will make this decision, who is the contact for each department and organization, and who will serve as a representative for the EOC or Joint Information Center?
- What message will need to be scripted for release to the media, who does this need to be coordinated with inside and outside the hospital, and how often and when will this information need to be updated?
- What scripted pubic information messages can be drafted beforehand, and who can do this?
- To whom will the media release be addressed, and how will this information be delivered (television, radio, door to door, etc.)?
- Who needs to be alerted and notified that there will be a media release, and what are the regulatory and planning requirements for notification?
- Who is responsible for public information, what message will need to be scripted for release to the media, who does this need to be coordinated with inside and outside the hospital, and how often and when will this information need to be updated?

Outcomes

This workshop design allows for experts in particular arenas (e.g., response, planning, recovery, public information) to work together to find new changes and solutions in a short period. Other outcomes include the following:

- Provides the opportunity for corrections and updates to be made to the existing plan.
- Provides the opportunity for experts to develop an outline of a public information and media plan.
- Provides the opportunity to define the roles and responsibilities of the participating organizations including the new water and wastewater utilities.
- Prepares and trains the participants for more advanced exercises.

Variations on a Theme

For breakout groups, consider having poster boards outlining the issues placed near each group. Have the players document their observations and suggestions on the poster boards, whiteboards, or flip charts along with the number or description of the question that is addressed and the initial of the person making the comment. Throughout the workshop, have the breakout groups rotate to another table to address a new set of questions and consider the findings documented by the previous group. This methods works well if there is a short list of questions for each breakout group, where face-to-face contact and relationships need to be developed, and where the breakout group topics are general enough for various groups to consider. For example, if the breakout group subject is transportation and ice/snow removal procedures, the hospital breakout group will not be capable of providing much input into this breakout group subject.

Tabletop Exercises

A TTX is a facilitated, stress-free analysis of plans, policies, and procedures within the context of a brief scenario. It is designed to elicit constructive discussion between participants as they examine and resolve problems based on draft or existing operational plans. The difference between the workshop and tabletop is that the tabletop is more focused on executive decision making and addresses issues at a much higher level (does not go into details of each step of a plan).

The success of the exercise is largely determined by group participation in the identification of problem areas and opportunities for improvement and refinement. Participants are typically those persons in decision-making positions and roles, persons responsible for directing staff or activities, elected or appointed officials, or other key personnel in an informal group discussion centered on one or more scenarios.

During a tabletop, equipment is not used, resources are not deployed, and time pressures are not introduced.

Level of Difficulty: moderate
Level of Effort: moderate
Development Time Frame: 2 weeks–6 months
Delivery Time Frame: 1 hour–3 days
Suggested Process
- Develop a design team.
- Review plans, policies, and procedures.
- Develop presentation materials.
- Establish meeting time, date, location, and attendees.
- Develop a SITMAN.
- Invite attendees and provide a copy of the SITMAN to them if they are evaluating themselves rather than using outside evaluators.
- Conduct exercise.
- Conduct Hotwash and/or after action review.
- Develop AAR, IP, and assignments.
- Follow up on assignments and changes.

Materials
- Presentation
- Messages or injects on paper
- SITMAN

Roles
- Facilitator and assistants
- Players
- Observers

Goals
- Identify strengths and shortfalls.
- Identify opportunities for improvement.
- Practice policies, procedures, and plans within one organization and between organizations.
- Enhance coordination, communication, and relationships between departments, agencies, and organizations.
- Do not assess or grade staff performance rather, evaluate plans, procedures, and policies, and identify ways to improve their application.

Products
- AAR and IP
- Develop and refine drafts of SOPs, policies, and plans
- Sharing of contact information, databases, and other materials

Conduct Characteristics
- Requires an experienced facilitator

- Allows participants to discuss the items of interest in-depth and skip through the issues that do not require review
- Nonstress and nonjudgmental environment
- In-depth discussion and problem solving

Format

The exercise begins with the reading of a short narrative of the scenario. The facilitator can lead the discussion in several ways: whether through problem statements or simulated messages to breakout groups. In either format, the discussion generated by the problem focuses on roles (how the participants would respond in a real emergency), plans, coordination, the effect of decisions on other organizations, and similar concerns. Often, maps, charts, and packets of materials are used to add to the realism of the exercise.

- *Problem statements.* The facilitator delivers problem statements that describe situations pertaining to the scenario(s) to the entire group. The facilitator then asks a participant or representative of a participating agency or department to discuss the activities or actions they would take given the background and scenario. All participants are grouped together so they can learn and understand the decisions and steps taken by the other participating groups and organizations. The best room arrangement is seating in a U shape that allows all participants to see each other. This takes a skilled facilitator to engage the participants, foresee the probable outcomes, listen and identify probable issues, and bring them up for discussion. The facilitator must also ensure that all parties have the opportunity to engage directly on addressing their particular issues from their perspective. The facilitator must also have diplomatic skills to know when a subject needs to be tabled for further analysis at another time, control the conversation and discussion, and make sure that a single participant does not monopolize the entire conversation.
- *Simulated messages/breakout groups.* These messages are more specific than problem statements and mirror the types of messages that would be expected in a real-world situation. The participants are broken into separate groups and discuss the responses. The group that received a message may find that one of their plan steps is to make a decision and notify another group. They would send a message to another group in the exercise. This process allows for groups to spend more time on issues that affect them directly. Using the arrangement of group breakouts allows for the practice and analysis of the lines of communication that would be expected in a real-world incident.

Examples and Outcomes

- Presentation to discuss a topic or problem to low-stress discussion of coordination and policy.
- Low-stress discussion to obtain input for changes and enhancements to existing or new policies, procedures, and plans.
- Discussion to build consensus and agreement and practice of coordination.

- Problem Statements: discuss activities as group at large, learn about other group policies and procedures, and identify enhancements for lines of communication and integration of steps across plans.
- Breakout groups: practice lines of communication between groups.
- Provide an opportunity for key agencies and stakeholders to become acquainted with one another, their interrelated roles, and their respective responsibilities.
- Practice, prepare, and review for more complex exercises.

It is important to take good notes of the possible changes and outcomes from your tabletop. If you are the facilitator, you might ask the participants to make notes in their SITMANS and provide the comments during the AAR and Hotwash comment period at the end of the exercise. Also, have some support to take notes for you and to capture the discussion. If information is not readily available or the discussion gets stalled on a particular subject, take control and have the topic tabled to be researched and settled at a later time.

Leadership

Tabletops require an experienced facilitator and an understanding of the roles and responsibilities of the participating groups. The facilitator must have leadership skills to control and guide the groups through their own processes and aid in encouraging frank and comfortable discussion of the issues. This person decides who gets a message or problem statement, calls on others to participate, asks questions, and guides the participants toward sound decisions.

As the facilitator, understand that you will never know all the answers or issues and you never will. The experts are in the room and it is your job to guide and help them find the answers based on their expertise and knowledge, not on what you assume to be the answer. If participants ask you what you think should happen, ask them what they think it should be and what they would need to get the topic addressed. If they cannot answer, ask everyone in the room what they think should be done in regards to the question and the scenario at hand.

Participants

The objectives of the exercise dictate who should participate. The exercise can involve many people and many organizations: essentially anyone who can learn from or contribute to the planned discussion items. This may include all entities that have a policy, planning, or response role whether or not they have a plan in place.

Facilities

The type of facility is defined to some degree by the number of participants and the seating arrangement required by your exercise design. For a problem statement format, you may need a room large enough to accommodate seating arrangements in a U shape. If your exercise design is based on a simulated statement with a breakout group format, the facility will need to be able to accommodate work groups at separate tables or have work groups in separate rooms.

Time

A TTX usually lasts from one hour to three days. Depending on the type and quantity of material you are going to cover and the level of discussion you are going to elicit, most comprehensive tabletops run four hours to two days. The time for each discussion is open-ended and participants are encouraged to take the time necessary to reach in-depth decisions: without time pressure. Although the facilitator maintains an awareness of time allocation for each area of discussion, the group does not have to complete every item in order for the exercise to be a success.

Preparation

It typically takes anywhere from two weeks to three months to prepare for a TTX. If you establish an exercise design team, you will need to have an initial planning conference (IPC) to kick off the effort, and three to four more design team meetings to revise and finalize the exercise objectives and materials. Ideally, participants will have participated in at least one orientation and one or more drills on plan specifics.

Examples of Tabletop Exercises

1. **Example Plenary Tabletop Exercise**

   The Concoran Corporation is concerned about business interruption (Figure 10.3). The emergency manager and head executives have determined that a tabletop would be a good tool to determine their capacity and capability to withstand a major disaster. The emergency/risk manager develops a design team consisting of executives from key departments including facilities, operations, management, and information technology. During the IPC, they identify objectives including coordination, communication, continuity of operations, and information backup and reconstitution. The design team also determines that the scenario that best captures all the objectives would be a fire that destroys all or part of the headquarters and adjoining manufacturing plant. The exercise design team determines that they will use a problem statement format and have the seating arranged in a U shape.

Figure 10.3 Tabletop exercise: plenary design.

Under each objective, the design team defines specific tasks and items that they want addressed during the tabletop. Their draft includes answering Who, What, Where, When, and How for each objective.

The design team develops a SITMAN that present the scenario, objectives, and questions that will be presented during the TTX. The design teams holds a second design team meeting to discuss changes and revisions to the draft SITMAN. The changes to the SITMAN are made, and a final version is sent out to all participants. Below is the outline of their SITMAN, followed by a brief sample from the Scenario and Questions section. An example of a SITMAN is available in Appendix A.

Concoran Corporation Tabletop SITMAN Table of Contents

- Date
- Introduction
- Associated Documents
- Agenda
- List of Participants
- List of Objectives
- Scenario and Questions
- Action Items and Implementation Plan

Scenario and Questions

Over the weekend of February 11, a fire breaks out at the manufacturing plant. By the time the fire department arrives, the fire has spread to the headquarters building. The president and several managers are notified and arrive at the scene. Once the fire is extinguished, it is determined that the manufacturing plant is completely destroyed, whereas the headquarters reports complete damage to the IT rooms and equipment. The human resources department has suffered some damage to records and files, whereas all other departments and materials only have smoke damage. Manufacturing processes can be relocated to a secondary site outside the town, whereas management operations can be started up from scratch at an available office space near the original site.

Coordination

- Who will be in charge?
- Who will make the decision to return to work or stay at home?
- What leave policies are in place, and should the staff that is waiting to return to work be paid?
- When and at what point will these steps be carried out?
- What departments are critical for operations and immediately need to be started up in the new location?
- What equipment and resources are needed?
- What can be done to ensure that those departments can resume operations after a disaster and how much time do they require to become operational again?

- Who among the people within those departments are critical and which persons can be called in at a later time?
- What are the arrangements for temporary operations, what departments will need to be there, and who will be in charge?
- What is included in the corporate plan, and is this different from what is being discussed?

Communication

- Who will make the appropriate notifications?
- How will the different departments communicate with each other if they are temporarily housed at different temporary locations?
- How will communications with outside clients, vendors, and other organizations be handled?
- Who is responsible for moving the phone and Internet services to another location?

Continuity of Operations

- What needs to be done, and what resources are needed to ensure that the company can be up and running in 12 hours, 24 hours, one week, two weeks?
- How will the different departments communicate with each other if they are temporarily housed at different temporary locations?
- How will communications with outside clients, vendors, and other organizations be handled?
- Who is responsible for moving the phone and Internet services to another location?

Information backup and reconstitution

- What needs to be done and what resources are needed to ensure that the company can be up and running in 12 hours, 24 hours, one week, two weeks? Does the company have a cold or hot site?
- When was the last information backup, and how can this information be restored?
- What equipment is needed and how can this be procured?
- How can persons and operations located offsite at the new manufacturing location access the information and databases?
- Can the new system withstand a large influx of people accessing the data offsite or from home?
- When and how will operations be stood up and reconstituted?

Concoran Corporation Tabletop Conduct

The design team decides that the best way to present the material is to present it to the entire group. The SITMAN is delivered to all participants in advance of the exercise so that the participants have the opportunity to think through the questions, identify additional issues, and prepare in advance of the exercise. Providing the SITMAN in advance also alleviates any anxiety and nervousness that participants may feel.

The facilitator arranges the room in a horseshoe shape so that all the participants can face each other and discuss the issues. Additional copies of the SITMAN are provided along with pens so that participants can take notes in their SITMANs.

On the day of the exercise, the facilitator provides a Powerpoint presentation describing the objectives, the rules that will be followed in the exercise (when questions will be asked, who will be able to speak, when breaks and lunch will be taken), the agenda, and the outcomes and goals of the exercise. The facilitator asks the participants to make notes in the SITMAN and to document key items and tasks that need to be addressed in the future. The facilitator then presents the scenario.

Using the questions from the SITMAN, the facilitator walks the participants through the issues. In this example, the facilitator starts with the question of who is in charge given the scenario. The facilitator expects the CEO to answer that he is in charge. The facilitator would then ask the CEO who would be in charge if he is not in town or unavailable. The CEO might answer that the Vice President of Operations would be in charge if the CEO is unavailable. The facilitator would then ask if the CEO and Vice President coordinate their availability to ensure that one or the other is always available. The facilitator might then ask the Vice President who he would need to assist him during this type of scenario.

Going through each of the questions, the facilitator then asks different participants in the exercise what they would do, what their responsibility would be, how they will coordinate with other groups, how they will communicate with other parties, and what resources they might need to carry out their operations.

Hotwash

At the end of the discussion portion of the exercise (whether all the questions have been addressed), the facilitator calls a close to the question portion of the exercise and announces that they will now begin the Hotwash portion of the exercise. The facilitator then asks the participants to consider the notes that they took during the exercise and to present three key findings, observations, and tasks that need to be done. The facilitator will then ask each participant to provide three main issues that they observed.

After Action Report and Implementation Plan

After the exercise is completed, the design team compiles the findings from the exercise into an AAR. This is presented to the executive management of the Concoran Corporation who will prioritize the tasks and determines which tasks need to be completed, who will complete them, and by what date. This prioritization and assignment become the IP of the exercise. The IP is then tracked to ensure that the tasks are carried out within the appropriate deadline, and whether there are any additional items or issues that are encountered. (An example of this table can be located in the Sample SITMAN in Appendix A.)

TIP BOX 10.3

It is important that the facilitator controls the conversations and ensures that the exercise is not about finding fault or error but rather to identify opportunities for improvement. The participants are not being tested; rather, the plans and procedures are being evaluated.

2. **Example TTX with Breakout Groups**

   A state emergency manager has asked you to develop a statewide TTX (Figure 10.4). The requirements include the involvement of four key agencies, addressing objectives including coordination and communication between the agencies and a scenario that addresses response to a major storm with flood damages. The state also requests that the tabletop be presented in breakout sessions with the different agencies arranged at separate tables or in separate rooms so that a simulated statement process can be utilized.

   You pull together a design team that includes representatives from the various agencies and hold a kickoff meeting. During this meeting, you present the general objectives and goals, ask for their particular goals and objectives, and identify whether any additional participants should be invited to join the design team. You also identify the potential locations for holding the exercise and sketch out a timeline of goals that need to be reached in order to accomplish the exercise. Several assignments are given out and assigned to the design team members including locating and inviting a representative from another agency, locating potential locations for the exercise, and several other items. You set the date for the next meeting in two weeks.

   At the second design team meeting, you present an outline of the scenario and ask for input from the design team. The team then refines the issues and topics that will be presented in the scenario. You then present ideas for the objectives and the potential tasks that could be addressed under each objective. The design team adds their topics and tasks that they would like to have

Figure 10.4 Tabletop exercise: breakout groups with injects and messaging.

addressed during the exercise. The design team also addresses the location, agenda, invitees, and other topics. The design team decides it would be helpful to have several presenters provide a short lecture of their operations and procedures. The presenters are identified and later invited to speak. You establish the date for the next meeting and announce that you will be sending out a draft of the SITMAN for their review and input.

During the next month, you develop a SITMAN addressing the topics described by the state and the input provided by the design team. Before the third design team meeting, you send out a draft of the SITMAN for the design team to review and provide comments. At the third design team meeting, you address the comments and any additional changes that the group wishes to make to the SITMAN.

TIP BOX 10.4

Do not expect to complete all the questions and topics in the SITMAN. As the participants work through particular issues, it may be found that a particular task is extremely important and deserves the time and discussion to address it. It is impossible to determine in advance which items and tasks will need more time and attention since this is something that will be driven by the participants. Just make sure to keep them on track and address the issues.

You also present a list of injects that may be used during the exercise. These injects will include specific messages and scenario information that will be provided to the particular breakout groups. Injects are messages of the status or operations from an outside source that provoke or steer exercise activities in a particular direction.

The final list of invitees is defined, an invitation letter is crafted, and a public information release is designed in the event that officials need to describe the exercise to outside parties. All participating agencies procure approval from management to participate.

An invitation is sent out to the invitees asking them to respond whether they can attend or not. The invitations includes a description of the exercise and the objectives, potential participants, and a statement that this is not a regulatory and enforcement activity but rather a nonjudgmental exercise to improve plans and procedures.

Before the exercise, the design team holds a final meeting to ensure that all issues have been addressed and that the SITMAN can be sent out to all the participants. (An example of a State Level Tabletop Exercise SITMAN is located in Appendix A.)

Exercise Conduct

At the day of the exercise, participants are asked to sign in, use name cards, and sit at tables labeled for their particular agency, including fire and police, state government and emergency management, public health, environmental and critical infrastructure (communications, transportation, and utilities).

As described in the agenda, you introduce the exercise, the objectives and goals, the rules that will be followed in the exercise (when questions will be asked, who will be able to speak, when breaks and lunch will be taken), how injects will be used, how messages will be passed from one group to another, and the goals of the exercise. The participants may address the injects either through sending written messages to other tables or by making a general announcement to all the groups and simulate a conference call.

The facilitator asks the participants to make notes in the SITMAN and to document key items and tasks that need to be addressed in the future. The facilitator then presents the scenario to all the participants and then asks the group to address the first question in the SITMAN, "What will your agency do in this situation?"

Sample Injects

You allow the participants some discussion time and then hand an inject to the Government and Emergency Management table. The inject is a written note that you prepared beforehand to provoke particular groups to take specific actions.

The first inject you provide states that local emergency managers and local cities and counties are overwhelmed with storm response and damages and need help from the state. There may be several outcomes and activities depending on how you want this to play out. For example, the inject may state that they need to establish an EOC at an empty table with representatives from the other tables. Another option is that the inject may state that notifications must be made to the other agencies in the room and request their support. The Government and Emergency Management table may request that they make a general announcement to the entire room or they may choose to write a message on note pads and hand them out to all the groups.

As the groups work through the issues of the first SITMAN question and inject, you may present the second question from the SITMAN, allowing the groups time to address and respond to the question accordingly, including documenting their response and/or sending out a message to another group.

Hotwash

At the end of the time frame, regardless of whether all the SITMAN questions have been addressed, stop all exercise activities and introduce the Hotwash activities. You might ask the group to take 10 minutes to summarize the top 5 (or 10) issues that have they identified during the exercise. They will need to select someone from their group to present their findings to all the participants. Remind them that the goal of the exercise is to find methods and procedures to improve their activities and to keep the activities at a very high level. Instead of saying "Jeff at the Government and Emergency Management table never notified us that groups were responding to the disaster and an EOC had been opened," the suggestion should be, "Notification and involvement of response groups needs to be evaluated and improved. The power utilities would like to be invited to the EOC during storm responses and the water and wastewater utilities would like to be invited to the EOC during flood situations."

Once the groups have completed summarizing their observations, you then have a representative from each group present their findings. It is important to take notes of the findings as well as any comments and discussion that are provided from the other participants.

After Action Report and Implementation Plan

After the exercise, the design team will compile the findings and comments into an AAR and include a matrix or master list of activities and improvements that should be considered for implementation. Using the task matrix at the back of the SITMAN, each agency is asked to consider and prioritize the after action items and tasks, decide which tasks they are capable of taking and are willing to take, and the date that this can be accomplished.

TIP BOX 10.5

Not all implementation items and tasks will be taken on or accomplished. Typically, you may ask to have the most important and feasible topics adopted by the participating groups and leave the remaining items for the next exercise. For example, buying backup generators for all government operations may be a nice solution for a particular problem but the feasibility and expense of doing so may be restrictive. Instead, rephrase the task to include defining the high priority agencies and obtaining backup generators over time or establish a way to loan equipment from one agency to another.

You and the design team will evaluate the comments to the AAR and the tasks that each agency has decided to address. The changes and updates are made to the AAR, which is sent out to the participants again. The design team decides that a teleconference meeting will be held in four months to get an update on the implementation of the tasks and what possible issues there may be.

Variations on Tabletop Designs and Conduct

- Another format can involve not providing the scenario but only providing portions of the information to particular breakout groups. It will become their responsibility to make announcements to the room and other breakout groups that will have to remain in the dark until they are included by the other groups. This format emphasizes the need for communication and coordination as well as a practice of roles and responsibilities. If one breakout group should have contacted another group and has not done so, it would be wise to give them a clue or a comment to do so in order to keep the exercise moving forward.
- Remote tabletop design allows for agencies and department to participate without interrupting daily operations. The tabletop design remains the same except that communication and release of information and issues is conducted through teleconferencing, phones, and e-mails. This style of exercise requires advance experience and strong control of the activities and exercise play to ensure that all participants are addressing the issues and topics and that they are included in the process.
- Use video presentations of news flashes or announcements from governors or officials providing new information for the exercise participants to act on.

Appendix A. Sample SITMAN and AAR/IP

Table of Contents

Subject	*Page*

Exercise Schedule

8:00 A.M. CDT	**Welcome and Introduction**

	• Review of Administrative Details and Exercise Overview
	  • Purpose and Objectives
	  • Roles of Participants
	  • Expected Outcomes

9:00 A.M. CDT	**Background**
	• Orientation to the New Madrid Seismic Zone (NMSZ) • Local government activities • State government activities • Federal government activities
10:30 A.M. CDT	**Break**

10:45 A.M. CDT	**New Madrid Earthquake—Phase 1**

- Introduction
    - Explores 0 hours to 4 days after earthquake, including emergency response actions of county and state
- Scenario
- Facilitated Discussion
- Transition and Wrap-Up

12:15 P.M. CDT	**Working Lunch**
12:45 P.M. CDT	**New Madrid Earthquake—Phase 2**

- Introduction
    - Explores 4 days to 2 weeks after earthquake, including arrival of some federal assets, damage assessments, and beginning of the recovery phase
- Scenario
- Facilitated Discussion
- Transition and Wrap-Up

1:45 P.M. CDT	**New Madrid Earthquake—Phase 3**

- Introduction
    - Explores 6 months after earthquake, including the long-term community recovery phase
- Scenario
- Facilitated Discussion
- Transition and Wrap-Up

2:45 P.M. CDT	**Break**
3:00 P.M. CDT	**Action-Planning Session**
	Review and Conclusion
4:00 P.M. CDT	**Closing Comments**
5:00 P.M. CDT	**Adjourn**

Introduction

Purpose of the Exercise

The primary purpose of this tabletop exercise (a type of discussion-based exercise) is to enhance the ability of local, state, and federal stakeholders to prepare for, manage, and respond to infrastructure consequences resulting from any major disaster or emergency.

Additional purposes of this exercise include

- Providing an opportunity to augment emergency response planning.
- Developing a list of action items to support refinement of emergency response plans.
- Building relationships between utility, local, state, federal, and other stakeholders and response partners.

Scope of the Exercise

This exercise emphasizes the roles, responsibilities, and relationships of local, state, and federal stakeholders during a response to and management of consequences resulting from any major disaster or emergency.

Processes and decision-making are more important than minute details.

Objectives

At the conclusion of this exercise, participants should be able to do the following:

- Improve coordination and communication between response partners and delineate their roles and responsibilities.
- Delineate how data management will occur between response partners, including types of data and procedures.
- Discuss how local and state organizations coordinate response with federal organizations.

This exercise will not be a success unless you as a participant follow through.

Exercise Structure

This will be a multimedia, facilitated tabletop exercise. Participants will discuss issues related to the following phases of an earthquake scenario:

Phase 1: **0 to 4 days after an NMSZ earthquake**
Phase 2: **5 days to 2 weeks after an NMSZ earthquake**
Phase 3: **6 months after an NMSZ earthquake**

Exercise Format

This is a tabletop exercise, a type of discussion-based exercise (no field activities will occur during the exercise) during which the facilitator provides a brief overview of the scenario and then initiates a plenary discussion relating to the response for that scenario.

In advance of the exercise, participants are expected to read this Situation Manual, which includes short narratives and supporting graphics outlining the three phases of the exercise scenario. Participants should develop individual responses to the discussion questions in each phase.

Exercise discussions are intended to cover the topics listed below as well as provide participants an opportunity to share answers they developed in advance of the exercise. Before transitioning to the next phase, the facilitator will remind participants to capture action items they identified during the discussion using an action-planning matrix.

Following the discussion of all three phases, the facilitator will lead a "brainstorming" session to highlight key ideas and develop a list of action items drawn from participants' action-planning matrices.

The scenario and related information presented in this exercise are based on various NMSZ studies provided through the Mid-America Earthquake (MAE) Center at the University of Illinois and Central United States Earthquake Consortium (CUSEC). The scenario in this exercise is consistent with the analysis provided by the MAE Center and CUSEC. This scenario does not represent the worst-case scenario and only serves to provide the background for discussion during the exercise.

Discussion Topics

Anticipated discussion topics during the scenarios include

- Communication and Information Management
  - Notification Process—local to state, state to federal level
  - Situation reports—development, content, information coordination
  - Data management—methodology for utilizing and managing analytical data
  - Public information and community outreach—sharing and reporting
- Activation Process and Operations
  - Coordination between local and state agencies
  - Transitioning from response to recovery and demobilization
- Resource Management
  - Resource requests

- Resource typing
- Anticipating requests
- Procedures for making and fulfilling requests
- Tracking and reporting on resource status
- Recovering resources

– Personnel requests—credentialing, qualifications, certification, training, expertise

Participant Roles and Responsibilities

Players discuss the situation presented based on expert knowledge of response procedures, current plans and procedures, and insights derived from training and experience.
Facilitators provide situation updates and moderate discussions. They also provide additional information or resolve questions as required. Key planning team members will also assist with facilitation as Subject Matter Experts (SMEs) during the discussion.
Recorders capture the discussions of the exercise in written form.

Exercise Assumptions and Artificialities

In any tabletop exercise, a number of assumptions and artificialities may be necessary to complete play in the time allotted. Participation in the discussion is in accordance with the assumptions and guidelines below:

- The scenario is plausible, and events occur as presented.
- There are no "hidden agendas" or trick questions.
- The scenario is based on earthquake response capability and hazard assessments completed for each of the eight NMSZ states evaluated by the MAE Center and CUSEC. The scenario in this exercise is not based on worst-case scenarios developed from the models but represents a likely outcome of an earthquake of this size at this location.
- All participants receive information at the same time.

Exercise Rules of Conduct

A successful tabletop exercise also depends on following the rules below, which have been proven to ensure effective discussion.

- **There is no "school solution."** Varying viewpoints, even disagreements, are expected. This exercise is intended to be a **safe, open, stress-free environment.**
- Respond based on your knowledge of current plans and capabilities (i.e., you may use only existing assets) and insights derived from training and experience.
- Your organization's positions or policies do not limit you. Make your best decision, based on the circumstances presented.
- Decisions are not precedent-setting and may not always reflect your organization's final position on a given issue. This is an opportunity to discuss and present multiple options and possible solutions.
- Issue identification is not as valuable as suggestions and recommended actions that could improve response and preparedness efforts. Problem-solving efforts should be the focus.
- Assume there will be cooperation and support from other responders and agencies.
- The basis for discussion consists of the scenario narratives, your experience, your understanding of your organization, your intuition, and other resources included as part of this material or that you brought with you. There are no situational injects.
- Treat every scenario as if it might affect your area.

Additional Resources

The following additional resources are included as appendices at the end of this SITMAN:

- **Appendix 1: Acronyms and Key Definitions**
- **Appendix 2: List of Participants**
- **Appendix 3: Participant Evaluation Form**—the form allows for the player to provide feedback in regards to the exercise and exercise materials
- **Appendix 4: Supporting Documents and References**

SCENARIO—NEW MADRID EARTHQUAKE

Background Tuesday, April 12, 2011

At 9:00 A.M. (CST) on Tuesday, April 12, a major earthquake along the NMSZ, outside of Blytheville, Arkansas (35°55′38″N/89°55′8″W), brings the region to a shaking stop (Figure 10.5). Local seismologists measure the earthquake at magnitude 6.5 on the Richter scale. Ground shaking occurs for approximately 50 seconds. Shaking from the main shock is the result of a rupture of the southern arm of the NMSZ at a depth of 6.1 miles below the Earth's surface. The path of the rupture runs from south of Paducah, Kentucky, southwest to Arkansas. Aftershocks of varying intensity are felt throughout the region over the course of the next few days. Soil liquefaction is extensive throughout the area.

Figure 1 Scenario Fault Location

Area of Impact

The significantly damaged area extends from the southwestern portion of Kentucky, the southeastern section of Missouri, and the northeastern portion of Arkansas. The heaviest impact to Kentucky encompasses 14 counties including the following:

1 Ballard	6 Daviess	11 Hopkins
2 Caldwell	7 Fulton	12 Livingston
3 Calloway	8 Graves	13 Lyon
4 Carlisle	9 Henderson	14 McCracken
5 Crittenden	10 Hickman	

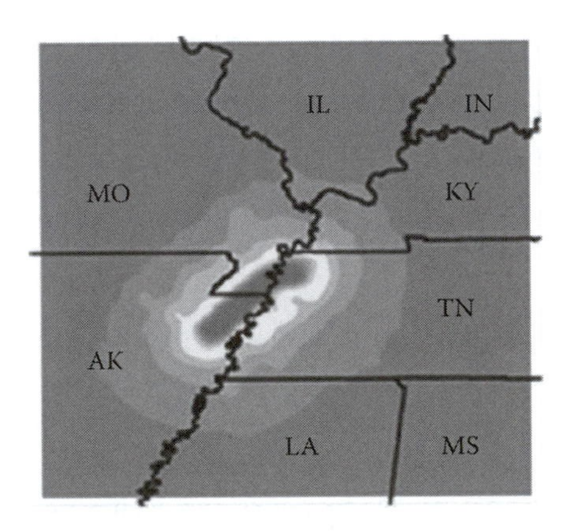

Figure 10.5 Scenario fault location.

Direct Earthquake Damage

Initial reports for Kentucky indicate more than 300 fatalities; 1200 injuries, with 1000 of these people requiring hospitalization. Thousands are missing or separated from their families. The preliminary estimate of displaced persons is 20,500 but this number is expected to increase within the first three days. Thousands of homes and possessions are destroyed or damaged. There are major shortages of essential supplies, such as fresh food, bottled water, fuel, and generators.

Collapsed bridges and roadways and sunken shipping vessels cause oil and hazardous material (HazMat) contamination along the Mississippi River. The river itself experiences an uplifting of its riverbed, causing temporary reverse flow for 12 hours and wave-surge events causing damage to the levee systems. Various levee systems and dams throughout the Kentucky area have failed resulting in extensive flooding and damages to critical infrastructures including wastewater facilities and telecommunications. Initial reports indicate significant damage to bridges including complete destruction of the Interstate 24 Bridge and the Irvin S. Cobb Bridge in Paducah, Kentucky.

Summary of Specific Damage

Shaking and soil liquefaction of land surrounding the fault line in Kentucky cause the following:

- Severe infrastructure damage in 14 counties
- Localized flooding in low-lying areas along the Mississippi River
- Destroyed or compromised roads, railroad tracks, buildings, and other infrastructure along the fault line
- Release of hazardous materials due to damaged structures and flooding from the river and levee and dam breaks
- Structural damage to levees and dams resulting in damage to wastewater and telecommunications systems
- Structural damage to drinking water treatment facilities and distribution systems
- Power outages as a result of transmission line damage
- Communication failure because of cell phone tower and phone line damage.

Phase 1: 0 to 4 Days after Earthquake Tuesday, April 12–Friday, April 15

Aftershocks continue during the days following the earthquake, several of them more than magnitude 6.5, dislodging debris and weakening already damaged structures. Aftershocks contribute to soil liquefaction and shifts in the Mississippi and Ohio rivers along western Kentucky.

Direct Earthquake Damage

As search and rescue operations continue, the death toll rises to more than 2500 throughout the various states. In Kentucky, more than 300 fatalities, 1500 injuries requiring hospital attention, and 5000 minor injuries are reported as a result of collapsed buildings and infrastructure damages. The primary locations of the casualties are located in McCracken, Graves, Ballard, Daviess, and Marshall Counties.

Hundreds of people are missing. Area hospitals and clinics are overwhelmed, especially in the major and medium impact zones. The Kentucky National Guard is setting up emergency tent hospitals to address the surge in patients. Communities within the major impact zone are completely without electricity, as are almost half of the communities in the medium impact zone.

Utility companies work to restore service, but efforts are hindered by damaged roadways and access restrictions. Various natural gas pipelines have suffered significant ruptures in segments that cross through the affected area. As the response teams for the pipeline companies attempt to isolate those segments of the pipeline, an explosion occurs along the section in Paducah, Kentucky, that severely damages a main water line and a sewer line. The resulting gas flare ends after a few hours.

Telephonic communications are virtually nonexistent due to major damage to towers and lines. Broken fuel lines and interrupted travel routes are creating fuel shortages, leaving very little fuel for backup generators. Meanwhile, hundreds of critical facilities are without power. Damages to pipelines throughout the area and manufacturing facilities located in Ballard and McCracken Counties has created a release of petroleum products (crude oil, gasoline, diesel, etc.) and HazMat (anhydrous ammonia, vinyl chloride, etc.). The earthquake and resulting water surges have broken portions of the levee systems and dams, allowing for HazMat-contaminated water to overflow into surrounding agricultural areas. Widespread environmental and economic impacts are expected. The contaminated water creates a significant risk to drinking water groundwater sources. More than 50 dams are damaged and all are located in Carlisle, Ballard, Hickman, McCracken, and Graves Counties. Levee damage is confined mostly to Fulton and McCracken Counties.

Federal Response Efforts

Due to the magnitude of the disaster, the state of Kentucky requested federal assistance under the Stafford Act for the 15 affected counties. Immediately, the President issued a disaster declaration for each of the affected states, including Kentucky. Representatives from federal agencies have arrived and EPA Region VI has deployed staff to provide support and guidance. Other than personnel supporting coordination, the full contingent of federal resources will not begin arriving for another week.

Local and state responders are overwhelmed. Many responders were injured while performing response operations in damaged, unsound structures, which has made the need for additional staffing paramount. In response to the widespread earthquake-caused damage or destruction of supply sources, resources from around the country are mobilized; many, however, have yet to arrive due to damaged transportation links. There are also significant problems in identifying where resources are most needed. Private sector organizations are offering resources and assets to aid in response efforts. There is a major shortage of qualified responders, especially for structural assessment, environmental assessment, and monitoring. Responders face additional stress due to lodging shortages.

Phase 1: 0–4 Days Response Questions

These questions serve to focus your thoughts on the issues associated with the scenario. In advance of the exercise, participants are expected to read the scenario, review each question, and develop individual responses to each question.

1. How do communication and information flow during this initial operational phase? What mechanisms are in place to help you establish situational awareness?
2. What are your initial concerns and priorities during this phase of your response based on the information that is provided? How would these priorities change over this four-day period?
3. What is your operational role? What roles would local, state, and federal agencies play?
4. At what point will you begin evaluating the impact on your operations and how will this information be managed?
5. How can Geographic Information Systems (GIS) be used in support of this scenario?
6. What triggers the request for support from the next higher level (county, state, federal)? What is your role in deciding to request, and making the request for support?
7. Once they arrive, how are response resources from outside your area integrated into your activities? What mechanisms for coordination are or should be in place?
8. How do you determine the type and quantity of resources needed? What resource requests are preplanned or prescripted?
9. How would state, regional, and federal resources be mobilized? How would these resources be tracked?
10. Where else can you obtain resources? What are the procedures for obtaining that support?

Phase 2: Day 5 to 2 Weeks after Earthquake Friday, April 15–Friday, April 29

During the two and a half weeks since the earthquake, aftershocks continue throughout the area, further damaging weakened structures and hampering operations.

Direct Earthquake Damage

Emergency shelters have reached overflow capacity and have operated with limited water or wastewater services for several days. The state has brought in portable showers and toilets to the shelters. Hospitals and clinics continue to be overwhelmed beyond capacity but no longer require emergency tents. Law enforcement officials are blocking off heavily damaged areas, so access is limited. Crews are clearing debris from roadways, and damage to roads and communications systems continue to hinder response efforts. Fuel shortages remain a major problem even though an emergency fuel depot has been established for use by responders.

Field teams consisting of local, county, and state specialists begin to assess operating status and needs assessments to address the changing needs and demands resulting from the earthquake. As federal support comes online, they join in to assist the field teams. Teams are sending assessments and reports to the State Emergency Operations Center (SEOC). This is enabling a more effective response, better use of resources, and more accurate incident action planning at the local, state, regional, and federal levels. The assessments and reported information also enable senior government officials to understand the extent of damage.

The SEOC begins to assign missions but it is immediately apparent that federal support will be required. The SEOC is coordinating efforts through the Emergency Management Assistance Compact (EMAC) to assess needs and serve as a vehicle for procuring out-of-state equipment and specialized personnel for utilities. Meanwhile, volunteer resources arrive unannounced, which has the potential to complicate the response efforts.

Federal Response Efforts

An Area Command is established outside of the impact zone. The Area Command provides strategic resource allocation and conflict resolution to the multiple Incident Command Posts. FEMA Region IV activates its Regional Response Coordination Center (RRCC) and is in the process of establishing a Joint Field Office (JFO) close to the Area Command to coordinate initial regional and field activities. Federal responders from activated Emergency Support Functions (ESFs) begin arriving to support operations throughout the impacted communities. The Multi-Agency Coordination Center (MACC) ESF #3 of the JFO is established to perform the same function for the Area Command, but on a national level.

Area Command receives pressure from the Governor, industry, Congress, and the President for information about physical damage, emergency operations, and safety issues. Requests for information updates and reporting requirements overwhelm the command center and become a source of frustration for responders. Planners begin forecasting response and recovery activities beyond the initial 30 days.

FEMA has staff and contractors in the field inspecting homes and neighborhoods for damage, using hand-held devices and relaying damage information back to headquarters. Volunteers and donations flood in, but a shortage of administrative staff and volunteer management programs makes it difficult to utilize these resources.

As the response continues, there is still a major shortage of qualified responders, partly due to overworked staffing needing relief breaks. In particular, the response effort needs support for environmental assessment and monitoring. Reporting requirements and data transfer challenges plague responders; highly specialized staff are finding themselves spending large amounts of time attempting to submit damage reports.

Phase 2: 5 Days to 2 Weeks Response Questions

These questions serve to focus your thoughts on the issues associated with the scenario. In advance of the exercise, participants are expected to read the scenario, review each question, and develop individual responses to each question.

11. How do communication and information flow change during continued operations? How do your requirements to maintain situational awareness change during this phase of the response?
12. What are your concerns and priorities during this phase of your response? How do they differ from your earlier concerns and priorities?
13. How does resource management during this operational phase differ from the initial assessment phase?
14. What financial and administrative procedures will you use to track reimbursement costs?
15. How can personnel from unaffected areas travel into the affected areas? What type of access control system is used?
16. At this stage, it is easier to review requests and identify resource needs. What can you do to anticipate these requests and stage these resources?

Phase 3: 6 Months after Earthquake Tuesday, September 13, 2011

It has been six months since the earthquake. FEMA has requested each ESF to begin planning for transition to remediation and recovery efforts under ESF #14, Long-term Community Recovery. Discussions are underway between agency representatives to begin the process of planning the transition, bringing all the parties together and establishing long-term goals. Recovery specialists will be requesting the coordination and integration of information, databases, and personnel. Although much damage still exists, critical infrastructure is operational as a result of repairs or work-arounds. Repairs and construction on residential and business properties continues, and more residents and workers are able to return to the area.

Direct Earthquake Damage

In Kentucky, most of the roadways are opened and the population has returned home. As more homes and business are repaired and occupied, the demand for utility services increases. However, debris from residential repairs hinders the ability of utility vehicles to access power lines and rights of way to make repairs. Access to fuel continues to be a problem, because many storage tanks were damaged. Although the accidental release of petroleum products and HazMat has greatly diminished, identification and decontamination efforts associated with exotic released chemicals remain an issue.

Federal Response Efforts

While response repair activities are ongoing, long-term recovery and reconstruction efforts are initiated and coordinated as more local communities regain the ability to address their issues. Outstanding issues for local communities include decontamination, remediation, construction, and financial record keeping. State personnel no longer are working three shifts, and many field support teams have returned home. Other states are requesting the return of their assets, but local and state entities may still need resources to support decontamination and cleanup efforts.

Phase 3—6 Months Response Questions

These questions serve to focus your thoughts on the issues associated with the scenario. In advance of the exercise, participants are expected to read the scenario, review each question, and develop individual responses to each question.

17. How do communication and information flow change during this phase of the scenario?
18. What are your concerns and priorities during this phase of your response? How do they differ from your earlier concerns and priorities?
19. What is the threshold and at what point does planning for the transition to recovery operations start?
20. What is your role in remediating contamination hazards in source, distribution, and collection systems? What hazard mitigation and reduction procedures and standards are in place?
21. What are the potential issues in a transition from response to recovery? What are the existing mitigation and preparedness activities already in place that you can leverage for recovery?
22. How does resource management differ during this phase of the operation compared to the previous phase? What complications do you anticipate as the transition to recovery operations approaches?
23. Looking forward and planning for the recovery phase, would anything change in the tracking and reporting of assets and resources used in the recovery? When should you begin planning for recovery operations?
24. What will be the process for the evaluation of continued support and the return of assets from other regions within your organization? What are the possible complications?

Action-Planning Session

Instructions

- *During the Action-Planning Session of today's exercise, you should focus on identifying the next steps to enhance your emergency response plans. During this session, you will discuss the answers to the following questions and complete an Action-Planning Guide that supports developing both short- and long-term actions or program capability development goals and objectives. After answering the following questions individually, categorize the indicated actions on the Action-Planning Matrix. Once the majority of participants have completed their questions and matrix, the group will complete the Action-Planning Guide.*

Action-Planning Questions

1. List the plans, policies, procedures, mission assignments, associated subtasking and other delegations, and other preparedness elements that you think should be further reviewed, supplemented, or developed. Which are the highest priorities?

2. Identify the actions you think should be taken to deal with the variety of possible infrastructure impacts and consequences associated with any major disaster or emergency.

3. What types of training do utility and local, state, and federal government personnel need to more effectively support response/recovery efforts? Other than Incident Command System (ICS) training, what training would help you be fully prepared?

4. Describe the personal action steps you plan to take to improve your level of readiness.

Action-Planning Matrix

Instructions: *For each issue that you identified during the facilitated discussion period, as well as your answers to the action-planning questions above, identify a corrective action/task/follow up that addresses that issue (Table 10.2). Put the issues and corrective actions on the Action-Planning Matrix into one of the four categories: Planning, Resources, Personal Action Steps, and/or Training.*

Table 10.2 Action-Planning Matrix

Planning		Resources	
Issue	Action	Issue	Action
Personal Action Steps		**Training**	
Issue	Action	Issue	Action

Action-Planning Guide

Instructions: *Put each categorized action from the Action-Planning Matrix into the Action-Planning Guide (Table 10.3). Each action should include an individual responsible for the action, people who will support the effort, resources (and possible sources) that will support the effort, and a timeline for completing the action, including short- and long-term milestones.*

Table 10.3 Action-Planning Guide

Action/ Task/ Follow-Up	Lead Individual or Agency Responsibility	Supporting Individual or Agency	Resources and Possible Sources	Timeline	
				Short-Term	Long-Term

Appendices

Appendix 1—Acronyms and Key Definitions

Note: Be sure to include Appendices with acronyms and definitions as well as any supporting documents and references.

Appendix 2—List of Participants

The following is a list of those that registered as of 00/00/0000.

First Name	Last Name	Organization

Appendix 3—Participant Evaluation Form

Name (optional) ______________ **Organization (optional)** __________
Position: Utility____ **Emergency Management** ___ **Gov (Loc/St/Fed)**___
Other_______
Role: Player_____ **Facilitator** ______________ **Support** _______________

1. How would you rate the training and exercise overall?

Excellent	Very Good	Good	Fair	Poor
5	4	3	2	1

2. The exercise identified how the loss of utility services affect other response and recovery efforts.

Explained Well		Agree		Not Explained
5	4	3	2	1

3. The training and exercise increased awareness of response partners' roles, responsibilities, authorities, and interdependencies.

Explained Well		Agree		Not Explained
5	4	3	2	1

4. The training and exercise defined how local, state, and federal governments can coordinate logistics of prioritizing critical resources following a catastrophic incident.

Explained Well		Agree		Not Explained
5	4	3	2	1

5. The exercise facilitator effectively engaged all of the participants and kept the discussion moving.

Strongly Agree		Agree		Strongly Disagree
5	4	3	2	1

6. Overall, participation in the training and exercise was a valuable use of your time.

Strongly Agree		Agree		Strongly Disagree
5	4	3	2	1

Additional Comments:

____

Chapter 11

Planning for Terrorism

Michael J. Fagel

Contents

Given the creativity of those committed to carrying out acts of terrorism, it is more important than ever that jurisdictions are equipped to respond to terrorist events. Planners must consider a broad range of incidents, including assaults on infrastructure and electronic information systems that could result in consequence affecting human, life, health, and safety.

State and local governments have the primary responsibility in planning for and managing the consequences of a terrorist incident using available resources in the critical hours before federal assistance can arrive.

The Planning Process

The process of developing a terrorism annex to your Emergency Operations Plan (EOP) is similar to that used to develop other EOPs. As is the case for these other plans, the terrorism planning process must begin before an emergency and before any planned special event that could be subject to a terrorist attack.

Traditionally, the planning process has consisted of six phases:

1. Initiation
2. Concept development
3. Plan development
4. Plan review
5. Development of supporting plans, procedures, and materials
6. Validation of plans using tabletop, functional and full-scale exercises

However, given the unusually intense and multifaceted nature of terrorist attacks, a seventh phase is recommended: Thorough coordination of plans, internally and externally.

You should carefully compare plans for the various response functions within that agency and revise the plans, if necessary, to remove any discrepancies. This will help prevent disconnects between vital functions that support one another and help ensure that each does what the others expect on a timely basis. Similarly, the various departments and agencies within a local jurisdiction should also compare their plans, focusing on issues of consistency and coordination. Again, this review will ensure that each organization does what the other expects, when it is expected.

Such reviews are especially important in planning for response to a major terrorist incident, since local jurisdictions are likely to be aided during the response by neighboring communities, its own and neighboring counties, and its own and possible neighboring states.

Terrorism Hazards

The terrorism annex to your EOP should identify and discuss the nature of the terrorist hazard. The hazard may be a weapon of mass destruction (WMD), including conventional explosives; secondary devices; and combined hazards, or it may be another means of attack, including low-tech devices; attacks on infrastructure; and cyber terrorism.

WMD Hazard Agents

A WMD is defined as any weapon designed or intended to cause death or serious bodily injury through the release, dissemination, or impact of toxic or poisonous chemicals, disease organisms, radiation or radioactivity, or explosion or fire.

Two important considerations distinguish these hazards from other types of terrorist tools. First, in the case of chemical, biological, and radioactive agents, their presence may not be immediately obvious, making it difficult to determine when and where they have been released, who has been exposed, and what danger is present for first responders and medical technicians. Second, although there is a sizable body of research on battlefield exposures to WMD agents, there is limited scientific understanding of how these agents affect civilian populations.

Chemical Agents

Chemical agents are intended to kill, seriously injure, or incapacitate people through physiological effects. A terrorist incident involving a chemical agent will demand immediate reactions from emergency responders—fire, police, hazardous materials

(HazMat) teams, emergency medical services (EMS), and emergency room staff—who will need adequate training and equipment.

Hazardous chemicals, including industrial chemicals and agents, can be introduced via aerosol devices (e.g., munitions, sprayers, or aerosol generators), breaking containers, or covert dissemination. Such an attack might involve the release of a chemical warfare agent, such as nerve or blister agent or an industrial chemical, which may have serious consequences.

Table 11.1 lists some indicators of the possible use of chemical agents.

Early in an investigation, it may not be obvious whether an outbreak was caused by an infectious agent or a hazardous chemical; however, most chemical attacks will be localized, and their effects will be evident within a few minutes. There are both persistent and nonpersistent chemical agents. Persistent agents remain in the affected area for hours, days, or weeks. Nonpersistent agents have high evaporation rates, are lighter than air, and disperse rapidly, thereby losing their ability to cause casualties after 10 to 15 minutes, although they may be more persistent in small, unventilated areas.

Table 11.1 Indicators of Possible Use of Chemical Agents

Stated Threat to Release a Chemical Agent
Unusual Occurrence of Dead or Dying Animals • For example, lack of insects, dead birds
Unexplained Casualties • Multiple victims • Surge of similar 911 calls • Serious illnesses • Nausea, disorientation, difficulty breathing, or convulsions • Definite casualty patterns
Unusual Liquid, Spray, Vapor, or Powder • Droplets, oily film • Unexplained odor • Low-lying clouds/fog unrelated to weather
Suspicious Devices, Packages, or Letters • Unusual metal debris • Abandoned spray devices • Unexplained munitions

Biological Agents

Recognition of a biological hazard can occur through several methods, including identification of a credible threat, discovery of bioterrorism evidence, diagnosis, and detection.

When people are exposed to a pathogen, such as anthrax or smallpox, they may not know that they have been exposed, and those who are infected, or subsequently become infected, may not feel sick for some time. This delay between exposure and onset of illness, the incubation period, is characteristic of infectious diseases. The incubation period may range from several hours to a few weeks, depending on the exposure and pathogen.

Unlike acute incidents involving explosives or some hazardous chemicals, the initial detection and response to a biological attack on civilians is likely to be made by direct patient care providers and the public health community.

Terrorists could also use a biological agent that would affect agricultural commodities over a large area (e.g., wheat rust or a virus affecting livestock), potentially devastating the local or even national economy. The response to agricultural bioterrorism should also be considered during the planning process.

Responders should be familiar with the characteristics of biological agents. Table 11.2 lists some indicators of the possible use of biological agents.

Nuclear/Radiological Agents

The difficulty of responding to a nuclear or radiological incident is compounded by the nature of radiation itself. In an explosion, the fact that radioactive material was involved may or may not be obvious, depending on the nature of the explosive device used.

The presence of a radiation hazard is difficult to ascertain, unless the responders have the proper detection equipment and have been trained to use it properly. Although many detection devices exist, most are designed to detect specific types

Table 11.2 Indicators of Possible Use of Biological Agents

Stated Threat to Release a Biological Agent
Unusual Occurrence of Dead or Dying Animals
Unusual Casualties • Unusual illness for region/area • Definite pattern inconsistent with natural disease
Unusual Liquid, Spray, Vapor, or Powder • Spraying; suspicious devices, packages, or letters

Table 11.3 Indicators of Possible Use of Chemical Agents

Stated Threat to Deploy a Nuclear or Radiological Device
Presence of Nuclear or Radiological Equipment • Spent fuel canisters of nuclear transport vehicles
Nuclear Placards/Warning Materials along with Otherwise Unexplained Casualties

and levels of radiation and may not be appropriate for measuring or ruling out the presence of radiological hazards.

Table 11.3 lists some indicators of the possible uses of nuclear or radiological agents.

The scenarios constituting an intentional nuclear/radiological emergency include the following:

- Use of an improvised nuclear device includes any explosive device designed to cause a nuclear yield. Depending on the type of trigger device used, either uranium or plutonium isotopes can fuel these devices. Although "weapons-grade" material increases the efficiency of a given device, material of less than weapons grade can still be used.
- Use of radiological dispersal device (RDD) includes any explosive device utilized to spread radioactive material upon detonation. Any improvised explosive device (IED) could be using by placing it in close proximity to radioactive material.
- Use of a simple RDD spreads radiological material without the use of an explosive. Any nuclear material (including medical isotopes or waste) can be used in this manner.

Conventional Explosives and Secondary Devices

The easiest to obtain and use of all weapons is still a conventional explosive device, or improvised bomb, which may be used to cause massive local destruction or to disperse a chemical, biological, or radiological agent. The components are readily available, as are detailed instructions on constructing such a device.

IEDs are categorized as being explosive or incendiary, using high or low filler explosive materials to explode and/or cause fires. Explosions and fires also can be caused by projectiles and missiles, including aircraft used against high-profile targets, such as buildings, as was the case for the September 11 attacks.

Bombs are firebombs are cheap and easily constructed, involve low technology, and are the terrorist weapon most likely to be encountered. Large, powerful devices can be outfitted with timed or remotely triggered detonators and can be designed to be activated by light, pressure, movement, or radio transmission. The potential exists for single or multiple bombing incidents in single or multiple municipalities.

Historically, less than 5% of actual or attempted bombings were preceded by a threat.

Combined Hazards

WMD agents can be combined to achieve a synergistic effect—greater in total effect than the sum of their individual effects. They may be combined to achieve both immediate and delayed consequences. Mixed infections or toxic exposures may occur, thereby complicating or delaying diagnosis. Casualties of multiple agents may exist; casualties may also suffer from multiple effects, such as trauma and burns from an explosion, which may exacerbate the likelihood of agent contamination.

Attacks may be planned and executed so as to take advantage of the reduced effectiveness of protective measures produced by employment of an initial WMD agent. Finally, the potential exists for multiple incidents in single or multiple municipalities.

Other Terrorism Hazards

Planners also need to consider the possibility of unusual or unique types of terrorist attacks previously not considered likely. For example, prior to the attacks of September 11, 2001, the use of multiple commercial airliners with full fuel loads as explosive incendiary devices in well-coordinated attacks on public and private targets, was not considered a likely terrorist scenario.

Although it is not realistically possible to plan for and prevent every conceivable type of terrorist attack, planners should anticipate that future terrorism attempts could range from simple, isolated attacks to complex, sophisticated, highly coordinated acts of destruction using multiple agents aimed at one or multiple targets. Therefore, the plans developed for terrorist incidents must be broad in scope, yet flexible enough to deal with the unexpected.

These considerations are particularly important in planning to handle the consequences of attacks using low-technology devices and delivery, assaults on public infrastructure, and cyber terrorism. In these cases, the training and experience of the responders may be more important than detailed procedures.

Low-Technology Devices and Delivery

Planning for possibility of terrorist attacks must consider the fact that explosives can be delivered by a variety of methods. Most explosive and incendiary devices used by terrorists would be expected to fall outside the definition of a WMD. Small explosive devices can be left in packages or bags in public areas for later detonation at a time and place when and where the terrorist feels that the maximum damage can be inflicted.

The relatively small size of these explosive devices can also be brought onto planes, trains, ships, or buses, within checked bags or hand carried. Although present airline security procedures minimize the possibility of explosives being brought on board airliners, planners will need to consider the level of security presently employed on ships, trains, and buses within their jurisdictions.

Larger quantities of explosive materials can be delivered to their intended target area via car or truck bombs. Planners need to consider the possible need to restrict or prohibit vehicular traffic within certain distances of key facilities identified as potential terrorist targets. Planners may also need to consider the possible use of concrete barriers to prevent the forced entry of vehicles into restricted areas.

Infrastructure Attacks

Potential attacks on elements of the nation's infrastructure require protective considerations. Infrastructure protection involves risk management actions taken to prevent destruction of or incapacitating damage to networks and systems that serve society.

Infrastructure protection is more often focused on security, deterrence, and law enforcement, than on emergency consequence management preparedness and response. Nevertheless, planners must develop contingencies and plans in the event critical infrastructures are brought down as the result of a terrorist incident.

Cyberterrorism

Cyberterrorism involves the malicious use of electronic information technology to commit or threaten to commit acts dangerous to human life, or against a nation's critical infrastructures in order to intimidate or coerce a government or civilian population to further political objectives. As with other infrastructure guidance, most cyber protection guidance focuses on security measures to protect computer systems against intruders, denial of service attacks, and other forms of attack, rather than addressing issues related to contingency and consequence management planning.

Emergency Operations Plan: Situation

The situation section of a terrorism annex to your EOP should discuss what constitutes a potential or actual terrorist incident. It needs to present a clear, concise, and accurate overview of potential events and discuss a general concept of operation for response. Any information already included in the EOP need not be duplicated in the terrorism annex, but should be referenced. The situation overview should include as much information as possible that is unique to terrorism response actions, including the suggested elements listed in Table 11.4.

As a state or local emergency manager or planner, you need to consider the possibility of unusual or unique types of terrorist attacks in addition to those that have occurred in the past. You need to think creatively about possible scenarios and response needs. The plans developed for terrorist incidents must be comprehensive in scope, yet flexible enough to deal with the unexpected.

Terrorism emergency response planning should include provisions for working with federal crisis and consequence management agencies. The key to successful emergency response involves smooth coordination among multiple agencies and officials from various jurisdictions regarding all aspects of the response.

Because of the need to interact with a wide range of organizations and individuals within these organizations, up-to-date directories of the points of contact must be maintained in the course of the planning process. It is important that these directories be updated to reflect changes in personnel and telephone numbers.

While assistance from federal agencies will be needed in the event of a terrorist incident, planning by state and local jurisdictions should take into account the difficulty that can be experienced in incorporating the federal resources into the initial local response. Coordination among state, local, and federal officials should take place well in advance of events that could be targeted so that all response organizations clearly understand the responsibilities of each organization, and how they will be integrated.

Potential Targets

In determining the risk areas within your jurisdiction, the vulnerabilities of potential targets should be identified, and the targets themselves should be prepared to response to a WMD incident. In-depth vulnerability assessments are needed for determining a response to such an incident.

Areas of vulnerability may be determined by several factors:

- Population
- Accessibility
- Criticality
- Economic impact

It may be beneficial to coordinate vulnerable areas with the Federal Bureau of Investigation (FBI).

Table 11.4 Details That Should Be Included in a Situation Overview

Suggested Emergency Operations Plan Elements	
Maps	• Use detailed, current maps or charts. • Include demographic information. • Use natural and man-made boundaries and structures to identify risk areas. • Annotate evacuation routes and alternatives. • Annotate in-place sheltering locations. • Use geographic information and analytical tools, as appropriate.
Environment	• Determine response routes and times. • Include bodies of water with damn or levees (these could become contaminated). • Specify special weather and climate features that could alter the effects of a WMD.
Population	• Identify those most susceptible to WMD effects or otherwise hindered or unable to care for themselves. • Identify areas where large concentrations of the population might be located, such as sports arenas and major transportation center. • List areas that may include retirement communities. • Note locations of correctional facilities. • Note locations of hospitals, medical centers, schools, daycare centers, or any other locations where multiple evacuees may require assistance. • Identify non-English-speaking populations.
Regional	• Identify multijurisdictional perimeters and boundaries. • Identify potentially overlapping areas for response. • Identify rural, urban, suburban, and city mutual-risk areas. • Identify mutual aid resources from adjoining municipalities. • Identify terrorism-specific or unique characteristics, such as interchanges, choke points, traffic lights, traffic schemes and patterns, access roads, tunnels, bridges, railroad crossings, and overpasses or cloverleafs.
Resources	• Identify mutual aid resources. • Identify terrorism-specific resources.

Table 11.5 gives you an idea of vulnerable areas in your jurisdiction that should be included in the planning process.

Various criteria may be used in determining the vulnerability of facilities to terrorist attack. In evaluating the vulnerability of facilities, state and local planners need to consider the existing security measures in place and the need, if any, to upgrade security.

In addition, the FBI has a standard vulnerability assessment paradigm that can be used for evaluating the vulnerabilities of potential targets. As a planner, you should also be aware that once target lists and vulnerability information are developed, careful decisions must be made regarding security considerations for handling this information based on applicable state and federal law regarding confidentiality and public information. Even when laws do not require strict confidentiality, you should use common sense regarding whether information that could be useful to terrorists should be made available.

Initial Warning

While specific events may vary, the emergency response and protocol followed should remain consistent. When an overt WMD incident has occurred, the initial call for help will likely come through the local 911 center. This caller may or may not identify the incident as a terrorist incident, but may state only that there was an explosion, a major accident, or a mass casualty event. Information relayed through the dispatcher prior to the arrival of first responders on scene, as well as the initial assessment, will provide first responders with the basic information necessary to begin responding to the incident.

With increased awareness and training about terrorist incidents, first responders should recognize that a terrorist incident has occurred. The information provided in this chapter applies where it becomes obvious or strongly suspected that an incident has been intentionally perpetrated to harm people, compromise the public's safety and wellbeing, disrupt essential government services, or damage the area's economy or environment.

You need to be aware of the likely occurrence of false warnings. Since these cannot be ignored, they must be investigated, resulting in wasted resources and psychological stress. You should develop procedures and training to deal with such threats.

Initial Detection

The initial detection of a WMD terrorist attack will likely occur at the local level by either first responders or private entities (e.g., hospitals, corporations). Consequently, first responders, the business community—both public and private—should be trained to identify hazardous agents and to take appropriate actions. State and local health departments, as well as local emergency first responders, will be relied upon

Table 11.5 Potentially Vulnerable Areas

Potential Areas of Vulnerability	
Traffic	• Determine which roads, tunnels, or bridges carry large volumes of traffic. • Identify points of congestion that could impede response or place citizens in a vulnerable area. • Note time of day and day of week that this activity occurs.
Trucking and Transport Activity	• Note location of hazardous materials (HazMat) cargo loading and unloading facilities. • Note vulnerable areas such as weigh stations and rest areas this cargo may transit.
Waterways	• Map pipelines and process or treatment facilities (in addition to dams). • Note berths and ports for cruise ships, roll-on/roll-off cargo vessels, and container ships. • Note any international flagged vessels that conduct business in the area.[a]
Airports	• Note information on carriers, flight paths, airport layout, and types of aircraft that use this facility. • Annotate location of air traffic control towers, runways, passenger terminals, and parking areas.
Trains/ Subways	• Note location of rails and lines, interchanges, terminals, tunnels, and cargo/passenger terminals. • Note and HazMat material that may be transported via rail. • Note locations of subway stations and ventilation control systems.
Government Facilities	• Note location of federal, state, and local government offices. • Include locations of post office, law enforcement stations, fire/rescue, town/city hall, and local mayor/governor's residences. • Note judicial offices and courts. • Note locations of monuments memorial structures and prominent governmental symbols.
Recreation Facilities	• Map sports arenas, theaters, malls, special interest group facilities, and locations of special events.

(*continued*)

Table 11.5 (*Continued*) Potentially Vulnerable Areas

Potential Areas of Vulnerability	
Symbolic Buildings and Locations	• Note national monuments, internationally well known facilities and locations. • Note potential areas of congestion connected with such buildings and locations.
Other Facilites	• Map locations of financial institutions and the business district. • Make any notes on the schedule that the business district may follow. • Determine whether shopping centers or heavily populated downtown areas are congested at certain periods. • Note location of special events facilities that may have national importance. • Note location of prominent high-rise buildings.
Military Installations	• Note locations and type of military installations.
HazMat Facilities, Utilities, and Nuclear Facilities	• Map locations of facilities, such as electricity generating stations, oil refineries, spent nuclear fuel storage facilities, etc.
Water Supply Facilties	• Note the locations of water supply intakes from lakes or rivers. • Note the locations of water supply pipelines and holding areas, such as reservoirs and tanks. • Note the locations of water supply treatment plants.
Food and Agriculture	• Note the locations of key agricultural facilities, such as large grain elevators and livestock concentrations. • Note the locations of food processing and packaging facilities.
Computer Systems	• Identify governmental and business-related computer systems located within the jurisdiction and ascertain their level of protection against terrorist cyber attack.

Note: Security and emergency personnel representing all of these facilities should work closely with local and state personnel for planning and response.

[a] The Harbor and Port Authorities, normally involved in emergency planning, should be able to facilitate obtaining information on the type of vessels and the containers they carry.

to identify unusual symptom occurrence, and any additional cases of symptoms as the effects spread throughout the community and beyond.

The detection of a terrorism incident involving covert biological agents, as well as some chemical agents, will most likely occur through the recognition of similar symptoms or syndromes by clinicians in hospital or clinical settings. Detection of biological agents could occur days or weeks after exposed individuals have left the site of the release. Detection will occur at public health facilities receiving unusual numbers of patients, the majority of whom will self-transport. Similarly, a biological attack aimed at agricultural assets might first be detected by veterinarians or agricultural inspectors.

First responders must be protected from the hazard prior to treating victims. Planning for response to terrorist acts must include provisions for appropriate personal protective equipment (PPE) for emergency responders, specifically first responders. This equipment should include protective clothing and respirators, with high-efficiency particulate air filters. Detection equipment for chemical, biological, nuclear, or explosive materials will assist in identifying the nature of a potential hazard.

You need to determine the present availability of this protective and detection equipment within your jurisdiction, determine if additional resources would be needed to adequately protect your first responders, and identify the funding needs to upgrade your resources, if needed.

Release Area

Standard models are available for estimating the effects of a nuclear, chemical, or biological release, including the area affected and consequences to population, resources, and infrastructure. Some of these models include databases on infrastructure that may prove useful when developing your terrorism annex.

Analogous to the area affected by a nuclear, biological, or chemical release is the area impacted by an explosive device. Models are also available for estimating the blast effects at various distances for various quantities of explosive materials.

These models can be useful in preparing your terrorism annex, especially in regard to determining minimum setback distances from a potential vehicle bomb to a vulnerable facility or structure. If a specific minimum distance cannot be maintained, then the planning effort may need to consider the cost and effectiveness of facility hardening to mitigate the effects of an assumed blast impact. You may also want to consider the removal or modification of window areas.

Investigation and Containment of Hazards

Local first responders will provide initial assessment or scene surveillance of a hazard caused by an act of WMD terrorism. The proper local, state, and federal authorities capable of dealing with and containing the hazard should be alerted to a suspected WMD attack as soon as first responders recognize the occurrence of symptoms that are highly unusual or of an unknown cause. Consequently, state and local emergency responders must be able to assess the situation and request assistance as quickly as possible.

Emergency Operations Plan: Assumptions

Although situations may vary, planning assumptions remain the same.

- The first responder or health or medical personnel will, in most cases, initially detect and evaluate the potential or actual incident, assess casualties, and determine whether assistance is required. If so, state support will be requested and provided. This assessment will be based on warning or notification of a WMD incident that may be received from law enforcement, emergency response agencies, or the public.
- The incident may require federal support. To ensure that there is only one overall lead federal agency, the Federal Emergency Management Agency (FEMA) is authorized to support the Department of Justice (DOJ). In addition, FEMA is designated as the lead agency for consequence management within the United States and its territories. FEMA retains authority and responsibility to act as the lead agency for consequence management throughout the federal response. In this capacity, FEMA will coordinate federal assistance requested through state authorities using normal FRP mechanisms.
- Federal response will include experts in the identification, containment, and recovery of WMD (chemical, biological, nuclear/radiological, or explosive).
- Jurisdictional areas of responsibility and working perimeters defined by local, state, and federal departments and agencies may overlap. Perimeters may be used to control access to the affected area, target public information messages, assign operational sectors among responding organizations, and assess potential effects on the population and the environment. Control of these perimeters may be enforced by different authorities, which will impede the overall response if adequate coordination is not established.
- Response activities may continue for an extended period of day or weeks. Early emergency responders may be pushed beyond their capabilities, and regional and federal resources may be needed. The incident will be covered extensively by the media. There may be many volunteer responders and donations of food and material that will require management.

Emergency Operations Plan: Concept of Operations

Your terrorism annex should include a concept of operations section to explain the jurisdiction's overall concept for responding to a WMD incident. Topics should include division of local, state, federal, and any intermediate interjurisdictional responsibilities; activation of the EOP; and other elements set forth in State and Local Guide 101.

Direction and Control

Local government response organizations will respond to the incident scene(s) and make appropriate and rapid notifications to local and state authorities. Control of the incident scene(s) most likely will be established by local first responders from either fire or police.

To ensure continuity of operations, it is important that the Incident Command Post be established at a safe location and at a distance appropriate for response to a suspected or known terrorist incident. In addition, in severe terrorist attacks, response operations may last for very long periods, and there may be more leadership casualties because of secondary or tertiary attacks or events. Therefore, planning should provide for staffing key leadership positions in depth.

The Incident Command System (ICS) should be used by all responding local fire, police, and emergency management organizations, and all relevant responder personnel should be trained in ICS use to prevent security and coordination problems in a multiorganizational response.

The ICS that was initially established likely will transition into a Unified Command System as mutual-aid partners and state and federal responders arrive to augment the local responders. It is recommended that local, state, and federal regional law enforcement officials develop consensus "rules of engagement" early in the planning process to smooth the transition from ICS to Unified Command. The Unified Command structure will facilitate both crisis management and consequence management activities.

The Unified Command Structure used at the scene will expand as support units and agency representatives arrive to support crisis and consequence management operations. The site of a terrorist incident is a crime scene as well as a disaster scene, although the lives, health, and safety remains the top priority. Because of these considerations, as well as logistical control concerns, it is extremely important that this incident site and its perimeter be tightly controlled as soon as possible.

State and local planners must realize that the integration of the federal response into the local response efforts can be a difficult and awkward process. Whenever possible, each entity should involve the others in its planning process so as to facilitate a better understanding by all parties of the anticipated actions and responsibilities of each organization.

Planners should understand that integration of the federal response into an urban setting would be different from that into a rural setting. In an urban area, there will be substantial manpower and equipment resources, and the control of the emergency response. The rapid influx of federal resources can be a sensitive issue unless properly coordinated. The federal response should not overwhelm the local emergency response organization but should provide resources, as needed (Figure 11.1).

It is assumed that normal disaster coordination accomplished at state and local Emergency Operations Centers (EOCs) and other locations away from the scene would be addressed in the EOP. Any special concerns relating to state and local coordination with federal organization should be addressed in the terrorism annex.

Response to any terrorist incident requires direction and control. The planner must consider the unique characteristics of the event, identify the likely stage at which coordinated resources will be required, and tailor the direction and control process to merge these resources into an ongoing public health response.

With many organizations involved, there is the danger of key decisions being slowed down by too many layers of decision making. Planners should be aware of the need to streamline the decision-making process so that key decisions or authorizations regarding public health and safety can be obtained quickly.

A primary EOC is necessary to properly coordinate response actions within the jurisdiction and to liaise with other jurisdictions and federal agencies. The EOC of the City of New York's Office of Emergency Management, a new state-of-the-art facility, was located at the World Trade Center, and was destroyed during the attack

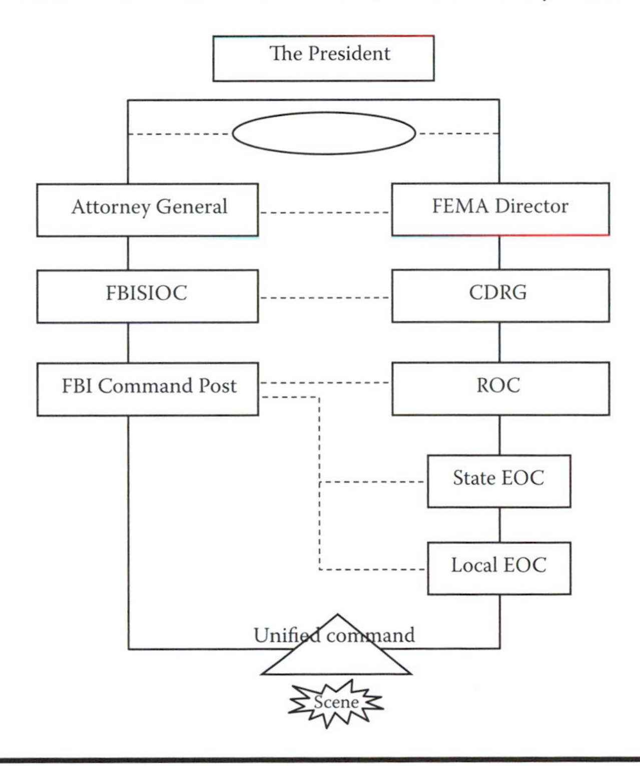

Figure 11.1 Incident Command from the federal level down.

of September 11, 2001. This necessitated the establishment of an alternate EOC. Therefore, planning should address the possibility that operations might have to be shifted to an alternative EOC or even a secondary alternative location.

When considering direction and control, as well as continuity of operations, planners must determine the availability of usable alternate EOC facility locations that can be brought up to operational levels within a reasonable period of time. In a large-scale terrorist attack, the local EOC might become uninhabitable, especially if it is not a hardened facility. In identifying and evaluating alternative EOC locations, planners will need to consider the availability of communications systems, space to accommodate all key staff, materials and supplies, backup power, kitchen, bathrooms, and the overall capability to maintain around-the-clock operations for an extended period of time.

Local, state, and federal interface with the FBI On-Scene Commander is coordinated through the Joint Operation Center (JOC). FEMA will recommend joint operational priorities to the FBI on the basis of consultation with the FEMA-led consequence management group in the JOC. The FBI, working with local and state officials in the command group at the JOC, will establish operational priorities.

Communications

In the event of a WMD incident, rapid and secure communications is important to ensure a prompt and coordinated response. Strengthening communications among first responders, clinicians, emergency rooms, hospitals, mass care providers, and emergency management personnel must be given top priority when planning. Planning should include adding 911 resources when an event requires extraordinary response.

In addition, terrorist attacks have been shown to overload nondedicated telephone line and cellular telephones. In these instances, the Internet has proven more reliable for making necessary communications connections, although it should be recognized that computers may be vulnerable to cyber attacks in the form of viruses. It is recommended that response organizations both establish relevant Internet connections with all coordinating emergency response organizations and have the use of these connections formalized in plans and practiced during training, drills, and exercises.

Responders with different functions within a jurisdiction or from different jurisdictions may use different radio frequencies, hindering communications. Use of 800 MHz radios alleviates this problem. Therefore, a backbone communications system to interconnect local, state, and federal responders is recommended. Also needed is the establishment of mutually agreed upon communications protocols during the planning and exercise stages so that all responding organizations will understand each other's codes and terminology during response to real events.

Planning should also consider the need for an integrated communications system for all key state agencies and local emergency response organizations. The interoperability of such a system would facilitate the integrated response to a terrorist incident. Planning also needs to consider the importance of reliable backup communications systems for emergency responders. Terrorist incidents may include the loss of radio transmission capabilities, and telephone land lines and cellular phone connections will be overwhelmed. Satellite telephones, which can operate when cellular and nondedicated land lines are overloaded, are another option for backup telephone systems.

State and local planners should consider the distribution of priority emergency access telephone cards to their emergency workers. Planners should determine the existing status of their emergency communications systems and identify the funding needed to upgrade them.

Warning

Every incident is different. There may or may not be warning of a potential WMD incident. Factors involved range from intelligence gathered from various law enforcement or intelligence agency sources to an actual notification from the terrorist organization or individual. The EOP should have HazMat facilities and transportation routes already mapped, along with emergency procedures necessary to respond.

The warning or notification of a potential WMD terrorist incident could come from many sources; therefore, open but secure communication among local, state, and federal law enforcement agencies and emergency response officials is essential. The local FBI Field Office must be notified of any suspected terrorist threats or incidents.

Similarly, the FBI informs state and local law enforcement officials regarding potential threats. An integrated communications system would be an aid in maintaining these communications channels and would expedite the dissemination of warnings about suspected terrorist threats. The interoperability of such a system would eliminate the need to switch back and forth between different communications systems for different organizations.

Pre-Event Readiness

The FBI operates with a four-tier threat level system that can be used as a basis for initiating precautionary actions when a WMD terrorist event is anticipated.

> Level 4 (Minimal Threat): Received threats do not warrant actions beyond normal liaison notifications or placing assets or resources on a heightened alert.

Level 3 (Potential Threat): Intelligence or an articulated threat indicates the potential for a terrorist incident; however, this threat has not yet been assessed as credible.
Level 2 (Credible Threat): A threat assessment indicates that a potential threat is credible and confirms the involvement of a WMD in a developing terrorist incident. The threat increases in significance when the presence of an explosive device or WMD capable of causing a significant destructive event.
Level 1 (WMD Incident): A WMD terrorism incident resulting in mass casualties has occurred that requires immediate federal planning and preparation to provide support to state and local authorities. The federal response is primarily directed toward the safety and welfare of the public and the preservation of human lives.

Emergency Public Information

Terrorism is designed to be catastrophic. The intent of a terrorist attack is to cause maximum destruction of lives and property; create chaos, confusion, and public panic; and stress local, state, and federal response resources. Accurate and timely information, disseminated to the public and media immediately and often over the course of the response, is vital to minimize the accomplishment of these terrorist objectives.

Crisis research and case studies show that accurate, consistent, and expedited information calms anxieties and reduces problematic public responses, such as panic and spontaneous evacuations that terrorist hope will hamper response efforts.

The news media will be the public's primary source of information, from both official sources and other nonofficial sources, during the course of the incident. Ensuring that the media will receive accurate, consistent, and expedited official information from the outset and over what may be rapidly changing and lengthy response requires careful planning and considerable advance planning and considerable advance preparations. It is important to build and maintain a strong working relationship with the media. This relationship should include a clear commitment that government representatives will be immediately available to provide information over the course of the emergency.

Local plans should reflect responsibility for emergency information operations during the crucial initial response until state and federal personnel and resources can arrive to provide support. Planning should also reflect:

- A mechanism for sharing and coordinating information among all responders' agencies and organizations
- Development and production of information materials
- Dissemination of information through various methods
- Monitoring and analysis of news media coverage with rapid response capabilities to address identified problems

A strong and ongoing public education program for terrorism response, built upon outreach and awareness programs for other types of emergencies, can enhance the response organization's credibility and benefit both members of the public and first responder efforts in the event of a terrorist attack.

Protective Actions

Evacuation may be required from inside the perimeter of the scene to guard against further casualties from contamination by primary release of a WMD agent, the possible release of additional WMDs, secondary devices, or additional attacks targeting emergency responders.

Temporary in-place sheltering may be appropriate if there is a short-duration release of hazardous materials or if it is determined to be safer for individuals to remain in place. Protection from biological threats may involve coercive or noncoercive protective actions, including isolation of individuals who pose an infection hazard, quarantine of affected locations, vaccinations, use of masks by the public, closing of public transportation, limiting public gatherings, and limiting intercity travel.

As with any emergency, state and local officials are primarily responsible for making protective action decisions affecting the public. Protocols should be established to ensure that important decisions are made by persons with the proper decision-making authority. The terrorism annex should include provisions for coordinating protective actions with other affected jurisdictions. Planning should also address ways of countering irrational public behavior that can hinder protective actions.

However, planning for evacuation should be flexible to account for difficult situations. After the attack on the Pentagon, federal buildings in Washington, DC, were evacuated as a precautionary measure. Evacuation was hampered by roads in the federal area being under the jurisdiction of several federal agencies and the roads in the rest of the District being under the authority of the District government. On the other hand, the Pentagon is located in Arlington County, Virginia, near a route persons from Washington, DC, were using to evacuate. Therefore, there was a situation where some evacuation was toward the incident site rather than away from it. Because of heavy traffic in the vicinity of the Pentagon, fire vehicles had difficulty reaching the incident site.

Mass Care

The location of mass care facilities will be based partly on the hazardous agent. Decontamination, if necessary, may need to precede sheltering and other needs of the victims to prevent further damage from the hazardous agent to either the victims themselves or the care providers.

The American Red Cross (the primary agency for mass care), the Department of Health and Human Services (HHS), and the Department of Veteran Affairs (VA) should be actively involved with the planning process to determine both in-place sheltering and mobile mass care systems for the terrorism annex.

A midpoint or intermediary station may be needed to move victims out of the way of immediate harm. This action would allow responders to provide critical attention (e.g., decontamination and medical services) and general lifesaving support, then evacuate victims to a mass location for further attention. Some general issues to consider for inclusion in you terrorism annex:

- Location, setup, and equipment for decontamination stations, if any
- Mobile triage support and qualified personnel
- Supplies and personnel to support in-place sheltering
- Evacuation to intermediary location to provide decontamination and medical attention
- Determination of safety perimeters (based on agent)
- Patient tracking/record keeping for augmentation of epidemiological services and support

Health and Medical

The basic EOP should already contain a Health and Medical Annex. Issues that may be different during a terrorist incident and that should be addressed in the terrorism annex include

✓ Decontamination
✓ Safety of victims and responders
✓ In-place sheltering and quarantine versus evacuation
✓ Multihazard and multiagent trage

Planning should anticipate the need to handle large numbers of people who may or may not be contaminated but who are fearful about their medical wellbeing. In addition, the terrorism annex should identify the locations and capacities of medical care facilities within the jurisdiction and in surrounding jurisdictions. The terrorism annex should also include a description of the capabilities of these medical care facilities, especially with regard to trauma care. Depending on the nature and extent of the terrorist attack, the most appropriate medical care facility may not necessarily be the closest facility.

In addition, first responders may be entering an environment rife with biological or chemical agents, radioactive materials, or hazardous pollutants from collapsed buildings, or collapsed buildings might be imminent. Other incidents may post environmental or physical risks to responders. Examples may be a structurally damaged and potentially deadly pipeline, tank car, tank truck, bridge, or tunnel.

In planning, you address the need for first responders to perform a risk assessment and to modify standard protocols (e.g., establish plans for inoculating first responders) if the risk assessment indicates.

The planning should also address how such assessments are made and what resources they may indicate are needed. The assessment may indicate monitoring and sampling resources before federal resources can arrive.

Responders will also need appropriate PPE, including respirators. The planning process needs to address the availability of regional monitoring and sampling capability and PPE.

A bioterrorism incident raises several other special issues. Such an incident may generate an influx of patients requiring specialized care. If an infectious agent is involved, it may be necessary to isolate the patients and use specials precautions to avoid transmission of the disease to staff and other patients.

State planning should also consider the need to obtain and integrate supplementary medical professionals and technicians who may be needed to respond to a terrorist incident. In addition to physiological health consideration, planners should take into account the need for mental health considerations in the consequence management planning. Support must be provided not only to those individuals directly affected by a terrorist attack, but also to those surviving family members experiencing emotional stress.

Planning issues to consider include

- Immunization and prophylaxis for biological agents
- Notification to and receipt of information from doctors and clinics
- Augmentation of medical facilities and personnel
- Management of medical supplies and equipment
- Patient tracking and record keeping for augmentation of epidemiological services and support
- Analytical laboratory support, including memorandums of understanding (MOUs), specifying special considerations, as appropriate
- Mental health support services, including clinical psychologists, psychiatrists, social workers, etc.

Resources Management

The following considerations are highly relevant to WMD incidents and should be addressed, if appropriate, in one or more appendices to your resource management annex:

- Nuclear, biological, and chemical response resources that are available through interjurisdictional agreements.

- Unique resources that are available through state authorities (e.g., National Guard units).
- Unique resources that are available to state and local jurisdictions through federal authorities (e.g., the National Pharmaceutical Stockpile, a national asset providing delivery of antibiotics, antidotes, and medical supplies to the scene of a WMD incident).
- Unique expertise that may be available through academic, research, or private organizations.
- Trained and untrained volunteer resources and unsolicited donated goods that arrive at the incident site.

Recovery

The basic EOP should already contain a Recovery Annex. However, different issues may arise during a WMD incident that should be addressed in the terrorism annex. A WMD incident is a criminal act, and its victims and their families may be eligible for assistance under your state's crime victim's assistance law, if one exists. Therefore, state crime victim's assistance staff should be included in the planning process. In addition, injured victims of a terrorist attack, those put at risk for injury, and the families of these persons may have suffered psychological trauma as a result of the attack and may be in need of crisis counseling.

In the event of an incident involving chemical or biological agents or radioactive materials, large areas or multiple locations may become contaminated. Decontamination may be required before buildings can be safely reoccupied. While decontamination is taking place, or until damaged areas are repaired or replaced, persons must be relocated from residences and offices, and office equipment must be relocated from office buildings.

Relocation after a terrorist incident tends to be of longer duration and entail greater costs than relocation following a natural disaster. You need to take these factors into consideration and make appropriate arrangements.

Urban Search and Rescue

Urban Search and Rescue (US&R) involves a rapid deployment of US&R task forces to provide specialized lifesaving support to state and local authorities, including locating and extricating, and providing on-site medical treatment to those trapped in collapsed structures.

There are several US&R task forces throughout the country. They have the ability to deploy within six hours and to sustain themselves for 36 hours. Currently, deployment plans rely on commercial air transport. Consequently, in incidents where air traffic is curtailed, arrival of remote US&R task forces may be delayed.

The capabilities of these US&R task forces are being enhanced to operate in a collapsed building environment contaminated with biological or chemical agents or radioactive materials. These enhanced task forces will have additional HazMat specialists and medical personnel and more monitoring and detection equipment.

Organization and Assignment of Responsibilities

As with hazard-specific emergency, the organization for management of local response will probably need to be tailored to address the special issues involved in managing the consequences of a terrorist incident. This organization should be defined in the terrorism annex or your EOP.

The consequences of a terrorist attack have the potential to overwhelm local resources, which may require assistance from state or federal governments. The response by state and local governments, as well as the types of support and assistance from the federal government may be different than the response and support received for a natural disaster. Because of this, not only must the plans be upgraded to include response to a terrorist incident, but training and exercising must also be expanded to ensure that the unique aspects of response to a terrorist incident can be carried out in a coordinated, effective manner.

Training needs to be planned for local, state, and federal staff involved in the response. You should identify their training needs, establish budgets for the training, and determine what funding resources will be required to implement the training. Periodic integrated exercises must be conducted to ensure that the emergency response at the local, state, and federal levels can be adequately coordinated. The following response roles should be articulated in your terrorism annex.

Local Emergency Responders

Local fire departments, law enforcement personnel, HazMat teams, and EMS will be among the first to respond to terrorist incidents, especially those involving WMDs. In incidents associated with public transportation, workers and officials from these transportation organizations may be among the first responders. As response efforts escalate, the local emergency management agency and health department will help coordinate needed services.

Interjurisdictional Responsibilities

The formal arrangements and agreements for emergency response to a terrorist incident among neighboring jurisdictions, state, local, and neighboring states (and those jurisdiction physically located in those states) should be made prior to an incident. When coordinating and planning the Risk Assessment and Risk Area sections of the terrorism annex, interjurisdictional responsibilities should be readily identifiable.

State Emergency Responders

If requested by the local officials, the state emergency management agency has the capabilities to support local emergency management authorities and the Incident Commander.

State and Local Public Health Authorities

State laws grant the state and local public health authorities emergency powers to combat communicable disease. The powers available, diseases that trigger them, and procedures for enforcement vary from state to state. Typical powers include the power to isolate or quarantine persons and places and the power to compel vaccination and other preventive measure, such as wearing masks. In some states, these measures may be taken whenever there is a threat of communicable disease; in others, the powers apply to only one or more specific, named diseases.

Medical Service Providers

Hospitals generally perform emergency planning, both to protect their own facilities and patients and to respond to disasters in the community. State licensing and accreditation standards require hospitals to meet certain criteria for emergency preparedness, which often include participation in local or regional medical planning for disasters. Hospitals accredited by the Joint Commission on Accreditation of Healthcare Organization must be prepared for a variety of disaster scenarios, including facilities for biological, radioactive, or chemical isolation and decontamination, where appropriate.

Local Emergency Planning Committees and State Emergency Response Commissions

These entities are established under the Superfund Amendment and Reauthorization Act of 1986 (SARA) Title III, and the implementing regulation of the Environmental Protection Agency (EPA). Local Emergency Planning Committees (LEPCs) develop and maintain local hazardous materials emergency plans and receive notifications of releases of hazardous materials. State Emergency Response Commissions (SERCs) supervise the operation of the LEPCs and administer the community right-to-know provisions of SARA Title III, including collection and distribution of information about facility inventories of hazardous materials, chemicals, and toxins. LEPCs will have detailed information about industrial chemicals within the community. It may be advisable for LEPCs and SERCs to establish MOUs with agencies and organizations to provide specialized resources and capabilities for response to WMD incidents.

Federal Emergency Responders

Upon determination of a credible terrorist threat, or if such an incident actually occurs, the federal government may respond through the appropriate departments and agencies. These departments and agencies may include FEMA, the DOJ and FBI, the Department of Defense, the Department of Energy, the HHS, the EPA, the Department of Agriculture, the Nuclear Regulatory Commission (NRC), and possible the American Red Cross and Department of Veteran Affairs.

Administration and Logistics

There are many factors that make consequence management response to a terrorist incident unique. Unlike some natural disasters, the administration and logistics for response to a terrorist incident require special considerations. For example, there may be little or no warning and because the release of a WMD may not be immediately apparent, caregivers, emergency response personnel, and first responders are in imminent danger of becoming casualties before the actual identification of the crime can be made. Incidents could quickly escalate quickly from one scene to multiple locations and jurisdictions.

The types of supplies needed to respond to a terrorist incident may differ from those needed for a natural disaster or other type of technological emergency. For example, the responders to the September 11 attacks incident needed hats, steel-toed shoes, respirators that were appropriate for the hazards, and other PPE. These were not stockpiled and had to be purchased.

Your planning should address administrative protocols to ensure that proper purchasing procedures are followed and that duplicate purchases are avoided.

On September 11, and the days that immediately followed, commercial airlines were not allowed to operate. The shutdown delayed the arrival of supplies and federal responders from distant locations. To avoid the inefficiencies of ad hoc purchasing of supplies and of delays in the arrival of supplies related to air traffic curtailments, you need to consider regional warehousing of supplies and equipment for emergency responders, including equipment for use of US&R task forces.

One of the key logistical problems in the initial stages of emergency response to a terrorist incident is the establishment of an Incident Command Post from which direction of response activities can be made.

In "routine" emergency, such as fire, small hazardous materials releases, or police actions, the Incident Command Post is established at a point that is close enough to observe the incident, but far enough away to maintain an overview perspective and a safe distance from the immediate hazards.

Because of the unique nature of terrorist activity and the inherent unpredictability of the incident, planners and emergency responders may need to rethink the protocol for locating the Incident Command Post.

One of the key administrative and logistical challenges in managing the emergency response to a terrorist incident is the successful integration of the federal response into the initial response by local and state emergency response organizations.

The very nature of a terrorist attack assumes federal response. Depending on the extent of the terrorist of the terrorist incident, the federal response could be swift and massive. The application, integration, and coordination of the federal resources into the existing local command and control structure can be a sensitive operation. Federal resources should not overwhelm the local response but should be made available as needed and requested.

You should involve federal agencies in your planning, to the extent possible, in order to develop a better understanding among all parties regarding the nature and extent of the federal response, including the logistical support needs of the federal agencies.

You should be aware that nonagricultural terrorist incidents are more likely to occur in urban areas than rural setting. Urban centers have significantly higher numbers of emergency personnel and material resources, and they routinely deal with emergency response. Local emergency response organizations will likely want to maintain the direction and control of the emergency response to a terrorist incident.

As a planner, you should be a aware of the potential logistical problems that may be caused by the unsolicited influx of volunteers and donated goods, as experienced post-September 11. Site and perimeter control is extremely important to avoid responder casualties and to prevent emergency operations from being disrupted by uncontrolled movement of such volunteers.

In developing plans for urban centers, you will need to identify potential staging areas for personnel and equipment and warehouses for materials, equipment, and supplies. Although these may not be needed for small-scale incidents, an inventory of available warehouse space and potential staging areas would assist in the response to a large-scale incident and/or a prolonged consequence management response and recovery effort.

Chapter 12

EOC Management during Terrorist Incidents

Michael J. Fagel

Contents

After action reports and studies of catastrophic disasters have identified the need to provide additional information in Emergency Operations Center (EOC) management. Although terrorist incidents have been less common than catastrophic natural disasters, effective EOC management is perhaps more critical for the management of terrorist incidents than for even the most catastrophic natural disasters because the stakes are, potentially, much greater.

By definition, terrorist incidents are Incidents of National Significance. Incidents of National Significance include those incidents that, under Homeland Security Presidential Directive 5, require a "coordinated and effective response by and appropriate combination of federal, state, local, tribal, nongovernmental, and/or private-sector entities ... to save lives and minimize damage, and to provide the basis for long-term community recovery and mitigation activities." One can assume, then, that terrorist incidents will cause coordination issues within the affected jurisdiction(s) and beyond.

The local EOC plays a critical coordination role during Incidents of National Significance. The ability of EOC personnel to work effectively with command personnel at the scene as well as state and federal personnel, public- and private-sector organizations, and others throughout the response and recovery phases is critical to saving lives, protecting property and the environment, and maintaining public confidence.

This chapter will describe the role that the Emergency Manager and other EOC personnel will play following a terrorist incident. To help ensure effective EOC operations following an incident, however, it is necessary to consider the steps of preincident preparedness.

Preparing the EOC for a Terrorist Incident

Preparing the EOC for a terrorist incident is not much different from preparing for other high-impact disasters. To be certain, however, the Emergency Manager should convene the planning team to revisit terrorist threats and targets identified during the hazard analysis process to determine additional needs and areas or tasks that might need to be performed differently.

Reviewing the Hazard and Vulnerability Analyses

All areas of the United States are vulnerable to a terrorist attack, because of the wide variety of means by which an attack could occur. When reviewing the hazard and vulnerability analyses, the planning team should determine the type(s) of incidents that could occur, potential targets of each type of incident, and the risk each type of incident poses to the jurisdiction and its citizens.

Note that jurisdictions can be affected by a terrorist attack, even when not attacked directly. For example, an attack on the power grid could affect a large area of the country. Rural areas that are less likely to experience a direct attack may be affected by a large influx of evacuees seeking shelter in a safe area. However, rural areas that are in close proximity to a military facility may be at risk of a direct terrorist attack.

The planning team, which may need to be expanded to analyze terrorist threats, should think expansively and consider all possibilities when reviewing the hazard and vulnerability analyses. Consider "worst-case" scenarios, but also consider other scenarios that are potentially high impact, even if they might not be high damage.

Incorporating Terror Analysis into Emergency Operations Plan

Use possible scenarios to determine the possible impact on the jurisdiction's Emergency Operations Plan (EOP). Then extrapolate from the EOP, especially the Concept of Operations and the Situation and Assumptions sections, to determine if alterations may be required in EOC operations. Then, play "what if" to change the parameters of a possible terrorist attack to determine how those changes affect necessary planning. (For example, what if a Timothy McVeigh would have used 2000 pounds of ammonium nitrate–fuel oil for his attack on the Murrah Federal Building, rather than 1000 pounds? What additional damage would have occurred? How many additional casualties could have occurred? What additional resources would be required for the response? What other issues would be affected (e.g., communication needs)? Finally, how does this information affect the coordination function and operations at the EOC?) Keep a careful record of these discussions for later use in revising the EOP.

Revisit Each EOP Annex

Next, revisit each EOP annex to identify gaps based on the terrorist threats, potential targets, and risks. The planning team should determine what portions of each annex should work in a terrorist situation and identify areas that need revision. For example, does the current Emergency Public Information Annex describe the huge number of uncertainties that will be felt by the public? Has a procedure been developed to ensure that citizens—and the media—receive critical information accurately and in a timely way? Continue this process with each annex, in turn, until all known planning gaps and needed revisions have been identified.

When completing revisions, keep in mind the key concepts and principles required in the National Incident Management System (NIMS) for all domestic responses, which focuses on command and control using the Incident Command System, resource management, including the typing of equipment according to capability, the training and credentialing of all response personnel, interoperable and redundant communications capability, and use of multiagency coordination systems to coordinate response activity from the incident scene to the federal level, including the Federal Bureau of Investigation. If your jurisdiction has not built NIMS requirements into its planning process, the terrorism review and revision process is a good time to do so.

Policies and Procedures

Policy and procedure review and revision is best done at the agency level, but after revisions have been completed, the planning team should review them to ensure that they are aware of other agencies' procedures and to identify areas that may not work in concert with other agencies' procedures.

Finally, extrapolate to EOC operations. What do changes in the EOP and response procedures mean for EOC operations? Have additional layers of communications systems been added? Does the jurisdiction need to identify and equip an alternate EOC? Examine each change made to the EOP for its possible effect on the EOC. Bear in mind that the EOC plays a coordination role during a response—not a command role—and that revisions to policies and procedures should focus on support needs at the incident scene.

Do *not* change the entire EOC organization unless it has been shown through exercises or actual incident responses that it does not work. If the EOC organization must be changed, change it for all responses, rather than incorporating a unique organization for terrorism incidents alone. A less effective organization can be overcome by the comfort level and enhanced performance of personnel who are familiar with it, which is certainly preferable to requiring staff to learn two different organizations based on the type of incident.

Several specific areas to look at are included in the following job aid. A "no" response to any of the questions on the job aid means that additional planning is necessary to prepare for postterrorism EOC operations. Use the checklist as a guide for identifying additional/different EOC operations during a terrorist response.

BOX 12.1 JOB AID 1: PLANNING FOR EOC MANAGEMENT DURING A TERRORIST RESPONSE

1. EOC Organization
   - Will additional organizations and/or agencies (e.g., public health) need to be included at the EOC following a terrorist attack?
   - What additional organizations should be included?
   - How will personnel from those organizations be notified?
   - What are their equipment needs? Do you have them?
   - With what other organizations and/or agencies will these personnel need to coordinate?
2. Communications
   - What will you do if the power grid goes down for an extended period of time? Are redundant communications systems in place?
   - Do all communications systems users know what the redundant communication systems are? Will they recognize when they need to switch to a backup system?
   - Do users know how to operate backup systems?
   - Have the systems been tested?
   - Will EOC personnel need to communicate with different people/agencies following a terrorist attack than during other operations? With whom will they need to communicate? Is a system in place to facilitate those communications?

- Are special systems and/or equipment required to facilitate communication among key personnel (e.g., between the EOC and state agencies, between the EOC and the mayor or governor, between the EOC and off-site experts)? Are these systems in place? Have they been tested?

3. Continuity of Operations
   - Do you have a backup facility for EOC operations in the event that the primary facility becomes inoperable or otherwise unusable?
   - How long will it take to shift operations to the backup facility? What will you do in the meantime?
   - Do personnel know where the backup facility is? Can they access it?
   - Are equipment and systems in place and operational at the backup facility (i.e., is the facility a "hot" site?)? If not, how long will it take to bring critical systems and equipment to an operational posture? How long will it take to assume full operations from the backup facility? How long can you maintain operations from the backup facility? How will you notify personnel that they should report to the backup facility?
4. Life Support Requirements
   - Do the EOC and the backup facility have independent power-generation capability?
   - Have arrangements been made to provide food, water, and a place to rest for EOC personnel?
   - Do personnel know what shift they work? Do they know when that shift begins and ends?
   - Can operations be maintained from the backup facility for an extended period?
5. Documentation
   - What additional documentation requirements exist for terrorist incidents?
   - What personnel are needed to support these additional requirements?
6. EOC Protocols
   - Has a policy/procedure review been completed to identify needed variances from typical EOC operations?
   - Have new terrorism-specific policies and procedures been developed? Have EOC personnel been trained and have exercises been conducted to test the procedures? Have lessons learned from the exercises been fed back into the jurisdiction's EOP?
7. Emergency Public Information
   - Has the jurisdiction established a Joint Information System (JIS)? Does the JIS include the Public Information Officers (PIOs) from all involved agencies?

- Has the Joint Information Center (JIC) location been identified? How long will it take to assume a fully operational status? What will you do in the meantime to ensure that citizens receive accurate, concise information about the incident and what they should do to protect themselves?
- Is the JIC collocated with the EOC? If so, are JIC operations clearly separated from EOC operations to avoid media interference in critical EOC operations?
- Have all JIC personnel been trained in the policies and procedures for coordinating the preparation and release of emergency public information?
- Has the JIS been tested and exercised? Have lessons learned been incorporated into the Emergency Public Information Annex?

8. EOC Organization
   - Will additional organizations and/or agencies (e.g., public health) need to be included at the EOC following a terrorist attack?
   - What additional organizations should be included?
   - How will personnel from those organizations be notified?
   - What are their equipment needs? Do you have them?
   - With what other organizations and/or agencies will these personnel need to coordinate?
9. Communications
   - What will you do if the power grid goes down for an extended period of time? Are redundant communications systems in place?
   - Do all communications systems users know what the redundant communication systems are? Will they recognize when they need to switch to a backup system?
   - Do users know how to operate backup systems?
   - Have the systems been tested?
   - Will EOC personnel need to communicate with different people/agencies following a terrorist attack than during other operations? With whom will they need to communicate? Is a system in place to facilitate those communications?
   - Are special systems and/or equipment required to facilitate communication among key personnel (e.g., between the EOC and state agencies, between the EOC and the mayor or governor, between the EOC and off-site experts)? Are these systems in place? Have they been tested?
10. Continuity of Operations
    - Do you have a backup facility for EOC operations in the event that the primary facility becomes inoperable or otherwise unusable?

   - How long will it take to shift operations to the backup facility? What will you do in the meantime?
   - Do personnel know where the backup facility is? Can they access it?
   - Are equipment and systems in place and operational at the backup facility (i.e., is the facility a "hot" site?)? If not, how long will it take to bring critical systems and equipment to an operational posture? How long will it take to assume full operations from the backup facility? How long can you maintain operations from the backup facility? How will you notify personnel that they should report to the backup facility?
11. Life Support Requirements
   - Do the EOC and the backup facility have independent power-generation capability?
   - Have arrangements been made to provide food, water, and a place to rest for EOC personnel?
   - Do personnel know what shift they work? Do they know when that shift begins and ends?
   - Can operations be maintained from the backup facility for an extended period?
12. Documentation
   - What additional documentation requirements exist for terrorist incidents?
   - What personnel are needed to support these additional requirements?
13. EOC Protocols
   - Has a policy/procedure review been completed to identify needed variances from typical EOC operations?
   - Have new terrorism-specific policies and procedures been developed? Have EOC personnel been trained and have exercises been conducted to test the procedures? Have lessons learned from the exercises been fed back into the jurisdiction's Emergency Operations Plan?
14. Emergency Public Information
   - Has the jurisdiction established a JIS? Does the JIS include the PIOs from all involved agencies?
   - Has the JIC location been identified? How long will it take to assume a fully operational status? What will you do in the meantime to ensure that citizens receive accurate, concise information about the incident and what they should do to protect themselves?
   - Is the JIC collocated with the EOC? If so, are JIC operations clearly separated from EOC operations to avoid media interference in critical EOC operations?

- Have all JIC personnel been trained in the policies and procedures for coordinating the preparation and release of emergency public information?
- Has the JIS been tested and exercised? Have lessons learned been incorporated into the Emergency Public Information Annex?

Test, Train, and Exercise

Finally, develop a progressive test, training, and exercise program to evaluate every aspect of a potential terrorism response. Begin with orientations that describe the changes required for a terrorism response. Then, proceed to tabletop exercises, functional exercises, and full-scale exercises.

Evaluate tests and exercises honestly, carefully, and thoroughly. Lessons learned from testing and exercises may point to additional EOP revisions or additional training needs. When evaluating tests and exercises, stay focused on the jurisdiction's goal, which is to improve incident response to a terrorism incident, including improved support offered through the EOC and the hierarchy of multiagency coordination entities.

Summary

Preparing your jurisdiction's EOC for a terrorism response may seem like a daunting task, but by building on the systems and procedures that are already in place and testing, training, and exercising the jurisdiction's plan, you can ensure an effective response.

Remember that resources are available to help you with your planning. Contact your State Training Officer for a complete schedule of course offerings. Other courses are offered numerous resources are available to you through the Department of Homeland Security training partners, including the Emergency Management Institute (EMI) to enhance U.S. emergency management practices through a nationwide program of instruction.

Chapter 13

The Active Shooter Incident

Rick C. Mathews

Contents

Over the years, emergency managers have had to activate emergency operations centers (EOC) for many different types of emergencies. Historically, these incidents could be categorized into one of two major groups: those caused by nature and those caused by accidents. Hurricanes, tornadoes, and snow storms are but three types of incidents that are caused by nature. Train derailments, spilling hazardous materials, and forest fires caused by an accidental spark are examples of emergency events caused by accidents. During the past decade, however, emergency managers have had to begin devoting more time and resources planning EOC operations for a third type of emergency, those incidents related to high consequence violent events, be they instigated by terrorists or others. The difference between the two distinctions is essentially whether the attack is consistent with the definition of terrorism as defined by the Homeland Security Act of 2002. Indeed, since the establishment of the U.S. Department of Homeland Security (DHS) in 2002, significant resources have been provided for state and local jurisdictions to plan and prepare for terrorist events. It is important to note, however, that even with the significantly increased focus on terrorism, emergency managers have also been encouraged to adopt a much more comprehensive approach to preparedness planning, one that is inclusive of "all hazards." With this concept in mind, this chapter will discuss certain planning and preparedness actions that should be considered by the emergency manager in terms of a particular violent incident, the "active shooter incident." These include

- Planning
- Training
- Exercises
- Emergency alerts/messaging and risk communications

According to the DHS, "an "active shooter" is an individual actively engaged in killing or attempting to kill people in a confined and populated area; in most cases, active shooters use firearms and there is no pattern or method to their selection of victims."[1] Most discussions about active shooter situations focus on the actual law enforcement response, whereas this chapter will explore this type of incident from the perspective of the emergency manager. One of the key aspects of the law enforcement response that the emergency manager should keep in mind is that this type of incident typically involves what can generally be thought of as an immediate action deployment in contrast with a typical tactical team or SWAT response. Many, if not most, tactical responses involve a well-trained and well-equipped team that will rapidly deploy, secure a perimeter, and will do everything possible to ensure that all subsequent actions are planned, rehearsed (if possible), all without firing a shot, if at all possible. The goal of the immediate action deployment is to stop the killing—usually by the first officers on the scene. In many cases, entry is made by the first two to five officers on the scene, depending on local protocols and procedures. Most active shooter situations are operationally over within 10 to 15 minutes.[1] In many cases, it may well be that the response and entry phase of the operations will be over before the emergency manager is notified.

An excellent report was recently published (2011) by the New York City Police Department (NYPD), entitled "Active Shooter: Recommendations and Analysis for Risk Mitigation" (NYPD Report).[3] This is perhaps one of the most comprehensive reports compiled on the topic, based on reviews of more than 280 active shooter cases between 1966 and 2010 (worldwide). Several interesting and relevant findings were presented:

- More than 98% of the active shooter incidents reviewed were carried out by a single individual.
- Active shooters targeted people they knew as well as those with whom they had no known prior relationship (22% of the cases). Figure 13.1 depicts this set of findings.
- The locations of the attacks were identified as shown in Table 13.1.

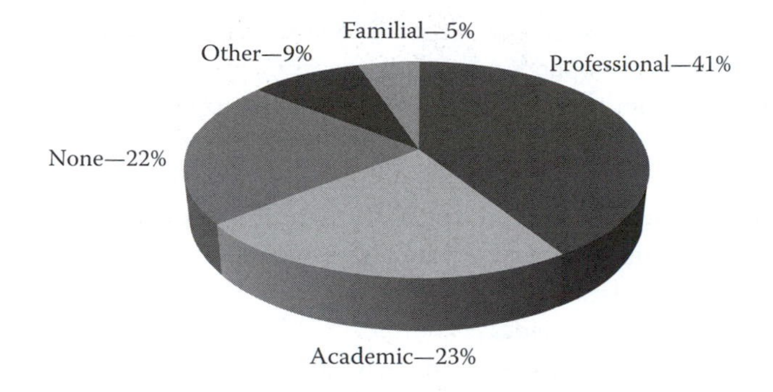

Figure 13.1 Settings and relations in active shooter incidents.

Table 13.1 Locations of Active Shooter Attacks

Location Type	*Number of Incidents*	*Percentage (%)*
School	64	29
Office Building	29	13
Open Commercial	52	23
Factory/Warehouse	30	13
Other	49	22
Total	224[a]	100

[a] The 202 cases in the active shooter data set occurred at 224 locations because several attacks involved more than one location.

Preattack Emergency Management Activities

Although responsibilities may vary from jurisdiction to jurisdiction and from agency to agency, there are generally three areas where an emergency manager can likely be most effective before an actual active shooter incident:

- Coordination of emergency alerting systems
- Coordination of preparedness planning
- Coordination of preparedness validation (exercises)

Across the nation, there are a variety of emergency alerting and messaging systems that can be made available to community leaders, emergency managers, and others that will enable the broadcast of emergency alerts or notifications. Some systems span entire communities, counties, or states such as the New York Alert system. Others may focus on specific agencies, campuses, or subscribers. Regardless of the system or geographical region covered, all had their genesis in the Emergency Alert System (EAS), which is operated by the Federal Communications Commission.[2] The EAS evolved from the Cold War era Emergency Broadcast System. Whereas the sole responsibility of the activation of the EAS rests with the President of the United States (authority delegated to the Director of FEMA),[2] local emergency alert systems and activation protocols, procedures vary widely.

There are several factors that should be assessed in terms of the local emergency alert system; all are relatively equal in importance. First, "how long does it take for a message/alert to be broadcast and received by the systems subscribers?" The system available may reach all, some, or none of its subscribers within three or four minutes. The longer the system takes for the alert to be *received* by its subscribers, the less it can be used, in most cases, to warn potential victims of an ongoing active shooter attack. On the other hand, the system is extremely useful to alert others in the attack vicinity and to provide other appropriate risk communications. Many organizations, especially educational institutions, also utilize other alerting mechanisms such as PA announcements, sirens, and horns. All of these can be used to immediately notify those within a particular facility or zone. It is important to thoroughly train these individuals as to what specific alerts might mean, unless they are PA announcements in clear language. For example, a long sustained siren might mean "evacuate," whereas a series of short siren blasts could mean "lock down and stay where you are."

The second question that must be answered is, "Who has the responsibility and/or authority to send an alert?" In some organizations, this might require the concurrence of a small committee; in others any one of a select group (4 or 5) may be able to authorize the alert. The process can impact the length of time needed to get an alert out. The third factor is the audience that will receive the alert. Is it everyone in an organization or only those that have voluntarily subscribed? What about nonemployees and visitors, are they alerted as well? The final factor is the actual message that will be sent. In many systems, there is a limitation on how many characters can be in any one message. Does the organization have predrafted, "canned" messages that can be quickly sent on the alert system?

As can be seen, there are several factors that can influence the potential effectiveness of an emergency alert system, depending on its purpose. Given an active shooter incident, there are basically three alert messaging purposes that should be given consideration. The first is to warn those in the immediate vicinity of where an attack is occurring. Given the warning, they can either evacuate the vicinity or "lock it down." Whether individuals should evacuate or lock down and stay put is based on the location of the shooter and the nearest evacuation route.[1] In many organizations, the general practice is to hide out and lock down the facility. Locking down a facility hinders the ability of the shooter to move about freely; it also hinders evacuations. In addition, the usefulness of an alert advising individuals to evacuate or "stay put/lock down" is dependent on how quickly the alert can be sounded and the knowledge of where the shooter is.

The second purpose of an alert messaging system is to advise individuals *not* in the immediate attack/threat area so that they will stay clear of that area. A message that issues a "general lock down order" can accomplish this as it advises all that receive the alert to stay where they are and to lock their doors. In this case, it is not critical to advise them as to where the actual shooting has occurred. On the other hand, most will agree that being better informed is the better way to go, meaning that the lock down order should be accompanied by the location of where the attack is occurring. The third purpose of the emergency alert system is to provide one (of several, in most instances) means for issuing usual risk and related public information messages. This includes informing the audience of the general situation and advising where to turn to or go (website, radio station, etc.) for more in-depth official messages. Coordination of the jurisdiction's or organizations emergency alerting system is an important part of the emergency manager's preattack preparedness responsibilities, in most organizations.

One of the key duties of any emergency manager is the coordination of a jurisdiction or organization's emergency preparedness planning efforts. Clearly, this planning should encompass the potential for an active shooter attack. A large part of this effort should focus on ensuring that organizations provide awareness training for everyone within its facilities on the active shooter threat as well as actions they should take should they be confronted with such an emergency. This effort can be imbedded within existing workplace violence training, and it should be made a part of the organizations fire drill, bomb scare, and storm emergency procedures. The emergency manager should consider the following as being important elements in any such training:

- Emergency alert system
- Procedures to evacuate or lock down a facility
- Indicators of potential violence
- Expected emergency response guidance from local law enforcement and security agencies as applicable
- Assurance that the training being provided is consistent with the community's established active shooter response procedures

The third preattack mission for the emergency manager is the coordination of drills and exercises intended to assess procedures and/or to validate preparedness efforts. In other words, has the organization or community achieved the level of preparedness to which it has planned and expended resources? In this area, the exercise might focus on the emergency alert system or how quickly a facility can "lock down" once it is given an alert or a message to do so. At the other end of the spectrum, the emergency manager may be tasked with the coordination of a large, full-scale exercise intended to measure the response of local law enforcement to a suspected active shooter incident. The emergency manager will need to tailor the drills or exercises to the locations participating as well as the objectives being assessed. The NYPD has suggested the following preparedness activities for facilities that may be useful to the emergency manager in terms of setting objectives for exercises:[3]

- Conduct a realistic security assessment to determine the facility's vulnerability to an active shooter attack.
- Identify multiple evacuation routes and practice evacuations under varying conditions; postevacuation routes in conspicuous locations throughout the facility; ensure that evacuation routes account for individuals with special needs and disabilities.
- Designate shelter locations with thick walls, solid doors with locks, minimal interior windows, first-aid emergency kits, communication devices, and duress alarms.
- Designate a point-of-contact with knowledge of the facility's security procedures and floor plan to liaise with police and other emergency agencies in the event of an attack.
- Incorporate an active shooter drill into the organization's emergency preparedness procedures.
- Vary security guards' patrols and patterns of operation.
- Limit access to blueprints, floor plans, and other documents containing sensitive security information, but make sure these documents are available to law enforcement responding to an incident.
- Establish a central command station for building security.

The DHS recommends that training for facilities and organizations should, to the extent possible, include "training exercises."[1] Within that context, DHS recommends that the "training exercises" incorporate the following:[1]

- Recognizing the sound of gunshots
- Reacting quickly when gunshots are heard and/or when a shooting is witnessed
- Evacuating the area
- Hiding out
- Acting against the shooter as a last resort
- Calling 911
- Reacting when law enforcement arrives
- Adopting the survival mind set during times of crisis

Activities that can or should be accomplished before an attack vary based on the agency or institution involved, and the emergency manager may also have responsibility for the provision or facilitation of training. Clearly, emergency responders must receive appropriate training in dealing with the response to an active shooter incident. The training will vary depending on the type of agency and their respective duties in such a response. The emergency manager should be familiar with this training but will generally not be responsible for the conduct or coordination of the training. Accordingly, this area of training will not be discussed in detail in this chapter. The emergency manager, however, may need to facilitate training for other disciplines and or audiences. For example, tenants in offices and businesses should receive "what to do" type training as part of the organization's regular "work place violence" training. The same is true for schools, universities, government facilities, and other public establishments. The NYPD recommends the following in terms of nonemergency responder training:[3]

- **Evacuate**: Building occupants should evacuate the facility if safe to do so; evacuees should leave behind their belongings, visualize their entire escape route before beginning to move, and avoid using elevators or escalators.
- **Hide**: If evacuating the facility is not possible, building occupants should hide in a secure area (preferably a designated shelter location), lock the door, blockade the door with heavy furniture, cover all windows, turn off all lights, silence any electronic devices, lie on the floor, and remain silent.
- **Take Action**: If neither evacuating the facility nor seeking shelter is possible, building occupants should attempt to disrupt and/or incapacitate the active shooter by throwing objects, using aggressive force, and yelling.
  - Train building occupants to call 911 as soon as it is safe to do so.
  - Train building occupants on how to respond when law enforcement arrives on scene.
  - Follow all official instructions, remain calm, keep hands empty and visible at all times, and avoid making sudden or alarming movements.

Attack Response Emergency Management Activities

If an active shooter attack occurs, the emergency manager's role will vary based on how long the attack ensues, the number of casualties (e.g., is this a "Multiple Casualty Incident?"), location of the incident (e.g., did this occur at a high school?), as well as other factors that may impact the role of the emergency manager. In the main, the primary determinant will be the type of support and resources required that the emergency manager will need to facilitate. If the number of casualties crosses the jurisdiction's threshold for being a Multiple Casualty Incident (MCI), then the emergency manager will need to perform those duties and respond according to the established MCI plan, keeping in mind that there will likely be other needs driven by the active shooter incident.

One of the first decisions that the emergency manager may need to make will be whether to activate the jurisdiction's EOC. Will it need to be fully staffed or will some partial staffing be sufficient? The answer will be determined in large part based on the resources that the emergency manager believes will need to be provided and the scope of the response and recovery phases that will need to be supported. The active shooter incident can be viewed as an iceberg, meaning that the response phase may be over in a relatively short period with the recovery phase comprising the bigger share in the event. It is very important for the emergency manager to be mindful that the recovery phase may last for many days or weeks. The psychological impact of the attack on the survivors, responders, as well as the families and friends of the victims will be significant and may require both time and compassionate expert counseling. The emergency manager may be called upon to support these crisis management activities for the duration of the recovery. In addition, there will likely be a prolonged need crisis and risk communications that may need to be supported by the emergency manager. The decision to open an EOC, at least at a minimal level, will often be made based on the recovery phase activities and not on the actual response operations.

The emergency manager will likely be the individual tasked with the facilitation of a comprehensive after action review (AAR) for the incident. The AAR will usually be compiled some time after the actual attack occurred, in order to capture the events comprising the recovery phase and to enable the various organizations and agencies involved to conduct their internal "hot wash assessments." During the actual response and early recovery phase activities, most organizations, agencies, and individuals involved in the response and recovery operations will likely not be focusing significant, if any, efforts on collecting necessary information and data that will become very useful when the AAR is begun. For this reason, one of the more useful activities for emergency management personnel may well be the active observation, documentation, and data collection during the response and recovery operations. To be effective, the emergency manager will need to consider a combination of two strategies. First, (s)he will need to encourage, where and when possible, the active collection of relevant data by those directly involved or in supervisory roles. It is important to emphasize that the intent of this activity is collection only, not analysis—that will come later. It may be useful for the emergency manager to provide a repository for this information that can be accessed by various agencies. In addition to the facilitation of information/data collection, the emergency manager may also engage in actual data collection. This activity involves the documentation of many activities and the compilation of a timeline. Another critical activity will be the compilation of names and contact information of agency points of contacts and leaders involved in the incident response and recovery, which will be useful in the AAR.

Important: It was noted that the emergency manager should consider the establishment of a data repository for the incident for the explicit purpose of the AAR. With this is mind, it is extremely important for the repository to *not* contain original records and reports dealing with the actual attack and response that could be construed as being evidentiary in the criminal investigation and proceedings that may follow. In addition, all data collected should be labeled "For Official Use Only—Work in Progress—For Assessment Purposes" or some similar label intended to afford some degree of protection to the release of the data collected before the production of the formal AAR. The emergency manager should follow the procedures and protocols provided by law and legal counsel guidance.

Personal Actions When Confronting an Active Shooter

The DHS has compiled a set of recommendations or good practices to follow in the event an individual is confronted with an active shooter scenario.

BOX 13.1 GOOD PRACTICES FOR COPING WITH AN ACTIVE SHOOTER SITUATION

- Be aware of your environment and any possible dangers.
- Take note of the two nearest exits in any facility you visit.
- If you are in an office, stay there and secure the door.
- If you are in a hallway, get into a room and secure the door.
- As a last resort, attempt to take the active shooter down. When the shooter is at close range and you cannot flee, your chance of survival is much greater if you try to incapacitate him/her.
- CALL 911 WHEN IT IS SAFE TO DO SO!

Source: "Active Shooter: How to Respond," U.S. Department of Homeland Security, 2008.

Summary

The emergency manager should play an essential role in the jurisdiction or organization's preparedness, preattack active shooter attack efforts. These activities include planning, training, and exercises. In the response and recovery phases of an active shooter attack, the emergency manager will play an important supportive and coordinating role including the alerting/emergency messaging and risk communications actions, resource supply coordination, and AAR responsibilities. Although a great deal has been written about active shooter incidents, most have either been descriptive of actual incidents or focused on law enforcement response actions. With the number of active shooter incidents seeming to increase over the past few years, the DHS determined the need to publish concise guidelines regarding recommended facility and jurisdictional preparedness efforts as well as individual "best practices" in the event the person is confronted with an active shooter. There have been many papers written about specific active shooter attacks with most focusing on either the Columbine School or Virginia Tech incidents. The U.S. Secret Service has published some analytical reports dealing with active shooter profiles. Arguably, the most comprehensive analysis of active shooter incidents, from a planning and preparedness perspective, is the one published in January 2011 by the NYPD. Accordingly, the DHS and NYPD publications served as essential resources for this chapter.

References

1. "Active Shooter: How to Respond;" U.S. Department of Homeland Security. Accessed on October 1, 2010.
2. http://www.fcc.gov/pshs/services/eas/ (accessed January 2011).
3. "Active Shooter: Recommendation and Analysis for Risk Mitigation," NYPD. Accessed on February 2011.

Chapter 14

Terrorist Tradecraft I: The Attack Cycle

Raymond M. McPartland

Contents

At its most rudimentary level, conducting a terrorist operation consists of following a generically created, yet ultimately effective, cycle. This cycle is known as the *terrorist attack cycle* and has been analyzed and reworked by countless counterterrorism officials since the inception of the threat. Although some discuss the possibility of it consisting of six steps and another, four, the reality is that at the cycle's core lies a very general yet germane set of actions that are required to launch and accomplish an effective attack. As an emergency manager and responder, it is imperative to not only know, but also understand, this cycle. Doing so helps a discipline better prepare, respond, and possibly interdict the next terrorist attack.

Successful terrorist operations are seldom self-contained undertakings. This holds true for large international terrorist groups that leave a much longer logistics trail and for so-called "lone wolves" who act alone or within a very limited circle of witting or unwitting accomplices. Even "lone wolves" are not free from the burden of applying the tools of their trade. Very few individuals have the requisite technical, physical, financial, and logistical skills to mount an operation without reaching out to like-minded individuals or sympathizers for assistance. In the age in which we live, considerable assistance can, of course, be gleaned online, but surfing the Internet for information can take one only so far, and it carries with it its own set of perils and vulnerabilities. Furthermore, it is probably safe to say that no one has carried out an attack in the real world solely by using materials and techniques acquired in the cyber world. At some point, this online knowledge and technical skills must be translated into real-world applications and actions.

In looking at individual terrorist operations, we sometimes exhibit a bit of myopia in that we tend to think that any method of attack that does not follow the pattern of recent attacks must be new. A closer look, however, reveals a set of recurring themes. And although the methods of attack are constantly being refined in an attempt to adapt to security countermeasures, the terrorists' basic repertoire remains pretty much fixed.

For example, the November 2008 attacks in Mumbai, India, were carried out by individuals highly trained in light infantry tactics and equipped with assault rifles, grenades, and relatively small explosive charges. Such tactics had been discarded in recent times in favor of large-scale, mass casualty attacks, but they can be traced back to the very beginnings of the current wave of terrorism that dates at least from the second half of the twentieth century. Although the Mumbai attackers took advantage of some newly available commercial technologies—global positioning systems, voice over Internet protocol, personal data assistants, etc.—the basic tactics they used were not all that different from those used by members of the Japanese Red Army in an attack in May 1972 on Tel Aviv's Lod Airport. In this attack, the perpetrators hid their assault rifles with removable stocks in violin cases and, upon entering the airport, removed the weapons from the cases, assembled them, and started firing indiscriminately, ultimately killing 26 people.

This same scenario was reprised more than 13 years later, when on December 27, 1985, the Abu Nidal group, using Libyan-supplied assault rifles and grenades, launched virtually simultaneous attacks on the El Al and TWA ticket counters in Rome's Leonardo da Vinci–Fiumicino Airport and on passengers waiting to board a flight to Tel Aviv at Vienna's Schwechat Airport. It was even reported that the Abu Nidal attackers were doped on amphetamines, mirroring similar reports about the Mumbai attackers. The cycle replicates itself because it works and is effective. A terrorist group will continue to do what is effective and is deemed to be nearly 100% accomplishable.

Aside from understanding the cycle and all its phases, the emergency practitioner must also be cognizant of the motivation and goals of the terrorist or extremist. Although ideologies differ between various terrorist groups, their motivation and goals need to be examined at an elementary level. Knowing what the attack planners set out to accomplish helps in determining their next target location, target group, and potential method of attack. Before one can begin looking at specific plot mechanics, it is important to recognize why a terrorist group, or possibly a single extremist, is motivated to undertake such a planned yet despicable act. By motivation, we will not discuss the various group ideologies but instead, group-sought goals in general. It is also critical to understand basic cell structure and operational parameters of a terrorist group as well as the terrorist attack cycle itself.

Terrorist Motivations and Goals

All terrorist groups have both primary and secondary goals motivating them to conduct a terrorist attack. Oftentimes these goals are intertwined and parlay into each other, allowing the consequences of each attack the ability to multiply. It is important for the emergency manager to understand the reason in which a terrorist group or individual commits an act of terrorism in order to better prepare for such an event, distinguish possible targets, and easily recognize an event as it occurs. Motivations can originate in a number of ways but often fall in a few central categories. These categories are explained below.

Primary Motivations/Goals
- Instill fear in the targeted population as well as those populations unsympathetic to their cause.
- Create mass causalities as well as mass confusion in the targeted population.

Secondary Motivation/Goals
- Cause financial distress in a nation, region, or population.
- Create political unrest for an organization, regime, or government entity.
- Create recognition for their cause through action and exploitation of resulting media.
- Coerce a government to make political and even operational changes in matters relating or unrelated to the event at hand.

BOX 14.1

Although initially planned as an attack on America, its way of life, and on targets of great symbolic importance, the September 11, 2001, attacks attained not only a tactical and psychological impact, but also a financial one. Recovery from the attack required millions of dollars of initial recovery, the mobilization of military forces to Afghanistan, and the increased expenditure and upgrade of modern-day security and law enforcement. The success of that on the financial stability of the United States has motivated other extremists to strike targets connected economically either directly or indirectly economically.

Terrorist Cells

From within, virtually all groups use variants of cellular organizations at the tactical level to enhance security and to task organize for operations. Historically, terrorist organizations displayed an organized, hierarchical approach to organization and operation (Figure 14.1). This approach allows for top-down direction normally led by an overall organization or individual. In a hierarchical approach, terrorist cells may be identified by five main categorizations all led by both an overall command cell and a smaller, more unit-specific column command. Those five categories are as follows:

1. Intelligence—The intelligence cell is responsible for reconnaissance of the target location and intelligence gathering.
2. Financial—The financial cell is responsible for the funding all aspects of the operation.
3. Logistical—The logistical cell is responsible for gathering the necessary resources and personnel.
4. Operational—The operational cell is responsible for planning the attack.
5. Attack—The attack cell is responsible for carrying out the attack.

The smallest elements of terrorist organizations are the cells that serve as building blocks for the terrorist organization. One of the primary reasons for a cellular or compartmentalized structure is security. The compromise or loss of one cell should not compromise the identity, location, or actions of the other cells. A

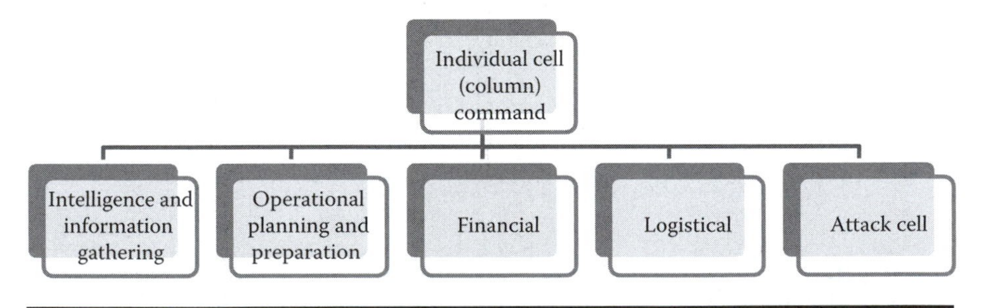

Figure 14.1 Terrorists historically operate within hierarchical organizations.

cellular organizational structure makes it difficult for an adversary to penetrate the entire organization. Personnel within one cell are often unaware of the existence of other cells and, therefore, cannot divulge sensitive information to infiltrators.

With the face of modern terrorism changing from an organized group with a discernible leader to a more ideological approach undertaken by inspired followers, so does the operational structure change. Although the hierarchical approach is still in place today in a more fragmented fashion, the groups or individuals are using more of a networked approach connected not so much in a linear way but in a less tangible and indirect manner. There may still be a command cell or an individual or group dictating actions of followers but the line of communication is not direct. Very often current groups are self organized without any direction other than the ideology itself. The structure may seem slightly different but the roles are still exact. Instead of each cell function containing multiple players, there might only be one. Some operators may even play dual roles. And in the case of a lone individual, he/she plays planner, preparer, and executioner. The method is still the same and the cycle of preparation is still composed of the necessary components.

The Terrorist Attack Cycle

The way in which a single terrorist or terrorist group plans for and executes an attack is known as the "terrorist attack cycle." The cycle is a predetermined group of steps undertaken before, during, and possibly after the event that, if followed properly, can lead to a more successful operation. The necessity to understand and recognize the terrorist attack cycle is no longer limited to members of the intelligence community (IC). Today, members of the emergency response community from all disciplines need to know the way in which an attack is planned and executed in order to better prepare and respond.

Contrary to popular belief, the execution of a terrorist attack is not random nor is it solely dependent on a moment of opportunity. Rather, there is a process that can be generically applied to all terrorist attacks, regardless of their scope. Attacks are often meticulously planned with inception to delivery times ranging from six months to years. The planning is often segmented and each phase is often carried out by separate individuals allowing for maximum operational security. Overall, operational planners of the attack look to exploit their adversary's weaknesses while avoiding all their strengths often leading to the selection of "soft targets," or those areas, locations, or people, carrying a weak security profile.

> On February 26, 1993, at midday, a truck bomb exploded on Level B-2 of the World Trade Center (WTC) garage in New York City. The act, meant to topple the Twin Towers, caused significant structural damage totaling over

$500 million. Six people were killed; more than a thousand were injured. Extremist Ramzi Ahmed Yousef led a small group of Islamic extremists who executed the attack.

The group conducted visual surveillance. Twice in February 1993, they drove to the WTC and inspected Level B-2 of the parking garage. The last trip was two days before the attack (Weiser, 1997). Additionally, Niddal Ayyad rented two separate cars one week apart to conduct surveillance of the WTC and listed Mahmud Salameh as an additional driver.

The cycle consists of 5 stages, each with its own unique set of requirements, and is normally carried out by specific member(s) of the overall terrorist cell. Not all phases are necessary for every mission. All is dependent on the overall goal of the attackers. *For example, not all events require an escape plan. Attacks involving the suicide method of attack allow the operational planners the luxury of not being weighed down with the often complex plan of escaping the incident.* Depending on the size of the group and the complexity of the operation, the cycle itself is scalable and will often appropriately match the planner's level of experience, expertise and resources.

The 5 general phases are normally allotted time constraints and often come with organizational limitations (Figure 14.2). Understanding this cycle and nuances within is critical for any emergency responder in any discipline to know and understand in order to better prepare for and mitigate against.

Phase I—Target Selection

Considering that most targets of terrorism tend to be utilities of public use (e.g., hotels, office buildings, shopping areas), the targeting aspect of the cycle is usually not easily predicted. Groups tend to overlay their own ideology with specific goals in order to choose the target that best suits their message. They often look to targets of symbolic value or those that can garner the most media attention and then combine it with a consideration for specific target groups in order to increase the impact of the attack and possibly avoid collateral damage.

They tend to avoid "hard targets," or those locations, people, or areas with a higher, more reinforced level of security. The softer targets consisting of civilian populations and civilian infrastructure will often result in a more successful mission and a higher number of casualties and deaths. This intended result, coupled

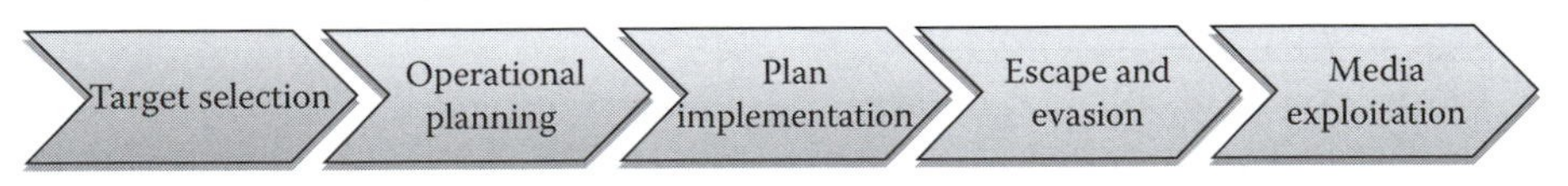

Figure 14.2 Five phases of the terrorist attack cycle.

with the impact both psychologically and financially on the targeted population, makes the choice of choosing a softer target much more enticing.

During the Target Selection Phase, operatives' research as much about the potential target as possible. The depth and detail of the collection depends on the group's resources and the overall scope of the operation. Oftentimes, there might be multiple collection processes placed in motion to look at multiple potential targets at once in order to prioritize and have various options to consider. Historically, target research was conducted on-site by operatives using average collection methods such as videotaping, taking photographs, and simply writing down notes about a particular area. Today, however, the methods used to collect information are different and more remote. The reason for this is twofold. First, with the invention and overall depth of the Internet, operatives need not physically see a location in order to collect nearly 85% of all information needed. Through remote portals such as live street and traffic cameras and typographical sites such as Googleearth.com, selection members are allowed to collect necessary information remotely and from the safety of their own living room. The second reason is the increased law enforcement awareness and the greater likelihood of discovery. Using the Internet and more remote methods of information collection is not only the easier method but is also a necessity. Law enforcement agencies around the country are more aware of information collection methods than ever before. Many are trained to look for signs of preoperational surveillance and security probing.

- **Preoperational surveillance** is the collection of information through various means conducted over a certain period of time before commencement of the operation.
- **Security probing** is when operatives test current security methods or procedures through legal means such as turning doorknobs or possibly unlocked doors or asking sensitive questions of assigned security personnel.

Target selection is the beginning of the planning cycle and is the first in a series of steps in the preincident phase of the actual attack. Understanding the methods makes interdiction a more viable option for any first responder, emergency manager or public/private sector director.

In preparation for the 1993 WTC attack, Salameh, Abouhalima, and Ayyad assisted with logistics, planning, and operational details. Their responsibilities included:

- Salameh renting a safe house and opening bank accounts
- Abouhalima mixing and transporting materials
- Ayyad advising technical procedures and acquiring chemicals for the weapon (Tyre, 1993)

Phase II—Operational Planning and Preparation

After the selection of the target is completed, the operation and method of the attack becomes the next issue. Terrorist attacks often require meticulous planning and preparation. Historically, this is done by a separate cell member(s) but as of late, given that many of the potential attacks domestically have come from radicalized, homegrown extremists, the function of planning may originate from those that conducted the original surveillance and selection.

During the selection process, through the use of on-site or remote surveillance, operatives have not only found a suitable target but also the possible "Achilles heel" of the said target. By pinpointing possible weak areas in security or even the physical façade itself allows the planners to then formulate an attack plan and decide upon an effective method. The planning process involves multiple steps including

- Conducting additional surveillance on the target location
- Selection of an attack method selection and possible weapon to use
- Careful consideration of funding issues and personnel needed
- Gathering of materials and weapons and possibly building the device
- Selection and formation of an attack team
- Rehearsals, or dry runs, of the attack to finalize the decided-upon method

Depending on the quality of the initial target selection process, the planning phase can be done more secretly and oftentimes more remotely. The amount of work needed is entirely dependent on how well the target information was collected and processed. If the information is vague, incorrect, or incomplete, the planning team has no choice but to revisit the scene and begin the collection process again. Planning and preparation can take months to years to complete. All depends on the size of the operation and the skill of the planners. Something of the magnitude of the September 11, 2001, attacks or the coordinated, armed assaults in Mumbai, India, in 2008 were elaborate, requiring various skill sets of the attack team members as well as an abundance of resources. Other smaller planned attacks may involve less people, less sophisticated skill sets, and minimal resources, allowing for a faster turnaround time from selection to planning completion.

Phase III—Plan Implementation

After carefully selecting the target and deciding upon the best method of attack, the third and most imminent part of the cycle is the Implementation Phase, or the actual deployment of attack members and the attack itself.

The last chance for interdiction by any member of the law enforcement community is during the implementation phase. During this phase, attack team members are mobilized from their respective safe houses and deployed to the attack area. Members are asked to collect any necessary resources and construct any needed devices, potentially explosive in nature, and begin the journey to their intended target. Interdiction by law enforcement and interaction by any member at this point is possible but solely dependent on the level of secrecy the group has attained as well as the amount of planning and safety methods put in place during the planning process.

> In the 1993 World Trade Center attack, the day before the attack, the conspirators finalized preparations and loaded a 1000-pound bomb consisting of urea nitrate into the cargo area of a rented Ryder truck.
>
> This was the same truck that, the night before, Salameh falsely reported to the police to have been stolen. On the report, he gave the police officers the wrong license plate number (Reeve, 1999).
>
> At approximately 0400 hours on the day of the attack, the Ryder van was filled with gas in Jersey City, New Jersey, and then later driven into Manhattan, New York.
>
> At approximately 0900 hours, Yousef picked up Ismoil at a hotel in midtown Manhattan and began to prepare for delivery. Around noon, the van was parked on Level B-2 of the WTC garage and then abandoned after Yousef lit the fuses, which took approximately 12 minutes to burn (Reeve, 1999).

Phase IV—Escape and Evasion

Even though members of the first responder community may have missed the initial three phases and are now handling the aftermath of the attack, there is still a good chance that capture is possible depending on the ability of the attack members to escape. The fourth phase of the terrorist attack cycle involves possible escape from the scene and then overall capture. Not all plans will include an escape plan and those that do not become much easier to accomplish and prepare for. Oftentimes the escape plan is more complicated and unpredictable than the attack itself.

Members will first need to escape the scene among multiple members of the first responder community. Once that is accomplished, they will still need to avoid detection and capture once the investigation begins. Both of these problems will be compounded by stress and the unpredictable nature of what is encountered on the route to refuge.

Members of the emergency response community need to be aware of the critical nature of this phase while still conducting lifesaving or investigative methods immediately after the incident. A possible victim could be a terrorist and that piece of debris, a critical piece of evidence. Those in the surrounding areas must also be aware that although they may not be responding to the actual incident, those that they may come in contact with who are acting suspiciously could be those involved in the incident. Given the suicidal nature of many recent attacks, the attack team members may not escape or be aware of any escape plan. The planners, however, or those involved in the other aspects of the preparation may.

Ramzi Yousef did not organize an escape plan for his associates, only himself. He stayed to witness the immediate aftermath and then left for JFK International Airport. Before the attack, he purchased a first class ticket under the name Abdul Basit Mahmud Abdul on Pakistani International Airlines ticket to Karachi.

Shortly after the incident one of his associates, Mahmud Salameh, planned to fly from New York to Amsterdam. He was captured when he tried to retrieve the security deposit for the rental van. Salameh was arrested when he returned to the Ryder agency for his money (Duffy, 1995).

Phase V—Media Exploitation

The fifth and final phase of the attack cycle comprises media exploitation immediately after the attack. Understanding that most groups want to take recognition for the incident to further gain support for their cause and notoriety of members, attackers will attempt to use either the main stream media or other nonconventional means (Internet, web postings, etc.) to send this message. Some groups may use their own branch of media collection and dissemination, whereas others will rely on national broadcast centers to show the fruits of their labors.

Given the changing face of modern terrorism and the increasing switch from operational groups to lone individuals, these five phases, at their core, are the framework for any terrorist attack. Some incidents may be more elaborate than others involving longer durations between each cycle, whereas others may not. The fact remains that the emergency response community must be in tune with the nuances of each in order to prepare for and possibly interdict the next attack.

Intelligence-Driven Counterterrorism

The law enforcement and intelligence communities always have had very different ways of looking at the problem posed by terrorism. Law enforcement's traditional approach to any criminal act was to conduct an investigation, establish "probable cause," make an arrest, and get a conviction. This was entirely in keeping with the procedures prescribed by our system of laws. The swift completion of this process was expected to have a deterrent effect on would-be criminals, who would know that any crimes they contemplated in the future would be dealt with similarly. Other than that, there were clearly defined limits as to how far law enforcement could go in taking preemptive steps to disrupt a potential crime.

The IC, on the other hand, operated under a different set of rules, especially with regard to collecting foreign intelligence. The only "probable cause" it needed was a reasonable expectation of gathering foreign intelligence. The IC, not being involved in the criminal process, was not concerned with such things as rules of evidence and chain of custody, since it never expected its findings to be presented in a court of law.

This does not mean that law enforcement and the IC did not, or could not, work together on terrorism cases. Urban legend has it that the two communities did not talk to each other before 9/11, but history tells a different story. The New York Joint Terrorism Task Force, for example, was formed in 1980, and successful law enforcement and intelligence collaboration resulting in disrupted terrorist acts and convicted terrorists followed shortly after. Examples of disrupted terrorist attacks were the Landmarks plot in New York City in 1993 and the multifaceted Millennium plot in 1999, which targeted Los Angeles airport and tourist hotels in Jordan. Examples of successful prosecutions were the sentences handed down on the perpetrators of the WTC attack in 1993, Shaikh Omar 'Abdul-Rahman, and the Al-Qaeda members involved in the bombings of the U.S. embassies in Nairobi and Dar es Salam. Since all of these cases were intelligence-driven, it is obvious that certain members in law enforcement were working with their counterparts in the IC, albeit not to the extent that would be the norm after 9/11, when the paradigm changed forever.

From the traditional criminal investigative perspective, cases are finite. There is a beginning (a criminal act), a middle (an investigation and arrest), and an end (a prosecution). From the intelligence perspective, cases just lead to other strands of information that must be followed, and this process in turn often produces other strands that must be investigated.

Intelligence is now expected to play an even greater role in supporting law enforcement and pointing it in the right direction so that its finite resources can be used in the most efficient manner. Collaboration must be as seamless as the law and prevailing authorities allow. Neither side can be allowed to work on its own world with stovepipes of proprietary information that is shared only on a case-by-case basis. Each side is expected to understand the authorities the other works under, and what the other side needs to do its job. Great strides have been taken since 9/11 to improve the flow of information in both directions. Whereas in the past, the JTTF structure was just about the only conduit for passing national intelligence to state, local, and tribal law enforcement agencies, we now find that the number of JTTFs has increased dramatically and the Department of Homeland Security has set up a number of intelligence fusion centers that facilitate the flow of information in both directions. This commitment on the part of the federal government has been matched by many state, local, and tribal governments, which have established, or added to, their own analytic cadres. Furthermore, this task is not limited to the intelligence and law enforcement communities. Many private sector security personnel across the country are engaged in securing their own facilities and interests. Increasing the collaboration with these private security professionals serves as a force multiplier, which complements what is being done in the public sector.

Terrorists study what is important to us and how we protect what is important to us. They also study our culture, habits, and ways of doing things in an attempt to find vulnerabilities that they can exploit to conduct operations aimed at exacting revenge, influencing policy, etc. As was pointed out earlier, however, each element of the terrorist tradecraft cycle is a potential vulnerability for the overall operation. Identifying and studying the elements of terrorist tradecraft is the job of the intelligence and analytic elements at every level of the government. If we are to "win" the struggle with terrorists, we must in turn be willing to study them with an eye toward uncovering these vulnerabilities in order to thwart their plots. Fighting terrorism is not an option, and the terrorists must be held accountable and forced to pay a price if we hope to protect our lives, economy, and way of life.

The New Terrorism—Radicalization and Growing Threat of Homegrown Terrorism

The 2003 Madrid attacks, followed by the July 2005 attacks against the London Underground and the follow-on failed attacks a few weeks later brought to the forefront the issue of radicalization. Since then, the study of radicalization has become its own industry inside the wider sphere of counterterrorism studies, with radicalization being solely focused on Islamist terrorist acts. In keeping with the Islamist angle, numerous books and reports have looked at the radicalization process, and in effect, have come to varying conclusions, some (e.g., New York City Police Department report on the subject) offering a simple formula for the prospective jihadist to follow.

Out of all this has come the notion of "homegrown terrorism," with the argument that prospective jihadists in Europe or Canada or the United States, who did not attend the Al-Qaeda training camps in Afghanistan are being radicalized inside their own homelands. Although this is true, it is misleading to say that the radicalization is homegrown since the ideas are not germane to the United States or to Europe. The ideas trace back to the ideology of Al-Qaeda, hence even if the Afghan camps have been disassembled, Al-Qaeda and its agents are still continuing to radicalize prospective jihadis. The Information Age has added information at a faster speed, which only accelerates the process, as evidenced with the Al-Qaeda online magazine, *Inspire*.

It was once thought that radicalization, in the Islamist sense, was a result of economics, that the poor were spurred on to conduct terrorist operations in order to change their economic situation. Abu Musab al-Zarqawi, was a career criminal with a limited future. His mother sent him to undergo religious instruction at the Al-Husayn Ben Ali Mosque in Amman, Jordan, during a period of the Soviet war in Afghanistan. From there, Zarqawi went to Afghanistan and met two men who would influence his life: Abu Mohammed al-Maqdisi and Saleh al-Hami. Al-Maqdisi, was a theologian and fighter who laid the Islamic foundations for "Zarqawism," before splitting with Zarqawi over his use of heavy-handed tactics in Iraq. Al-Hami became Zarqawi's brother-in-law, and was wounded in Afghanistan. The meetings of these two with Zarqawi showed the career criminal what life was like, and contributed to Zarqawi's development of a world in which Islam was triumphant.

For every Zarqawi, there are two Mohammed Atta's, European-educated, emotionally stable, and children of parents who left the Middle East for Europe and the United States and elsewhere. These individuals attend Western schools, but whereas Western students who attend institutions of higher learning and are more likely to become secularized, the Atta's of the world become more attuned with their religious upbringing. These feelings are mixed with societal maladies, or the actions of various governments, perceived by these individuals to be attacks against Muslims and Islam. Into this already volatile mix is the idea of *takfir*, the idea in which jihadists will condemn those who do not agree with their various worldviews, with apostasy, in essence, giving them, in their own minds, the license to conduct terrorist attacks.

Radicalization is a study unto itself, and contrary to past studies, there is no magic formula for law enforcement or counterterrorism agencies to follow. Although some lump these homegrown jihadists together, they themselves have followed different paths to their various end states. Step one does not immediately lead to step two on the path to jihadism; it is all personally driven. Below are the paths to radicalization of three different individuals, which by no means encompass the entire spectrum of radicalization studies, but are meant to sow the distinct pathways that were taken to get there.

Rashid Baz

On March 1, 1994, Rashid Baz, a Lebanese livery cab driver, opened fire on a van full of Hasidic students on the Brooklyn Bridge. The attack left several students wounded and 16-year-old Ari Halberstam dead. Baz was angered by a massacre of Muslim worshippers at an Islamic holy site in Hebron, and in the wake of the attack, Baz saw himself as defender of Islam.

Baz was born into a middle-class family in Lebanon in 1965, although his father was a Druze, a sect that broke from Islam and have since been persecuted by both Sunni and Shiites. Baz came to the United States in 1984 and began frequenting mosques in New York City, but he did not convert until 1992. On September 4, 1992, Baz was involved in an accident and told the driver of the other car, "I am a Muslim." Baz began identifying himself as a Palestinian Muslim, and in February 1994, when Baruch Goldstein killed 29 Muslims at the Cave of the Patriarchs, Baz began a rapidly accelerated radicalization process.

During his trial, Baz's psychiatrist told the court that after the Goldstein massacre that Baz "was enraged. He was absolutely furious. He was, I think, Hebron put him from condition yellow to condition red." Baz began attending the Islamic Center of Bay Ridge, in Brooklyn, New York, and was influenced by the rhetoric of the imam who talked about revenge. Days before the shooting, Baz began referring to himself as an "Arab soldier crusader" and wrote a poem about fighting Crusaders during one of the Crusades.

Major Nidal Malik Hassan

Twelve U.S. service members and one civilian Department of Defense employee were killed and 32 were wounded when Major Nidal Malik Hassan opened fire inside a meeting room inside the base at Fort Hood, Texas, in November 2009.

Hassan was born in Virginia in 1970, graduated from Virginia Tech in 1992, and joined the U.S. Army in 1995. He later entered medical school and became a resident in the psychiatric program at Walter Reed Army Medical Center where he began to show outward signs of his radicalization. At a post residency graduate program at the Uniformed Services University of the Health Sciences, he became even more radicalized to the point where he fully embraced violent Islamic thought. He gave several presentations espousing the Koran and stated that one of the risks of having Muslim Americans in the military was the possibility of fratricidal murder of fellow service members. In August 2007, instead of giving a lecture on a health care subject, Hassan gave a presentation entitled, "Is the War on Terror a War on Islam?: An Islamic Perspective," in which Hassan argued that the U.S. military operations are a war against Islam. He continued days and weeks later, openly supporting suicide bombers and saying that he put religion above the United States Constitution. Hassan was reported numerous times to superiors, who, in an effort to remain politically correct, did not act, resulting in the attack.

Taimour Abdulwahab al-Abdaly

On December 11, 2010, two explosive devices detonated in Stockholm, Sweden, killing one person and injuring two. The fatal injury was caused to Taimour Abdulwahab al-Abdaly, an Iraqi-born Swedish citizen (Figure 14.3). Abdaly moved from Sweden to Britain in 2001, where he moved to Luton, an area that was home to some of the July 2005 British bombers. He moved to Luton to study sports therapy, and in 2004, he married Mona Thwarny and had three children, including

Figure 14.3 Taimour Abdulwahab al-Abdaly.

one child named Osama. Abdalay traveled to Iraq and Jordan and moved his family from Luton to Leagrave, another enclave that is home to known jihadists in the U.K.

His radical views grew stronger around this period after an undisclosed event happened to him in England. At one time, a beer-drinking womanizer, Al-Abdaly grew a beard and repeatedly talked about Afghanistan and Islam and cut all contact with his friends. Qadeer Baksh, chairman of the Luton Islamic center where al-Abdaly worshipped, stated that al-Abdaly stormed out of the mosque after his views on jihad were challenged. Baksh said, "He was just supporting and propagating these incorrect foundations of Islam, so I stepped in and I left feeling he had changed.... He believed that the scholars of Islam were unreliable because they were in the pocket of the government. He proposed a physical jihad." That said, nobody from the mosque reported him to police. Baksh said, "You can't just inform on any Muslim having extreme views."

In an audio message recorded just before the bombing, Abdaly said, "So will your children, daughters, brothers and sisters die, like our brothers, sisters and children die. Now the Islamic state has been created. We now exist here in Europe and in Sweden. We are a reality." The audio message was joined by an email warning of an attack and talked about the Swedish cartoon controversy, in which a number of cartoonists drew cartoons of the Prophet Mohammed.

The preceding examples are just small snapshots of individuals who have conducted jihadist acts, but as mentioned earlier, there is no set blueprint of radicalization or what type of individual who will even go down that path. The cited individuals were lone actors, who pose an even greater threat for law enforcement. Although the focus has been put on large groups such as Al-Qaeda, the interaction between individuals, who support the Al-Qaeda ideology, and the group itself has grown closer due to technology. Online chat rooms espousing Islamic rhetoric, videotapes of radical clerics as well as webcastings of their sermons proliferate the Internet. For every Najibullah Zazi, there is a Rashid Baz; for every Mohammed Siddique Khan of the 7/7 cell, there is a Carlos Bledsoe, a follower of Al-Qaeda in the Arabian Peninsula who shot and killed one soldier and wounded another at a Little Rock, Arkansas, recruiting station.

In July 2010, Al-Qaeda in the Arabian Peninsula released the first issue of *Inspire Magazine*. The magazine is aimed at British and American readers and provides instructions in topics such as bomb assembly, surveillance, and other topics that would be taught in a training camp. The magazine is believed to have connections to an American citizen now in Yemen alongside AQAP, Anwar Aulaqui. Aulaqui, who was the former spiritual advisor for three of the 9/11 hijackers, has become one of the most famous purveyors of Islamic rhetoric in the world today. Although Aulaqui is not the only one, he is linked to a number of would-be jihadists that include Mahmud Brent, who admitted attending a Lashkar-e-Taiba camp in Pakistan; Hysen Sherafi, a member of a North Carolina cell that planned on attacking the Marines at Quantico, Virginia; and Zachary Chesser, a Muslim convert who wished to travel to Somalia to train with the Al-Qaeda–affiliated al-Shabaab as well as Umar Farouk Abdulmutallab.

Al-Qaeda as a group saw itself as the vanguard, a group that would last 20 years and lay the groundwork for the next generation. It saw itself as moving from a network to an ideology, and in this respect, it has. One of the senior leaders of al-Shabaab is a U.S. citizen from Alabama of Syrian descent, Omar Hammami. Hammami laid out a strategy for al-Shabaab to establish an Islamic state in Somalia, and he has become a deep thinker regarding jihadist issues. Hammami said, "They can't blame it on poverty or any of that stuff.... They will have to realize that it's an ideology and it's a way of life that makes people change."

Chapter 15

Terrorist Tradecraft II: Case Studies—Past, Present, and Future

Raymond M. McPartland

Contents

There are numerous sources of information for those wishing to study terrorism; however, to truly understand terrorist doctrine, it is important to study actual cases. This type of study will allow the student, the emergency planner, the law officer, and the intelligence professional and operator to learn how terrorists plan attacks; how they conduct surveillance, select targets, train, and finance the operation and obtain necessary materials; how materials are obtained; and how the attacks are executed. It should be noted that the tradecraft involved in terrorist attacks used by groups 30 to 40 years ago are still used today; the difference is that they are more refined. Instead of a bomb inside a suitcase, Al-Qaeda in the Arabian Peninsula (AQAP) operative Umar Farouk Abdulmutallab boarded a commercial airliner with an explosive device hidden inside his underwear. The attacker is looking for vulnerabilities to exploit in the defenses, whereas the defender is looking to exploit vulnerabilities within the tradecraft used.

> The July 2010 Kampala bombers used excellent tradecraft in planning and carrying out the attack, yet an unexploded device left at one of the targets led investigators to the Kenyan cell leaders, who made several calls to the cell phone located inside the unexploded device.[1]

Counterterrorism professionals are always studying the vulnerabilities of various targets as intelligence professionals continue to assess the capabilities of terrorist groups—made much more difficult now by the growth of lone actors—taking advantage of what military theorist John Robb calls the "Open Source Bazaar." Robb argues, "Rather than a single 9/11 style attack, we may see small attacks (less planning and training, fewer people, less support) against a plethora of targets."[2] Therefore, one of the keys of an effective defense, one that will lead to the mitigation of the threat, is the study of the tradecraft cycle and methods, with an understanding that everything that is old is once again new.

Terrorist operations are meant to be high-reward, low-risk propositions, as the planners look to hit the soft underbelly of the defenses. Soft, underdefended targets were preferred, but as of late, there is a shift in doctrine. Targets now thought to be fortified also continue to remain a target as shown in the October 2010 AQAP airline plot. The advantage is in favor of the terrorist planners as they only have to find the weak point, whereas the defender must continue to defend large areas and entities, not knowing the when, where, and how of the attack. The key on both sides is innovation, a concept that will be explained later in the discussion of the various modes of attack.

Planning

Successful tradecraft planning involves a healthy combination of **desire** and **capability**. A plan will be dictated by what the terrorists want to accomplish and their capability to carry out the attack.

In 1998, Osama bin Laden established the World Islamic Front for the Jihad against Jews and the Crusaders (*al-Jabhah al-Islamiyaah al-Alamiyaah Li-Qital al-Yahud Wal-Salibiyiin*). The aim was simple: since the United States declared war on Islam, it was therefore the duty of all Muslims to comply with the will of Allah and kill Americans and their allies, civilian or military, the world over. This is what drives Al-Qaeda and its affiliates. In terms of Timothy McVeigh, he developed a deep-seated anger toward the United States government and the Federal Bureau of Investigation (FBI) after its role in the Ruby Ridge standoff in 1992 and, later, the Bureau of Alcohol, Tobacco, and Firearms for the 1993 siege of the Branch Davidian compound in Waco, Texas. McVeigh's anger culminated in the April 19, 1995, Oklahoma City bombing of the Alfred P. Murrah Building in Oklahoma City.

THE ALFRED P. MURRAH BUILDING BOMBING

The details of the plan evolved from McVeigh's experience during and after his time in the U.S. military, which he joined when he was 20 years old. McVeigh (see photo) met his eventual coconspirator Terry Nichols during basic training. While serving with Nichols in an infantry division at Fort Riley, Kansas, McVeigh discovered that he and Nichols shared contempt for the U.S. government. After his discharge from the military, McVeigh joined the Ku Klux Klan, which led to his involvement in a militant separatist movement. It was through this involvement that McVeigh was exposed to *The Turner Diaries*, a seminal publication in far-right extremist circles that was written by National Alliance leader William Pierce. McVeigh sympathized with the book's protagonist, who believed that bombing a federal building would precipitate "the white revolution" to overthrow the U.S. government.

Mohamed Osman Mohamud, a 19-year-old Somali living in Portland, Oregon, designed a plan for a spectacular show as he looked to target a Christmas tree lighting ceremony in downtown Portland. Mohamud had the desire but lacked the capability and looked for help from others, who happened to be undercover agents. Mohamed, who said, "I want whoever is attending that event to leave, to leave dead or injured." He never got the chance as his lack of knowledge and need for outside help led to his arrest.

The 1993 World Trade Center Attack and the 1995 Bojinka Airline Plot

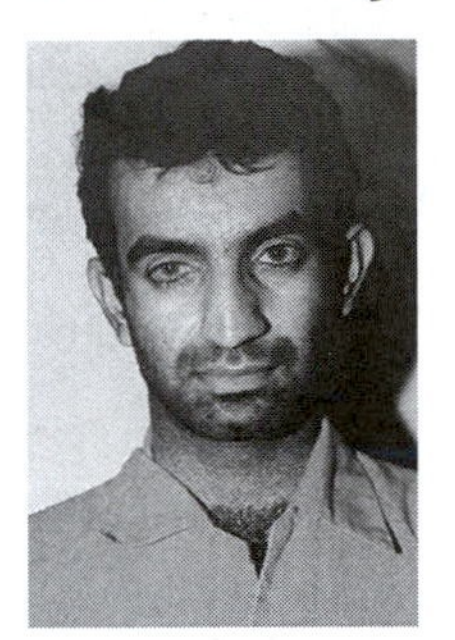

The planner of the 1993 World Trade Center attacks was Ramzi Yousef (see photo), who successfully detonated a vehicle-borne improvised explosive device (VBIED) in the North Tower's underground parking garage, using roughly 1200 pounds of nitrate-based explosives in an attempt to topple one tower into the other. Although the attack generated an explosion that caused significant damage and casualties, it failed to bring down the towers. In turn, the 9/11 plot stemmed in part from the strategic failure of the 1993 World Trade Center bombing. When conceptualizing the attacks that would take place on September 11, 2001, a select group of high-ranking Al-Qaeda officials wanted to overcome this past failure. When the brainstorming for the 9/11 "planes operation" began in 1999, the worldwide operations of Al-Qaeda were commanded and controlled by a central leadership headed by Osama Bin Laden. There were also subordinate commanders, including Khaled Sheikh Mohammed (KSM), the plot's operational mastermind. KSM, the uncle of 1993 World Trade Center mastermind Ramzi Yousef, first started discussing the use of planes to attack the United States in 1994 in the Philippines. He and his nephew conceived the Bojinka plot. The Bojinka plot was scheduled for delivery in early 1995 and aimed to bomb 12 American aircraft within a two-day span. However, Yousef was arrested in Pakistan in January 1995, and KSM continued on his own. After years of travel and a continued interest in planning a terrorist attack against the United States, KSM connected with bin Laden in Tora Bora. KSM took this opportunity to brief bin Laden on the first World Trade Center bombing and the Bojinka plot and proposed his idea to train pilots who could hijack and crash planes into buildings in the United States. Bin Laden agreed and ordered KSM to start planning the 9/11 operation sometime in late 1998 or early 1999.

Operation Crevice

Operation Crevice, British law enforcement's successful arrest of an Al-Qaeda–linked cell in the United Kingdom, reveals how an operation can be exposed in the planning stages. The Crevice operatives wanted to attack the United Kingdom because it was a strong supporter of the United States and had not yet been hit by jihadis.[3] The objective of the mission was not only to bring the jihad to the United Kingdom but to unite international jihadist planners with local individuals to execute the mission. Omar Khayyam, the cell's leader, was raised in the town of Crawley, south of London, in a relatively secular household. In 1998 he joined the radical Islamic group Al Muhajiroun. Khayyam was first noticed by law enforcement after he defrauded bankers and builders' merchants upward of £13,000, which he sent to the jihadists in Pakistan. By February 2004, Khayyam was under surveillance. The surveillance of Khayyam allowed MI5 personnel, United Kingdom's counterintelligence and security agency, to record 3500 hours of audio conversations that detailed a variety of potential attack scenarios.[4] Surveillance also allowed authorities to capture a web chat between Khayyam and Salahuddin Amin discussing fertilizer, an ingredient in a homemade explosive device, which tipped them off that the plot was underway. By the end of March 2004, Khayyam had booked a flight with his brother Shujah Mahmood to Pakistan for April 6 and withdrew more than £10,000 from his bank account. Also, the manager of the space Khayyam and his men were using to store the fertilizer had recently attended a briefing given by Scotland Yard detailing how self-storage units had been used in the past. The manager provided the names of Khayyam and Anthony Garcia, who were already under surveillance for terrorist financing.[5] Surveillance was enhanced and security cameras were set up both outside the storage unit and outside the Langely Green Mosque in Crawley. Recording devices were placed in the homes and cars of many of the operatives. On March 30, 750 officers brought Operation Crevice to an end by raiding 28 locations throughout England before an attack could be executed.

The 2004 Madrid Train Bombings

Moroccan Amer el-Azizi, one of the masterminds of the 2004 Madrid attacks, worked with seasoned operatives such as Tunisian Sarhane Ben Abdeljamid Fakhet, Allekema Lamari, Egyptian Rabei Osman Sayed Ahmed, and Jamal Zougam, among others, to carry out the attacks. This cell is of interest in that the cell had a local infrastructure in place to execute the bombings locally, targeting the rail system as it was a huge soft target. The timing was possibly chosen to influence the general elections, although that is disputed by the fact that other devices were found in the safe house to be used well after the elections were decided.

Target Selection

Targets of potential operations are chosen based on the objective, as well as taking into account the iconic nature of a target, the critical nature of the infrastructure targeted, and, of course, a target that will ensure mass casualties. John Robb has written extensively about terrorist planners targeting the key critical node in the infrastructure rather than the entire network. Attacking these nodes will lead to a cascading effect of the entire network. He writes, "A nonobvious approach to node failure is to attack the connections radiating from high-load nodes. The result of an attack on the connections between nodes will be the redistribution of the load carried by the damaged connection to the remaining connections. This will result in the failure of a high-load node when the remaining connections fail due to overloading."[6] This is what Al-Qaeda is aiming to do with its attacks: target economic targets in the hopes of cascading the larger network.

THE PLOT AGAINST THE USS *THE SULLIVANS*

Terrorist planners have the advantage of time. The collection of information on the target can span from a short time to years and years. Each plan can be updated and put back on the shelf and used at a later date. The bombers who conducted the attack on the USS *Cole* (see photo) originally attempted to attack the USS *The Sullivans* in Aden Harbor. On initial approach to the USS *The Sullivans*, the explosive-laden skiff was found to be too heavy to move into the harbor and nearly sank. This was noted and updated to become the plan used to attack the USS *Cole*.

Once a target is chosen, preoperational surveillance is conducted in order to develop a target package. The aim of such a crucial activity within the terrorist tradecraft cycle is to learn as much about a target as possible, particularly in terms of its strengths, weaknesses, and surroundings.

Would-be Times Square bomber Faizal Shazad monitored pedestrian traffic in Times Square remotely via live streaming video sites with cameras focused on the area. By conducting his surveillance, he was able to determine the best location to place his VBIED. Shazad illustrated the innovations in terrorist preoperational activities.

In the past, the terrorist needed to physically see the target through actual presence. With the proliferation of computer applications such as Google Earth and streaming webcams, surveillance can be conducted from one's own home computer. Maintaining "eyes on the target via technology" also lessens the risk that the terrorist operative will be caught.

One can also argue that intelligence collection emphasizes an institutional knowledge, in which the terrorist looks to understand how operations were conducted in the past and looking backward to see weaknesses and vulnerabilities in plans that are going forward.

One of the most detailed and eye-awakening acts of preoperational surveillance was conducted by Dhiren Barot aka Issa al-Hindi (see photo) against major business and financial targets along the East Coast in the summer of 2000. al-Hindi compiled detailed surveillance notes. He was able to do so because he was the epitome of the "Gray Man," someone who fails to stand out, who seems bland, almost boring, causing no thoughts or suspicions. He blended into crowds with ease and, as his actions showed, was difficult to spot. Not an engineer or an architect, al-Hindi, through his detailed research, was able to find the critical nodes of each building and considered a number of different ways to attack his targets. Below are a few examples of the notes he took regarding some of his targets.

CITIGROUP CENTER: NEW YORK

"These four giant external columns are nine stories high, raising the building some eight stories above ground level. Together with the concrete inner core, they uphold a 26.6 foot/8 meter deep trust platform. For your information, these trusts have been the subject of criticism amongst architects and building safety specialists who have coined a saying which states, 'don't trust the trust.'"[7]

THE NEW YORK STOCK EXCHANGE

Inside the New York Stock Exchange, al-Hindi noticed a number of cameras that led him to write, "There are round, tinted opaque (black) glass ones [cameras] thus allowing freedom of rotation without public knowledge of which direction they are turning."[7] al-Hindi was very concerned about these cameras as he thought every camera was a working camera; which makes the point that any type of deterrent may stave of a future attack.

PRUDENTIAL PLAZA, NEWARK, NEW JERSEY

It was the Prudential Plaza area that al-Hindi wished to target with his Gas Limo Project. The plot consisted of loading a black car—the type used to drive company executives—with gas cylinders, poison gas, and explosives. al-Hindi believed that the black car would blend in with the others. Also, by using a black car, it cut down on the awareness of the security force, as they were under the impression it was just another car for another executive. He wrote, "The most obvious technique to utilize...if you do not mind history repeating itself, would be a limousine in the V.I.P. underground car park with all except the front seats removed.... Moreover, an added benefit of utilizing this method is that the underground V.I.P. car park is situated directly underneath the center of main offices that rise up."[7]

At the conclusion of his 37-page report on the Prudential building and plaza alone, he summed up what was the overarching theme of his surveillance on all the targets. He wrote, "I have not left any stone unturned in my investigation, and have spent countless weeks, days, hours, and months poring over literature as well as exploration in order to finish this report/presentation to the highest possible standard."[7]

Rehearsals

The last aspect of intelligence collection is conducting rehearsals to familiarize oneself with the target area and also the mode of the attack. Terrorist agents will conduct a "dry run" or rehearsal of planned actions at the attack location in order to see if what was laid out on paper is deliverable in the real world.

Before conducting his Bojinka plot, Ramzi Yousef boarded a Philippine Airlines flight in Manila, assembled his device after successfully boarding the plane, and deplaned the aircraft during a layover, leaving the device under his seat. It detonated during the next leg of the flight, while the plane was in-transit, killing the Japanese national who occupied Yousef's seat and injuring several others.[8] On June 28, 2005, at least three of the 7/7 bombers (see photo)—two of whom were wearing backpacks—carried out their dry run of the London Underground network. This allowed them to test the vigilance of security personnel and civilians at their initial embarkation point

at the Luton railway station, as well as within the Underground system in central London.[9] It should be noted that one of the members of the cell that did not partake in the dry run, Hasib Husain, became confused on the day of the bombings. Closed-circuit television (CCTV) footage showed him repeatedly calling on his cell phone in the moments after his cohorts detonated their devices. Husain was waiting on a platform, not realizing that the attacks had shut down the train system. CCTV cameras followed him up the stairs to the street, where he is seen pacing back and forth. He happened to see the Tavistock Square bus pull up near him, which he boarded and soon after detonated his device.

Training

After planning and surveillance, the cell or cell members may undergo training. In the post-9/11 era, there have been numerous case studies involving military-trained operatives participating in terrorist activities. In 2005, authorities warned of the threat posed by local extremists infiltrating the Spanish military garrison located in the North African enclave of Ceuta, citing specific instances in which extremists sought to recruit soldiers.[10] There have also been several case studies of those with military backgrounds leading terrorist training camps, showcasing drills and tactics. Training can consist of formal military training, such as that received by Timothy McVeigh, training camp training such as that received by the 9/11 hijackers, and informal training.

The official report investigating the July 7, 2005, London bombings said:

> Camping, canoeing, white-water rafting, paintballing and other outward bound type of activities are of particular interest because they appear common factors for the 7 July bombers and other cells disrupted previously and since.... It is worth noting that for some extremist activities—e.g., fighting overseas—physical fitness and resilience are essential. They may also be used to help with bonding between members of cells already established, or more direct indoctrination or operational training and planning.[11]

INSTANCES OF TRAINING

- In 2009, Faisal Shazad left the United States for Pakistan with the intent of training with the Tehrik-e-Taliban Pakistan (TTP). He returned to the United States in February 2010. While in Pakistan, Shahzad underwent practical training as well as explosives training in Waziristan.

- Members of the 7/21 cell conducted training in the hills in Cumbria in the northwest of England. One British police officer conducting surveillance on the group said the men were conducting a military-like maneuver.
- In 1989, members of the World Trade Center cell went shooting a number of times at a Calverton, New York, shooting range.

Financing

According to one Al-Qaeda operative, "There are two things a brother must always have for jihad, the self and money."[12] It is undeniable that terrorism costs money, but funding is needed not only to wage jihad in the form of an attack. Terrorist organizations need money both to achieve their goals and run their organizations. Money is needed for every step in the tradecraft cycle, from recruiting members and feeding those recruits to acquiring weapons and transportation for those carrying out the attacks.

There are no strict rules or formulas for how terrorist organizations raise funds. Funding sources are often diversified, and come from a variety of means, even within one organization. Similarly, groups can finance their efforts from legitimate sources as well as illegitimate sources. The 1993 Landmarks Plot was financed primarily through speaking engagements by Omar Abdel Rahman (the Blind Sheikh) and Siddig Ali, which were held at local mosques and even at one public school in Brooklyn, New York. Those who attended his speaking engagements were told that the money raised was to go to Islamic charity groups and organizations, along with the families of soldiers slain in the "holy war" in Sudan. While some of the money did go to local mosques and charities, the rest would go to the terrorist cell.[13] Before 9/11, revenue streams for terrorist organizations primarily came from large-scale, top-down sources such as state sponsors, financial donors, and large charities. After the attacks on 9/11, the United States government responded with an aggressive and targeted response to attack the money trail of terrorist organizations, which has forced cells to finance their own operations.

The total cost of the 2004 Madrid attack is unknown, but the attack was financed through sales of illegal drugs. The cell financier, Jamal Ahmidan, was a known drug trafficker. He continually traded hashish and ecstasy in exchange for explosives in deals with known criminals. Abu Mezer and Lafi Khali, the two individuals who attempted to attack New York City subways in 1997, supported themselves by working full-time jobs at various delis and odd construction jobs. They eventually left New York for Greenville, North Carolina, where they worked in a supermarket owned by a relative of a connection they made in New York.

Faisal Shazad (see photo) was given $5000 to fund his attack by the TTP while he was in Pakistan. He also sued the hawala system when he returned to the United States to continue his funding streams. He received $12,000 from two individuals at the behest of the TTP.

Terrorist groups are increasingly turning to trafficking of drugs and tobacco products as a method of raising money. The drug trade is highly lucrative and is estimated to be a $400-billion-per-year industry.[14] According to the Drug Enforcement Agency, 19 of the 43 designated foreign terrorist organizations are linked definitively to the global drug trade, and up to 60% are connected in some way to the illegal narcotics trade.[15]

In 2002, Mohamad Hammoud became the first defendant convicted of providing material support and resources to a designated foreign terrorist organization, namely, Hizballah, a Muslim militant group based out of Lebanon.[16] For a period of six years, Hammoud and his associates profited from the trafficking of contraband cigarettes purchased in North Carolina, where tax rates on cigarettes were low, to Michigan, where tax rates were much higher. The profits procured from this activity were used to finance Hizballah directly, as well as to purchase dual-use military equipment that was sent to the terrorist organization. Each week, the cell would purchase approximately $13,000 worth of cigarettes from a discount tobacco wholesaler in North Carolina and transport them to Michigan for sale. At the time of the operation, "North Carolina taxed a carton of cigarettes at 50 cents, and no tax stamp was required on the pack or carton to show the permitted state of possession or resale. Michigan, on the other hand, taxed cigarettes at a rate of $7.50 per carton and no tax stamp was required."[17] Although it is not known exactly how much money was sent to Hizballah, it was estimated that the group had generated $8 million in revenue at the time of their arrest.[18] It is believed that Hizballah received more than $100,000.[18]

Terrorist organizations need money to maintain their organizations as well as to pursue their operational goals. Since 9/11 and the crackdown on the "money trail" of terrorist organizations, funding sources have trended away from externally generated macrofunding and toward self-generated microfunding. Terrorist cells—especially small, homegrown cells—are increasingly self-funded, using the proceeds of criminal activity, personal funds, or exploitation of government welfare benefits and insurance fraud. Terrorist organizations will innovate and raise funds through all feasible channels, including criminal activity. As one revenue stream is cut off, another will appear.

Acquiring the Weapons

Once the cell is in place, funding is secured, and the targets have been picked, the next logical step is the acquisition of weapons. Terrorists groups have used all types of acquisition methods, whether it is purchasing weapons legally, stealing them, or using the back market and third parties. Abu Eisa al-Hindi, who conducted detailed surveillance of East Coast financial centers for Al-Qaeda in the summer of 2000, advocated that others following in his footsteps "make use of that which is available at your disposal and bend it to suit your needs, improvise, rather than waste valuable time becoming despondent over that which is not within your reach."[19] Al-Qaeda in Iraq (AQI) did not have the same problems the Provisional Irish Republican Army (PIRA) had in acquiring weapons with the large cache of military-grade explosives available to it; yet the PIRA, operating in a more stringent environment did what it could to make use of what was available to it, continuing to improvise to make technically efficient weapons.

AL-QAEDA FI AL-JAZEERA

Forgotten in the wake of groups such as Al-Qaeda in Iraq was the tactical success of the Saudi branch of Al-Qaeda in the Arabian Peninsula (Al-Qaeda fi-al-Jazeera) from 2003 to 2006.

Although Al-Qaeda did not attack Saudi Arabia before 2003, the kingdom was regarded as an important target as it is home to the two holiest sites in Islam as well as huge oil reserves, which became a target for Al-Qaeda senior leadership's economic jihad. For three years, the group conducted a number of complex attacks, each aimed at finding the vulnerabilities in the Saudi defenses surrounding economic targets.

May 2003: On May 12, 2003, three foreign worker targets were attacked in Riyadh using small arms and explosives. Thirty-five were killed and 200 wounded, the majority of which were Muslims. A similar attack was conducted against another compound in Riyadh killing 18, mostly Muslims.

May 2004: In an attempt to reduce the number of Muslims killed, the group targeted the offices of ABB Lumus, an oil company in Yanbu. Attacking in the predawn hours, before Muslim workers arrived, cell members wearing Saudi Coast Guard uniforms killed four foreigners, including two Americans. Separate attacks targeted other office parks, killing 22 people, mostly Saudis.

February 2006: Saudi Al-Qaeda attacked the giant oil processing plant at Abqaiq with multiple VBIEDs and small units. Two VBIEDs in cars with Aramco logos penetrated the gate near the main gate. The vehicle detonated its charge, allowing the second VBIED to plow through. Security forces stopped the second vehicle. In the aftermath of the attack, the leaders of Saudi Al-Qaeda were hunted down and subsequently killed by security forces.

Fourth Generation Warfare

One of the number of ways that the terrorists have grasped the concept of fourth-generation warfare is their use of materials freely available in modern society to destroy that society. The terrorists do not have a need to defend bases, but they can rather be focused on offensive measures. It "also relieves him of the logistics burden of moving supplies long distances. Instead, he has to move only money and ideas, both of which can be digitized and moved instantly."[20] Acquisition methods reflect this fourth-generation mindset, exploiting society's openings and using what is available to decrease the chances of detection and also the cost. However, like all things involved in this type of operating environment, if and when the defenders put any type of countermeasures in place, life becomes more difficult for the attacker.

As mentioned earlier, the terrorist is going to use whatever methods that will allow him to gain access to the weapons he so desires. Niddal Ayyad, one of the members of the cell that conducted the attack against the World Trade Center in 1993, worked at Allied Signal. He used his position there to acquire restricted chemicals, such as lead nitrate, phenol, and methylamine, for the attack, along with an order of hydrogen tanks from ALG Welding Company.[21] He acquired the materials in February 1993, at times faxing purchase orders on the company letterhead. ALG delivered the final hydrogen shipment to the storage facility the day before the event.[22]

In 1997, Abu Mezer and Lafi Khalil, a Palestinian and Jordanian, respectively, were arrested as they prepared to detonate pipe bombs at the Atlantic Avenue subway station, a major transportation hub in downtown Brooklyn. The two had made their way to North Carolina and worked in the back of a supermarket owned by a contact of theirs in New York. It was there that they gathered the ingredients for their device of choice: pipe bombs. These ingredients were all obtained legally at a hardware store and a gun shop at Greenville, North Carolina, on July 10, 1997.[1] The gunpowder was purchased on the same date. Khalil and Mezer constructed the bombs in the trailer home they were renting in Ayden, North Carolina.[1] Mezer and Khalil used an open environment in 1997 New York to put their operation together, but it was thwarted by the NYPD.

When one thinks of weapons acquisition, thoughts immediately turn to the acquisition of larger weapons, such as nuclear devices or chemical weapons, yet in recent years, terrorists—specifically, those affiliated with Al-Qaeda—have effectively used simple homemade bombs built from everyday things that span the gamut from nail polish remover to fertilizer and black pepper, all used to concoct new innovations in lethality.

Purchasing Components

The best-case scenario is to legally purchase the components—meaning mostly explosive precursors—taking advantage of either a lax security environment or gaps in various countermeasures. Terrorists have used these holes to build homemade explosives, perhaps not as powerful as a VBIED in terms of sheer power, but as the 7/7 London bombings has illustrated, just as deadly. Homemade explosives are inexpensive to buy and difficult to detect.

Times Square bomber Faizal Shazad purchased the components for his explosive device from February to April 2010. He purchased the fertilizer and propane from a store in Connecticut and bought approximately $195 worth of fireworks from a Pennsylvania fireworks store.

The components used to destroy the Murrah Federal Building in Oklahoma City were acquired through both legal and illegal means. On September 30, 1994, Mike Havens entered the Mid-Kansas Cooperative Association in McPherson, Kansas, and purchased 2000 pounds of ammonium nitrate. A few weeks later, on October 18, 1994, a man calling himself Mike Havens returned to the Co-Op and purchased another 2000 pounds of ammonium nitrate. The Co-Op employees knew all their customers and did not recognize Havens, who said he was going to use the ammonium nitrate to sow his wheat fields, a practice that had once been popular among farmers in the community. In between the two purchases, a burglary at the Martin Marietta rock quarry just outside of Marion County, Kansas, was reported. One of the explosive magazines at the quarry had been burglarized.

The thieves stole 229 sticks of Tovex, aluminized water–gel dynamite, 93 rolls of Primadet nonelectric blasting caps, and 544 electric blasting caps.[23] Around the same time, a man calling himself Shawn Rivers rented a storage facility at the Herington Industrial Park in Herington, Kansas. A man calling himself Joe Kyle rented a storage unit in Council Grove, Kansas, on October 17, just a day before Mike Havens made his second purchase. On October 21, 1994, a man purchased three 55-gallon barrels of nitromethane at $925 a barrel at a race track in Ennis, Texas, from Tim Chambers of VP Racing Fuels. The purchaser told Chambers that he used nitromethane to fuel the motorcycles he raced in Oklahoma City. The buyer paid in cash. The investigation that took place after April 19, 1995, revealed that Mike Havens was Terry Nichols, and Shawn Rivers was actually Timothy McVeigh, who was also the unidentified buyer of the nitromethane in Ennis, Texas.[24]

In the wake of the Oklahoma City bombing, legislation was enacted to make buying large amounts of ammonium nitrate more difficult, and it has become more difficult to purchase large amounts of this material without arousing suspicion. Al-Qaeda looked to use fertilizer-based bombs in Operation Crevice in 2003. A cell member bought 1200 pounds of Kemira GrowHow from a gardening store in Great Britain. An employee at the store thought it was odd that the suspect was purchasing that much fertilizer off-season, and the cell was compromised when an employee at a storage facility tipped off the police. Both Oklahoma City and

Crevice brought renewed attention to this precursor. When Joel Hinrichs, who blew himself up outside a football stadium on the campus of Oklahoma University in 2005, tried to purchase ammonium nitrate, he was denied. In light of these countermeasures, Al-Qaeda has simply switched to liquid explosives. The 2006 Trans-Atlantic bombers were looking to use nexamethylene triperoxide diamine, which is produced by mixing hydrogen peroxide and hexamine with citrus acid.[25]

Theft

An interesting study for future researchers would be to find out how many weapons and explosives stolen across the United States each year are used in terrorist attacks.

- In 2008, 31,000 pounds of ammonium nitrate were stolen from various locations across the country.
- From January 1998 through December 2001, 27,000 pounds of high explosives were stolen from construction sites, mining quarries, and demolition companies.[26]
- In December 2005, 400 pounds of high-powered plastic explosives was stolen from Cherry Engineering, a company owned by a scientist used by Sandia National Laboratories. The stolen items included 150 pounds of C-4, 250 pounds of thin sheet explosives, and 2500 detonators.[27]

One of the most famous examples of weapons theft by terrorists revolved around the Goma-2 dynamite that was provided to the 3/11 Madrid bombers. Spaniard José Emilio Suárez Trashorras, who stole the Goma-2 dynamite from the Asturias coal mine where he worked, supplied the explosives. He was introduced to the groups through an intermediary he met in jail.[1]

State Sponsorship/Third Party Support

When one thinks of state sponsorship of terrorism, thoughts immediately turn to countries such as Iran and Syria, two countries, along with Cuba and Sudan, that populate the U.S. State Department's list of state sponsors of terrorism. In the 1970s through the 1990s, the list was longer, as Soviet Bloc countries supplied and supported terrorists such as Carlos the Jackal; Libya was a major sponsor of state terrorism as well supporting groups from the Abu Nidal group to the PIRA. The use of proxies to conduct attacks gives the state sponsor plausible deniability for responsibility for the attack.

Libyan support of the PIRA proved to be a major boost for PIRA operations in the 1980s. Now, in the wake of the release of Pan Am bomber, Abdelbaset al-Megrahi, a number of victims groups in the United Kingdom are seeking reparations from Libya for supplying weapons to the PIRA. Libyan support of the PIRA took place between two time periods that spanned between 1972 and 1975 and again in the 1980s. During the first period, it is believed that the Qaddafi government supplied over $10 million to the PIRA.[28] The Libyan shipments stopped in

1975 after the government became uneasy about the dealings with the PIRA. The shipments in the late 1980s exceeded those of the previous decade and were more effective in that these shipments gave the PIRA its "magic marble," Semtex. The importance of Semtex to PIRA arsenals cannot be forgotten.

The PIRA used Semtex as a booster for its giant ammonium nitrate VBIEDs that were used to strike the City of London's financial centers. The IRA and PIRA had received large quantities of weapons and support from the United States, yet the second round of Libyan shipments came at a time when the FBI was cracking down on U.S. support, creating a void in the amassing of its arsenal. Beginning in August 1985, Libya sent four shipments of weapons. The largest shipment was the 150-ton shipment on board the trawler *Eksund*, which was tracked by the Royal Navy and French customs. Authorities found two tons of Semtex, 1000 mortars, 1000 Romanian-made AK-47s, 600 Soviet F-1 grenades, 120 RPG-7s, 20 surface-to-air missiles (SAM-7s), 10 DShKs, 2000 electric detonators, 4700 fuses, and more than a million rounds of ammunition.[29]

Iran's role as a state sponsor is well known, but it goes beyond Hizbollah. The Islamic Republic has been very active in Iraq, facilitating the flow of weapons to Shiite militias that have then used those weapons to kill American soldiers. This is illustrated in its Ramazan Corps. In the spring of 2003, the Iranian Revolutionary Guards Corps, through its Qods Force, began acting in Iraq. The Qods Force worked to supply the various Shia militias with money, men, and weapons. It was a massive undertaking and forced the Qods Force to find a better method of managing its operations, so it created the Ramazan Corps. The Ramazan Corps divided its Iraqi area of operations into three commands: Nasr, Zafar, and Fajr. The Zafr Command was led by Mahmudi Farhadi, who was captured by U.S. forces and was recently released by the U.S. government. The Zafar command was spread out across the central regions of Iraq and served as the main point of entry for the Iranian-made explosively formed projectiles that have been responsible for the deaths of U.S. and Coalition servicemen.[30]

Building the Bomb

Although terrorists can look to acquire any type of weapon, the focus today is on explosive precursors. Once a group has all the materials it needs, the difficult part is building the device. The 1993 World Trade Center cell was not able to build an operable bomb until the arrival of Ramzi Yousef. Homemade explosives may be makeshift, but they can still be lethal. When the PIRA began utilizing homemade explosives, it faced a steep learning curve as it suffered a number of premature explosions that killed its bomb makers or operatives. Husein Mikdad was a Hizbollah operative recruited by Imad Mugniyah to bomb flights leaving Israel's Ben Gurion airport in 1996. The device Mikdad was constructing out of plastic explosives prematurely went off, blowing off his leg and arm as well as blinding him in the process.[31]

Others were more successful. Oklahoma City bombers McVeigh and Nichols were not bomb makers, but they learned how to make their device from reading

Andrew Macdonald's *The Turner Diaries* and *Hunter*. Macdonald's *Hunter* provided a detailed description of how to mix ammonium nitrate into a bomb:

> Then he bought himself a used Chevrolet pickup. With the pickup he drove to a large feed and fertilizer store on the edge of town and bought 15 bags of fertilizer-grade ammonium nitrate. He would have bought more, but 1,500 pounds was about as much as he estimated he could manage in one load... In a rented garage he removed several five-gallon cans of wallpaper adhesive and dozens of rolls of wallpaper from the back of the van and replaced them with four 40 gallon plastic trash barrels... he spent the next three hours emptying sacks of ammonium nitrate into the barrels and stirring a fuel-oil sensitizer into the white pellets.[32]

In May 1993, McVeigh ordered and received Ragnar Benson's *Homemade C-4*, which detailed how to mix ammonium nitrate with nitromethane to build a powerful bomb.[33]

Another point to mention in regard to building bombs is where to build them. Ahmed Ressam trained in Al-Qaeda's Khalden Camp in Afghanistan, where he received training in explosives construction. He was also tasked by Al-Qaeda senior leadership to launch a VBIED attack against Los Angeles International Airport in 2000. Ressam stated that when he left the training camps and returned to Canada, he brought with him hexamine and glycol as precursors to his bomb. He also brought a notebook of bomb-making instructions.[34] In Canada, he was visited by the major Al-Qaeda recruiter Mohamedou Ould Slahi, who gave him further instructions.[35] Ressam completed his bomb by stealing nitric and sulfuric acid from various fertilizer stores.

The capability to make explosives is greater today because of the large number of veteran jihadists who continue to operate. Experience gained in various theaters of the global jihad make these veterans indispensable to terror cells, especially those that are composed of individuals who did not train in the camps. However, while the experience of these veterans is important, it is not an absolute necessity. The role of the Internet in regard to online bomb making has been a new research topic. McVeigh was able to build his lethal bomb by reading a few books in conjunction with his military experience. Another interesting question to ask is, how many of today's current bomb makers who have built successful devices have trained in the terrorists' camps versus those who took the route of McVeigh?

CBRN

Although much of the focus of this chapter has been on the acquisition of conventional weapons, the realm of unconventional weapons cannot be dismissed. Terrorists have increasingly shown an interest in CBRN weapons. In 2002, an Al-Qaeda tape was found in Afghanistan showing militants killing dogs with what appeared to be nerve gas; in 2003, a Saudi cleric affiliated with Al-Qaeda issued a fatwa justifying the deployment of weapons of mass destruction against those he called "enemies of Islam;"[36] in 2008, a 39-minute Al-Qaeda video called for the use of CBRN materials against the West; and in February 2009, an Al-Qaeda video discusses the potential use of anthrax against the United States.

In 2004, Mustafa Setmarian Nasar, also known as Abu Musab al-Suri, published a 1600-page treatise titled "The Call for Global Islamic Resistance," which attempted to lay the groundwork for the generation of Al-Qaeda members. al-Suri's thoughts follow the lines laid out by the thinkers of fourth-generation warfare, discussed earlier, that there is a "decreasing dependence on centralized logistics," that the onus on supplying the materials needed for an operation is on the respective cell members. al-Suri's slogan is *nizam, la tanzim* or "system, not organization."[37] al-Suri's thoughts on the importance of Al-Qaeda central leadership aside, the system is already in place. Using the fourth-generation model, terrorists are taking it upon themselves to acquire their own weapons and build them in place without having to rely on a central hub not only for materials but also for the construction of the device, thus lowering the risk factor.

The July 2009 Jakarta hotel bombers reportedly used a florist, Ibrohim Muhharm, to smuggle in the bomb components. Using his position as an employee, Muhharm allegedly smuggled the components through the hotels loading dock in packages addressed to the florist. The bombs were later constructed in a room inside the hotel.[38]

As with all other elements of tradecraft, the terrorist is looking to exploit vulnerabilities in the defense, whether it is in deriving financing, acquiring weapons, or conducting the actual attack. Due diligence on the part of the defenders is a given, but so is the effort to build a public–private partnership. The tip that led to the arrests of the 2006 airline plotters was started by a phone call made by neighbors living in the same complex where the safe house was located.[39] Rick Schlender and Jerry Showalter were the employees on duty the two days that Terry Nichols bought the ammonium nitrate from the Mid-Kansas Co-Op. Although the two men did nothing wrong, they suffered anguish at the thought that they unknowingly supplied the key ingredient for the bomb. Educating and training not only law enforcement but also the multitude of public–private partners, hardening soft targets such as explosive storage sheds, and enacting legislation that will make it more difficult to acquire specific precursors will go a long way in thwarting this fourth generation of warfare—by making the key decisions needed to counter the chaos of the current threat environment.

Modes of Attack

One of the requirements for an effective counterterrorism strategy is an understanding of the enemy's tactics. The diffuse nature of terrorist groups, especially in how they plan their attacks, makes this requirement a great challenge. In looking at terrorist tactics, it is a dance of movement by the terrorist and countermove by the counterterrorism forces, or vice versa.

The great strategist Sun Tzu discussed the concepts of "cheng," to seek out and expose the vulnerability of the opposition, and "chi," the exploitation of those vulnerabilities by a decisive strike. This concept was built upon centuries later by the German General Staff with the development of "blitzkrieg." According to the Germans, the focal point of blitzkrieg was the concept "Schwerpunkt," which is the modern form of "chi," with the importance of getting inside the decision-making loop of the defender."[40] Expanding on this, General Gunther Blumentritt said, "The entire operational and tactical leadership method hinged on a rapid concise assessment of situations... quick decision and quick execution on the principle: 'each minute ahead of the enemy is an advantage'."[40]

Assuming the importance of the lightning strike, one would think that the enemy constantly invents new tactics with an eye toward the spectacular. In actuality, the enemy is a learning organization that constantly refines old tactics. By refining old tactics and adding a new twist, the enemy can gain not only tactical advantage but also psychological surprise, as illustrated in the Mumbai attacks, Umar Farouk Abdulmutallab, and others. The counterterror forces, or successful ones, will find ways to counter these new methods, as part of a continuous cycle between each side, with each side looking for an opening to exploit.

Factors and Constraints

It can be argued that since the defender has to defend as much as possible, the terrorist has the initial advantage, but as time goes on, the terrorist will more often than not reach back in time to use new twists on old tactics that may take the defender by surprise, with the 9/11 attacks being one such example. There may also be instances when the terrorist faces factors or constraints dissuading them from using one tactic over another. It was long rumored that the PIRA had amassed a large arsenal of SAMs but never used them. The decision not to use them was not to give the British forces an excuse for escalation of hostilities. Even though these weapons were never used, they did force the British to change their defensive measures in Britain and Northern Ireland.

The unavailability of weapons was compensated by an impressive amount of ingenuity. Al-Qaeda in Iraq (AQI) had the luxury of using stockpiles of military-grade weapons, but the PIRA was instrumental in building its own homemade explosives long before Al-Qaeda even existed. The PIRA moved from using military-grade weapons to building homemade devices by the early 1970s. British authorities were absolutely certain that the group was stealing weapons from British military bases, but the PIRA was using commercial explosives that were acquired under the auspices of the 1875 Explosives Act, which allowed the import of commercial explosives, such as gelignite, into Northern Ireland, shipped from English manufacturers. It was using ammonium nitrate-based bombs by 1972, bombs that were made even more lethal with experimentation in regards to dual switches and the acquisition of Semtex.

Terrorists also have to factor in security measures as seen in Iyman Faris' aborted attempt to attack the Brooklyn Bridge as a result of security being "too hot." In other cases, the symbolism of a possible target may result in the terrorist losing possible empathy or engendering a massive retaliation against them.

In 1994, the Landmarks Plot targeted multiple targets in New York City that included the United Nations, 26 Federal Plaza, the New York home of the FBI, and others. When cell leader Siddiq Ali asked the Blind Sheikh, Omar Abdel Rahman, for permission to target the UN, the Blind Sheikh replied, "It is not forbidden but it will put Muslims in a bad light. The UN is not a force of pressure. It will hurt Islam before the UN. Think of something else because the UN is considered to be a presence for peace. People will say that Muslims are against peace."

When told of plans to attack 26 Federal Plaza and asked for a blessing, Abdel Rahman told Siddig, "It doesn't matter. Slow down. Slow down a little bit. The one who killed Kennedy was trained for three years."

Source: Laurie Mylroie, "The War Against America," Regan Books, New York, 2001, p. 189.

Sometimes terrorist groups do not consider the effects of their strikes carefully enough. At the beginning of the 1970s the PIRA conducted a number of attacks in Northern Ireland aimed at Nationalists and British forces in an attempt to make Northern Ireland ungovernable and force the British to leave the territory. On July 21, 1972, in what was to be known as "Bloody Friday," the PIRA detonated 22 bombs within two and a half hours in a small area of Belfast. Two of the bombs killed two people and wounded 130 others. Many of the victims were wounded by multiple devices as the kill zone was a small area.[41] Although the PIRA phoned ahead of the attack, many of the warnings did not reach the proper authorities in time and the group miscalculated the impact of the operation.

In the wake of the Bloody Friday attacks in Northern Ireland, the PIRA began to target economic targets on the British mainland. In the early 1990s, as the British economy was weakening, the PIRA devised large VBIEDs as a way to leverage its position in negotiations.

April 1992: The PIRA attacked the Baltic Exchange with a VBIED composed of ammonium nitrate mixed with 99 pounds of Semtex and destroyed the target, the leading shipping market in London. The bomb killed three people and caused over $1.3 billion in damages in current U.S. dollars.

April 1993: The Bishopsgate attack saw the PIRA detonate a 1-ton ammonium nitrate device against a London financial target, which was the largest and most costly attack against a City of London target, resulting in roughly $2.3 billion dollars in damage. The device was primed by switching on the hazard lights of the vehicle the VBIED was housed in.

February 1996: The PIRA VBIEds become even more powerful as a 3000-pound ammonium nitrate device, mixed with icing sugar packed around booster tubes mixed with SEMTEX and packed with detcord made of PETN and RDX, detonated. The blast resulted in $2 billion in damages. What is interesting is that this bomb, although larger than others, caused less damage.

June 1996: The infamous Manchester bomb, which consisted of 3500 pounds of explosives, was placed inside a truck parked in the center of Manchester. The blast was enormous, causing over $800 million in damages and releasing a mushroom cloud that rose more than 1000 feet in the sky.

Preponderant Forms of Attacks

One of the few limiting factors facing a terrorist in regard to modes of attack is simply human imagination. Throughout the course of its history, the IRA and later the PIRA continued to improvise, using its own research and development capabilities to find new ways to kill—ranging from sniper rifles to the favored weapon of explosives. The IRA created such weapons as letter bombs—small devices with gelignite—destroyers that used ammonium nitrate with a Semtex booster.

Long before Al-Qaeda used VBIEDs, the PIRA was on the way to making VBIEDs as a lethal tactic. This practice not only allowed the PIRA to use new methods but allowed it to store large amounts of explosives in vehicles, making it easier to transport explosive devices. No longer did the group have to rely on dynamite but could now build larger bombs using ammonium nitrate and fuel oil and Semtex as a booster charge. Said one IRA man about the VBIED, "We could feel the rattle where we stood. Then we knew we were on to something, and it took off from there."

In his memoir, *Seven Pillars of Wisdom* (1922), Lawrence revealed his most effective tactic: "Mines were the best weapon yet discovered to make the regular working of their trains costly and uncertain for our Turkish enemy." Today's terrorists are not using mines but are using improvised explosive devices (IEDs) to attack everything from Humvees to trains to other pieces of infrastructure. In 2004, Chechen terrorists used the renovation of Grozny's Dynamo Stadium as an opportunity to plant a massive IED under the reviewing stand where then Chechen President Akhmad Kadyrov was seated during a Victory Day parade. The device killed Kadyrov and five others.

In Iraq, AQI and other Sunni groups used networks to build, place, and detonate the IEDs. The IED threat in Iraq was countered by up-armoring vehicles and using jamming devices, along with arresting members of the various IED networks. Because the Taliban in Afghanistan has learned about the countermeasures used in Iraq, it has littered Afghanistan's roads with IEDs. This is clearly an obstacle for U.S. and Coalition forces, yet it will be difficult for the enemy to keep up that pace as it is expensive and, in order to thwart the countermeasures, the Taliban will have to place more and more of these devices at great expense. However, the IEDs being used going forward, while perhaps not at the same rate, will be more powerful in nature.

In the United States, Eric Robert Rudolph developed IEDs that he used in the fatal bombing at Centennial Olympic Park on July 27, 1996, which killed Olympic spectator Alice Hawthorne and seriously injured more than 100 people; the bombing of a Sandy Springs, Georgia, family planning clinic on January 16, 1997, which injured more than 50 people; and the bombing of a Midtown Atlanta nightclub, the Otherside Lounge, on February 21, 1997, which injured five people. Rudolph used secondary devices to target responders, including the secondary device used at the Sandy Springs, Georgia, abortion clinic in 1997. Rudolph planted a secondary device aimed at first responders when it exploded an hour after the first and wounded seven people. He also used a similar device at the Otherside Lounge in Atlanta in 1997 but that device was discovered and rendered safe by police.

Although the IED does not have the cache of the VBIED in terms of flashiness, it is still very effective especially when using remote detonation. The technology is easy to obtain—anything from garage door openers to toy car remote controls can be easily disguised as something innocent if discovered—and for nonsuicide operations it allows the terrorist to be at a safe distance from the pursuer. In Iraq, IEDs were not only hidden in the ground but strung from poles and other locations off the ground, which increases the size of the killing zone. Even in suicide operations, the remote detonator can be used to successfully detonate a squeamish bomber, a tactic the Chechens have used to deadly effect.

The IED, unlike a VBIED, can be hidden under a number of different guises from backpacks to garbage debris. In 1982, the Fuerzas Armadas de Liberación Nacional planted three bombs in lower Manhattan. One bomb was secreted in a

Kentucky Fried Chicken box that severely injured NYPD officer Rocco Pascarella outside of One Police Plaza by blowing off his left leg. That same night, other bombs detonated at 26 Federal Plaza and the Metropolitan Correctional Facility, injuring Detectives Salvatore Pastorella and Anthony Senft from the NYPD Bomb Squad.

Small Unit Tactics

The events that began on November 26, 2008, in Mumbai were heralded by much of the media along with many analysts as the advent of a new form of terrorism. The West was taken aback by the use of small units of heavily armed individuals conducting sieges in urban areas. Small units in urban areas had already caused bloodshed long before Mumbai. In 1972, members of the Japanese Red Army killed 26 people at Israel's Lod Airport. On December 27, 1985, the Abu Nidal Group conducted simultaneous attacks against the El Al and TWA ticket counters inside the Leonardo da Vinci Airport in Rome and passengers waiting to board a flight to Tel Aviv at Schwechat International Airport in Vienna. Upon entering the airports, Abu Nidal gunmen threw hand grenades toward the large crowds and fired toward the ticket counters. In Rome, police and Israeli security guards killed three of the four attackers before capturing the fourth. In Austria, three attackers fled in a stolen car before being captured by police.

In 2007 in Karbala, Iraq, a 12-man team, dressed in U.S. military uniforms, attacked the Karbala Joint Provincial Coordination Center, where U.S. troops were holding a meeting. The attack cell drove past Iraqi checkpoints in a convoy, disguised in their uniforms. The attackers entered the compound, killed one American, wounded three, and captured three others. The three were driven somewhere and killed. The attack was quickly deemed the most sophisticated seen in that theater of operations.[42]

Another variation of the small unit attack is the use of snipers. From October 2 to 22, 2002, two snipers, John Muhammad and Lee Boyd Malvo, shot 13 people, killing 10 people with a Bushmaster XM15 E2S in Maryland, Virginia, and Washington, DC. The DC snipers got their idea from the Provisional IRA's lethal South Armagh Brigade.

The PIRA in South Armagh practiced the tactic of "one shot, one kill" sniping. According to the PIRA's former Chief of Staff, Sean MacStiofain, in his memoir *Memoirs of a Revolutionary*, said, "One-shot sniping was in fact, the theory of the guerilla rifle. It turned up once struck, and vanished, presenting no target in return."

The United States has also seen variations of small unit tactics. One came in 2007 in the form of the Fort Dix, New Jersey, plot. Its execution called for armed assaults against a military installation in the United States. In the summer of 2009, Australian authorities arrested four Australian citizens believed to have trained with Al-Qaeda affiliate, al-Shabaab, for planning an armed assault on the Holsworthy army barracks outside of Sydney. During the late 1960s and throughout the 1970s and into the 1980s, the Black Panther offshoot, the Black Liberation Army, killed a number of police officers across the country including New York City police officers Joseph Piagentini, Waverly Jones, Gregory Foster, and Rocco Laurie.

But what really changed the way we look at the terrorist use of small unit tactics was the November 2008 attacks in Mumbai, India, by the militant group Lashkar-e-toiba. What was so fascinating about Mumbai was that for 60 hours, the assault team brought a city of 20 million people to a standstill. The attackers outgunned the defenders and were trained down to the smallest detail. Unlike other instances, Mumbai was not an event that took place in a containable venue, changing to multiple venues over time. It was a hybrid attack using intermodal means (land and sea) and used small arms and small IEDs. A small armed cell was able to change the battle space not only by using technology as mentioned previously but also by winning the battle of time. The attackers were successful in penetrating the area of vulnerability, which slowed down an already-slowed response on the part of the police. Until challenged by the defenders, the attackers have the luxury to act freely, and this is what the Mumbai attackers managed to do. There were conflicting reports of whether the Mumbai attackers took hostages for the sake of taking hostages, yet in terms of the terrorism espoused by Al-Qaeda, death for the attackers and the hostages is the intended end result.

Abu Eisa al-Hindi thought of multiple attack scenarios during his surveillance of financial centers along the Eastern seaboard of the United States in the summer of 2000 that spanned from simple arson to the Gas Limo Project. The Gas Limo Project called for explosives, gas cylinders, and shrapnel to be placed inside a black car similar to those used by executives at one of the buildings cased, as they drew little security attention from security personnel. According to notes found during the investigation, "Perhaps the best example of how a building can be totally gutted by an inferno (blaze) and more was that of the WTC (World Trade Center)," the project said, referring to the September 11 attacks in 2001. Although al-Hindi was arrested before this plan could come to fruition, an element of this form of attack was seen in the 2007 failed VBIED attack outside a London nightclub.

CBRN

While IEDs and VBIEDs are clearly the weapons of choice for terrorists today, unconventional weapons cannot be discounted. Although terrorists have yet to successfully conduct a mass-casualty CBRN attack, a number of instances illustrate that they have a rudimentary framework for launching such attacks. Since chemical, biological, and radiological materials tend to be inherently dangerous, the key to a successful attack is to perfect the dispersal mode. In many CBRN attacks, terrorist groups have had difficulty in accomplishing this feat, which is why conventional weapons are more dangerous.

Refining the Modes of Attack

As discussed previously, successful terrorist groups are learning organizations that look for lessons learned not only in their own attacks but others as well, all looking for that gold nugget that will bring success. In regards to Al-Qaeda, a key group ideologue Abu Musab al-Suri wrote in his manifesto, *The Call to Global Islamic Resistance*, of the need to learn from past mistakes to help the operational perspective in order to devise a new perspective that he called *nazariyat al-ʿamal*, or "operative theories" for future operations. It also used the *Al Battar* magazine, an Al-Qaeda in-house journal that reiterated the lessons of the training camps for those who were there, and offered a baseline to start for those jihadists who were not.

In 2008, before its switch to Al-Qaeda in the Arabian Peninsula, Al-Qaeda in Yemen created its own magazine, *Sada al-Malahim* (*Echo of Battles*), followed now by *Inspire*, that discussed how to attack oil facilities properly. Prior to the attack on the USS *Cole*, the attack cell targeted the USS *The Sullivans* with an explosives-laden skiff, but the skiff was weighed down by too many pounds of explosives and sank. The plan was refitted and reworked and was successful a few months later. As mentioned earlier, the laboratories of Iraq and Afghanistan as well as Chechnya and elsewhere have given Al-Qaeda and its affiliates room to work with and experiment in new ways of killing. It is not just Al-Qaeda. The Unabomber, Ted Kaczynski, found a way new method to make his bombs more lethal by including ammonium nitrate and aluminum powder.

When Kaczynski admitted to the bomb placed at Northwestern University in 1979, he said, "The bomb used match-heads as an explosive. The bomb was in a cigar box and was arranged to go off when the box was opened. I did it this way instead of mailing the bomb to someone because an unexpected package in the mail might arouse suspicion, especially since a short while before there had been an incident in the news where cops in Alabama had been killed and maimed by a bomb sent them in the mail."

Simultaneity and secondary devices having been used with IEDs, and suicide bombers have also been used with VBIED attacks. Pioneered by the PIRA, Al-Qaeda, adding the suicide element, has elevated this tactic to deadly effect. On August 7, 1998, Al-Qaeda, borrowing from the PIRA's playbook regarding simultaneity, conducted two suicide VBIED attacks against the U.S. Embassy in Nairobi, Kenya, and Dar es Salaam, Tanzania, that killed 224 people and wounded 5000. The Nairobi bomb consisted of close to 2000 pounds of TNT and ammonium nitrate along with aluminum powder. Oxygen and acetylene tanks were included in the payload. The Tanzania bomb was wired in a similar manner. The Bali bombings in October 2002 were notable for its use of an initial explosion designed to drive a large number of victims into a "kill zone," where they would be cut down by a second blast.

AQAP bomb maker Ibrahim al-Asiri developed an IED that was inserted into the anal crevice of his brother Abdullah in an attempt to kill the Saudi Arabian Deputy Interior Minister Prince Mohammed Bin Nayef. The bomb detonated, killing Abdullah, but not Prince Nayef. Asiri was also believed to be the mastermind behind the October 2010 AQAP airline plot in which explosive devices were placed inside printer cartridges.

The hijacking of the 9/11 airliners evolved into what it became after a number of other attacks targeted commercial aviation. Hijacking became the favored tactic of terrorist groups starting in the late 1960s. In the 1980s, the Libyans were notable for finding ways to blow aircraft from the sky. In 1988, Abdelbaset Ali Mohamed Al Megrahi placed an explosive device hidden inside a RT-SF 16 BomBeat radio cassette player inside a suitcase that found its way into the luggage hold of Pan Am Flight 103. The device, consisting of PETN, RDX, and Semtex, exploded over Lockerbie, Scotland, killing all 259 people onboard, along with 11 people on the ground. The Libyans struck again in a similar vein in September 1989, when a bomb placed in the forward cargo hold of a Union de Transportes Aeriens (UTA) Flight 772 flying from Congo to Paris exploded over the Sahara Desert, killing all 170 onboard including the wife of the American ambassador to Chad.

New twists in targeting commercial airliners were on display on Christmas Eve 1994 when four members of the Algerian Armed Islamic Group boarded Air France Flight 8969 at Houari Boumediene Airport in Algiers dressed in Air Algerie uniforms. The hijackers immediately killed an Algerian policeman and a Vietnamese diplomat. They killed another French citizen before the plane was allowed to leave the airport and fly to Marseille, France.

THE BOJINKA PLOT

Ramzi Yousef, the nephew of Khalid Sheikh Mohamed, built upon the actions of the Libyans and worked at devising new methods to blow commercial airliners from the sky. Whereas the Libyans used radio cassette players in suitcases, Yousef looked to use liquid explosives using nitroglycerin to blow up at least 12 U.S.-flagged airliners over the Pacific as part of Op-Plan Bojinka. Bojinka focused not only on airliners but also called for the assassinations of Pope John Paul II and President Clinton during visits to the Philippines. A laptop recovered by authorities contained a plan to hijack a commercial flight in the United States, where the lone hijacker would seize controls of the plane and crash it into CIA headquarters in Langley, Virginia. The twist that Yousef added was that the attacker would board the plane, assemble the bomb, and depart the plane on a layover, circumventing any screening procedures. Yousef and associate Wali Khan Amin Shah worked on building these liquid devices and did a test run in a Manila theater that, when it exploded, wounded several people. Ten days later, Philippine Airlines Flight 434 left Manila for New Tokyo Airport. The flight had one stop in Cebu before heading to Japan. One of the passengers on this first leg was Ramzi Yousef. Yousef assembled the liquid device and placed it under seat 26K near the right fuselage. He set the timer for four hours later and departed the plane when it reached Cebu. As the plane approached Okinawa, the bomb exploded, killing Japanese tourist Haruki Ilkegami and tearing a hole in the plane.

Al-Qaeda's "Great Raid" of September 11, 2001, is just one example of how this group displays innovation by refining its tactics based on previous attacks and plots. The 9/11 attacks were so simple in nature, yet so complex. Using commercial airliners as weapons against simultaneous domestic targets was, according to one Al-Qaeda member, similar to "me tightly holding your finger, turning it toward you and poking it into your own eye." On the heels of 9/11, Richard Reid, another operative sent by Khalid Sheikh Mohammed, attempted to light an explosive device hidden in his shoe on board American Airlines Flight 63 traveling from Paris to Miami in December 2001. August 2006 saw an even further twist on Bojinka and 9/11 as British authorities arrested 20 men planning on using triacetone triperoxide placed in a sports drink bottles on flights leaving the United Kingdom for the United States and Canada. Whereas Yousef used timers, these cell members were geared to act as martyrs over the Atlantic or North America. The plot was foiled, and Abdulla Ahmed Ali, 28, the plot's ringleader, was given a minimum of 40 years in prison. Assad Sarwar, 29, and Tanvir Hussain, 28, were imprisoned for a minimum of 36 and 32 years, respectively.

In 1932, British MP Stanley Baldwin addressed the House of Commons in 1932 and made an astonishingly frank admission. "I think it is as well for the man in the street to realize that there is no power on earth that can protect him from being bombed. Whatever people may tell him, the bomber will always get through."[43] Baldwin was speaking of the threat of German bombers, yet his words have been echoed by others in relation to terrorism. The terrorists have and will continue to get through as they are unpredictable and have time on their side. They also have the advantage of not having to go on the defensive but remain on the offensive. Part of the reason as to why the terrorist has been successful is the reliance on effects-based responses, meaning that if a group obtains a particular type of weapon, he will use it—yet, if anything, the attacker remains unpredictable, reaching further and further back to find new modes of attacks. In the beginning, the advantage was on their side, yet the terrorist fighting an asymmetric war cannot sustain the pace in innovation. With each new innovation in attack, the defenders quickly work on a countermeasure to combat that innovation.

Using the example of IEDs in Iraq, the devices grew in size, forcing American forces to up-armor Humvees in the wake of the large number of casualties caused by this mode of attack. U.S. countermeasures did not drive down the number of IED attacks, but they drove down the number of casualties resulting from them. In conjunction with armor plating, radio jamming rendered some remotely detonated devices inoperable, forcing the terrorists to either find ways to build larger devices or find new tactics. In Iraq, this made AQI use the VBIED, a much more indiscriminate weapon, resulting in more civilian casualties. Modes of attack are part of a continuous cycle of move versus countermove, yet, like weapons systems, no mode of attack, with apologies to Mr. Baldwin, is a sure thing. Enacting countermeasures forces the attacker to keep up with a pace he cannot maintain. Vehicle checkpoints at sensitive locations have forced the attackers to find new ways to enter the target zone; bag screening of passengers boarding planes or trains have forced new shifts in tactics, this time putting the onus of time on the attacker. Public–private relationships also play a role. Learning terrorist tactics, techniques, and procedures is not for the defender alone but also for the web of private industries that help defend the target area. As al-Hindi was finding new ways to conduct attacks on his target set, he monitored the police presence and took note of each camera, forcing him to look elsewhere. Iyman Faris immediately saw that his target area was "too hot." It is true that the terrorists will attack again, but they cannot sustain the attacks; if they cannot, the advantage will be on the side of the defenders.

A British Special Air Services major operating against the Provisional Irish Republican Army (PIRA) snipers in South Armagh said, "To prevent attacks, you have to get inside the enemy's decision-making cycle and throw so many changes to the situation at such a rate that he has to break off his operation. He is then being purely reactive. That's something I don't believe we had ever achieved in South Armagh in 25 years. We'd always been reacting to incidents. We could just about contain them but we couldn't work out where and when the next one was going to take place."[44]

Conclusion

The ballet of move versus countermove remains fluid, but while the attackers have reached back to gain the short-term advantage, their actions also provide an opportunity for the defender to get inside the terrorists decision-making cycle. As General Blumentritt stated, "Each minute ahead of the enemy is an advantage." The onus is on the defender to win that battle just as much as it is on the attacker. The terrorists have been able to generate a number of different possibilities in terms of modes of attacks by shifting tactics that allow them to find vulnerabilities and exploit them. At the same time, the defenders have learned the hard lessons and have found new ways to respond to those innovations as each response is a defeat for the terrorist. It is true that terrorists will still conduct attacks, but by not learning their tactics, techniques, and procedures, the terrorists will be given a large window of opportunity to freely adapt to changing events.[45]

References

1. "Trail of Clues Helping Unravel Kampala Bombings," *Kampala New Vision*, July 31, 2010.
2. Robb, John, "The Bazaars of Open Source Platforms," September 24, 2004; http://globalguerrillas.typepad.com/globalguerrillas/2004/09/bazaar_dynamics.html.
3. Rayner, Gordon, "How the Preachers of Hate Turned Suburban Boys into Zealots Bent on Mass Murder," *The Daily Mail*, May 1, 2007.
4. MacDonald, Calum, "Four Crucia Sightings that Could Have Led to 7/7 Killers," *The Herald* (Glasgow), May 1, 2007.
5. Johnston, Philip, Gardham, Duncan, and Davies, Caroline, "How MI5 Got Lucky with Hotline Call," *Telegraph*, May 1, 2007.
6. Robb, John, "Cascading System Failure," Global Guerillas, May 24, 2004; http://globalguerrillas.typepad.com/globalguerrillas/2004/05/cascading_syste.html.
7. Information released by the Metropolitan Police Service; http://www.nefafoundation.org/miscellaneous/barot/.
8. Appeal: United States v. Ramzi Yousef et al. United States Court of Appeals for the Second Circuit, decided April 4, 2003, pp. 9–10.
9. O'Neill, Sean, "CCTV Shows 7/7 Bombers on Dummy Run in June," *Times Online*, September 21, 2005.
10. Jordán, Javier, "The Threat of Grassroots Jihadi Networks: A Case Study from Ceuta, Spain," *Jamestown Terrorism Monitor*, 5(3), February 21, 2007.
11. "Report of the Official Account of the Bombings in London on 7th July 2005," House of Commons, May 11, 2006.
12. National Commission on Terrorist Attacks Upon the United States, *Monograph on Terrorist Financing*, p. 17.
13. *NYPD Terrorist Modus Operandi*, Vol. 1, Tradecraft Book.
14. Levitt, Matthew, and Jacobson, Michael, "The Money Trail: Finding, Following, and Freezing Terrorist Finances," Washington, DC: The Washington Institute for Near East Policy, 2008, p. 9.
15. New Jersey Office of Homeland Security, "Financing Terrorism: Methods and Indicators for New Jersey," October 2008, p. 9 (U/FOUO).
16. Broyles, Scott D., and Rubio, Martha, "A Smokescreen for Terrorism," *United States Attorneys Bulletin*, January 2004, p. 34.

17. Broyles, Scott D., and Rubio, Martha, "A Smokescreen for Terrorism," *United States Attorneys Bulletin*, January 2004, pp. 32–33.
18. Kaplan, David E., "Homegrown terrorists," *U.S. News and Report*, March 2, 2003, http://www.usnews.com/usnews/news/articles/030310/10hez.htm.
19. Whitlock, Craig, "Homemade, Cheap and Dangerous," *Washington Post*, 7/8/07.
20. Ibid, p. 209.
21. United States of America, Appellee, v. Mohammed A. Salameh.
22. Statement by J. Gilmore Childers and Henry J. DePippo, Senate Judiciary Committee Subcommittee on Technology, Terrorism, and Government Information, "Foreign Terrorists in America: Five Years after the World Trade Center," 24 February 1998.
23. Hersley, John, Tongate, Larry, and Burke, Bob, "Simple Truths," Oklahoma Heritage Association, Oklahoma, 2002, p. 142.
24. Ibid, pp. 227–230.
25. Van Natta, Don Jr., Sciolino, Elaine, and Grey, Stephen, "Details Emerge in British Terror Case," *New York Times*, 8/27/09; http://www.nytimes.com/2006/08/28/world/europe/28plot.html?_r=1&pagewanted=all.
26. Thomas, Pierre, "Tons of Explosives Routinely Disappear," ABC News, June 19, 2009.
27. "Explosives Heist One of the Biggest in Recent History," ABC News, 12/20/05.
28. Families Acting for Innocent Relatives (FAIR); http:www.victims.org.uk/petition.html.
29. Oppenheimer, A.R., *IRA: The Bombs and the Bullets: A History of Deadly Ingenuity*, Irish Academic Press, Dublin, 2009, pp. 163–166.
30. Roggio, Bill, "Iran's Ramazan Corps and the Ratlines into Iraq," *The Long War Journal*, 12/5/07; http://www.longwarjournal.org/archives/2007/12/irans_ramazan_corps-print.php.
31. Miesels, Andrew, "FBI Grills Bomber in Tragedy," *NY Daily News*, 8/13/96.
32. Hersley, John, Tongate, Larry, and Burke, Bob, "Simple Truths," Oklahoma Heritage Association, Oklahoma, 2002, pp. 80–82.
33. Ibid, p. 125.
34. U.S.A. vs. Mohktar Haouari, 7/3/01.
35. Emerson Vermaat, "Homegrown Terrorism in Germany: The Case of Christian Ganczarski," *Militant Islam Monitor*, 10/8/07; http://www.militantislammonitor.org/article/id/3204.
36. Nasir bin Hamd al-Fahd, "A Treatise on the Legal Status of Using Weapons of Mass Destruction against Infidels," Carnegie Endowment for International Peace, May 2003, p. 8.
37. Black, Andrew, "Al-Suri's Adaptation of Fourth Generation Warfare Doctrine," *Jamestown Terrorism Monitor*, 4(18), 9/21/06.
38. Deutsch, Anthony, "Slain Jakarta Florist Said to Have Plotted Blasts," Associated Press, 8/13/09.
39. Van Natta, Don Jr., Sciolino, Elaine, and Grey, Stephen, "Details Emerge in British Terror Case," *New York Times*, 8/27/09.
40. Boyd, John R., "Patterns of Conflict," Reprinted by Defense and the National Interest, 1/07; http://www.d-n-i.net/boyd/patterns_ppt.pdf.
41. Oppenheimer, A.R., *IRA: The Bombs and the Bullets: A History of Deadly Ingenuity*, Irish Academic Press, Dublin, 2009, p. 64.
42. U.S.: "Soldiers Abducted in Karbala, Killed Elsewhere," Associated Press, 1/26/07; http://web.archive.org/web/20070202031215/http://www.cnn.com/2007/WORLD/meast/01/26/iraq.karbala.ap/index.html.

43. Bishop, Patrick, "The Bomber Will always Get Through, Be He ETA or Al-Qaeda," *Telegraph UK*, 3/12/04.
44. Harnden, Toby, *Bandit Country: The IRA and South Armagh*, Coronet Books, United Kingdom, 1999, p. 417.
45. Boyd, John R., "Patterns of Conflict," Reprinted by Defense and the National Interest, 1/07; http://www.d-n-i.net/boyd/patterns_ppt.pdf.

Chapter 16

Agroterrorism

Michael J. Fagel

Contents

The potential of terrorist attacks against agricultural targets (agroterrorism) is increasingly recognized as a national security threat, especially after the events of September 11, 2001. In this context, agroterrorism is defined as the deliberate introduction of an animal (Figure 16.1) or plant disease with the goal of generating fear, causing economic losses, and/or undermining stability.

Figure 16.1 Cattle and other livestock could be targeted through intentional disease-caused outbreaks.

Agriculture as a Target: Overview of Terrorist Threat

Agroterrorism is a subset of the more general issues of terrorism and bioterrorism. People more generally associate bioterrorism with outbreaks of human illness (such as from anthrax or smallpox), rather than diseases first affecting animals or plants. Agriculture has several characteristics that pose unique problems for managing the threat:

- Agricultural production is geographically disbursed in unsecured environments (e.g., open fields and pastures throughout the countryside). Although some livestock are housed in secure facilities, agriculture in general requires large expanses of land that are difficult to secure from intruders.
- Livestock are frequently concentrated in confined locations (e.g., feedlots with thousands of cattle in open-air pens, farms with tens of thousands of pigs, or barns with hundreds of thousands of poultry). Concentration in slaughter, processing, and distribution also makes large-scale contamination more likely.
- Live animals, grain, and processed food products are routinely transported and commingled in the production and processing system. These factors circumvent natural barriers that could slow pathogenic dissemination.
- The presence (or rumor) of certain pests or diseases in a country can quickly stop all exports of a commodity, and can take months or years to resume.
- The past success of keeping many diseases out of the United States means that many veterinarians and scientists lack direct experience with foreign diseases. This may delay recognition of symptoms in the case of an outbreak.
- The number of lethal and contagious biological agents is greater for plants and animals than for humans. Most of these diseases are environmentally resilient, endemic in foreign countries, and not harmful to humans—making it easier for terrorists to acquire, handle, and deploy the pathogens.

The general susceptibility of the agriculture and food industry to bioterrorism is difficult to address in a systematic manner because of the highly dispersed, yet concentrated nature of the industry, and the inherent biology of growing plants and raising animals.

The results of an agroterrorist attack may include major economic crises in the agricultural and food industries, loss of confidence in government, and possibly human casualties. Humans could be at risk in terms of food safety or public health, especially if the chosen disease is transmissible to humans (zoonotic). But an agroterrorist attack need not cause human casualties for it to be effective or to cause large-scale economic consequences.

The production agriculture sector (Figure 16.2) would suffer economically in terms of plant and animal health, and the supply of food and fiber may be reduced, especially in certain regions. The demand for certain types of food may decline

Figure 16.2 Food chain supply could be dramatically affected by targeted terrorist attacks.

based on which products are targeted in the attack (e.g., dairy, beef, pork, poultry, grains, fruit, or vegetables), whereas demand for other types of food may rise because of food substitutions.

An agroterrorism event would cause economic losses to individuals, businesses, and governments through costs to contain and eradicate the disease, and to dispose of contaminated products. Economic losses would accumulate throughout the farm-to-table continuum as the supply chain is disrupted, especially if domestic markets for food become unstable or if trade sanctions are imposed by other countries on U.S. exports. The economic impact can spread to farmers, input suppliers, food processors, transportation, retailers, and food service providers.

Public opinion may be particularly sensitive to a deliberate outbreak of disease affecting the food supply. Public confidence in government could be eroded if authorities appear unable to prevent such an attack or to protect the population's food supply. As the United States evolved away from an agrarian society during the twentieth century, food and the fear of inadequate food supplies moved further from the minds of most U.S. residents. However, because food remains an important part of everyone's daily routine and survival, significant threats to the currently held notion of food security in the United States could cause a reordering of people's priorities.

Because an agroterrorist attack may not necessarily cause human casualties, be immediately detected, or have the "shock factor" of an attack against the more visible public infrastructure or human populations, agriculture may not be a terrorist's first choice of targets. Nonetheless, some types of agroterrorism could be relatively easily achieved and have significant economic impacts. Thus, the possibilities are treated seriously, especially in the post-September 11 world.

Importance of Agriculture in the United States

Agriculture and the food industry are very important to the social, economic, and arguably, the political stability of the United States. Although farming employs less than 2% of the country's workforce, 16% of the workforce is involved in the food and fiber sector, ranging from farmers and input suppliers, to processors, shippers, grocers, and restaurateurs. In 2008, the food and fiber sector contributed $1.4 trillion, or 11%, to the gross domestic product (GDP), even though the farm sector itself contributed less than 1%. Gross farm sales exceeded $200 billion, and are relatively concentrated throughout the Midwest, parts of the East Coast, and California. Production is split nearly evenly between crops and livestock.

Agriculture in the United States is highly advanced and productive. This productivity allows Americans to spend less than 11% of their disposable income on food, compared with a global average of 20%–30%.[1]

Although the number of farms in the 2008 Census of Agriculture totaled 2.1 million, 75% of the value of production occurs on just 6.7%, or 143,500, of these farms. This subset of farms has average sales of $1 million annually, and averages 2000 acres in size.

The United States produces and exports a large share of the world's grain. In 2008, the United States exported $53 billion of agricultural products (8% of all U.S. exports), and imported $42 billion of agricultural products (4% of all U.S. imports), making agriculture a positive contributor to the balance of trade. The U.S. share of world production was 39% for corn, 38% for soybeans, and 8% for wheat. The United States accounted for 23% of global wheat exports, 54% of corn exports and 43% of soybean exports. If export markets were to decline following an agroterrorism event, U.S. markets could be severely disrupted, since 22% of U.S. agricultural production is exported (10% of livestock and 23% of crops).

The price of land is directly correlated to the productivity and marketability of agricultural products, along with federal farm income support payments. In 2008, farm assets exceeded $1.45 trillion, with $1.1 trillion in equity. Land and other real estate accounts for 80% of those assets. Of the 938 million acres of farm land in the United States, 46% are in crop land, 42% are pasture and range land, and 8% are wood land.

Livestock and poultry are concentrated in various regions of the country, and in large numbers. In 2009, the inventory included 95 million cattle and calves, and 60 million hogs and pigs. Farm sales of broilers and other meat-type chickens exceeded 8.5 billion birds.

Cattle are the most widely distributed given the prevalence of small cow–calf herds throughout the country and pockets of dairy on the West Coast, upper Midwest, and Northeast. However, beef cattle feedlots are particularly concentrated from northern Texas through Kansas, Nebraska, eastern Colorado, and western Iowa.

Hog inventories are concentrated in the Midwest, especially Iowa and southern Minnesota, and in North Carolina. The production of broilers for poultry meat is concentrated throughout the Southeast, ranging from the Oklahoma–Arkansas border up to the Delmarva peninsula (Delaware–Maryland–Virginia).

A Brief History of Agricultural Bioweapons

Attacks against agriculture are not new, and have been conducted both by nation-states and by substate organizations throughout history. At least nine countries had documented agricultural bioweapons programs during some part of the twentieth century (Canada, France, Germany, Iraq, Japan, South Africa, United Kingdom, United States, and the former USSR). Four other countries are believed to have or have had agricultural bioweapons programs (Egypt, North Korea, Rhodesia, and Syria).

Despite extensive research on the issue, however, biological weapons have been rarely used against crops or livestock, especially by state actors. Thus, in recent decades, using biological weapons against agricultural targets has remained mostly theoretical consideration. With the ratification of the Biological and Toxin Weapons Convention in 1972, many countries, including the United States, stopped military development of biological weapons and destroyed their stockpiles.[2]

Although individuals or substate groups have used bioweapons against agricultural or food targets, only a few can be considered terrorist in nature. In 1952, the Mau Mau (an insurgent organization in Kenya) killed 33 head of cattle at a mission station using African milk bush (a local plant toxin). In 1984, the Rajneeshee cult spread salmonella in salad bars at Oregon restaurants to influence a local election.[3]

Chemical weapons have been used somewhat more commonly against agricultural targets. During the Vietnam War, the United States used Agent Orange to destroy foliage, affecting some crops. Among possible terrorist events, chemical attacks against agricultural targets include a 1997 attack by Israeli settlers who sprayed pesticides on grapevines in two Palestinian villages, destroying up to 17,000 metric tons of grapes. In 1978, the Arab Revolutionary Council poisoned Israeli oranges with mercury, injuring at least 12 people and reducing orange exports by 40%.

Economic Consequences

Economic losses from an agroterrorist incident could be large and widespread.

- First, losses would include the value of lost production, the cost of destroying diseased or potentially diseased products, and the cost of containment (drugs, diagnostics, pesticides, and veterinary services).
- Second, export markets would be lost as importing countries place restrictions on U.S. products to prevent possibilities of the disease spreading.
- Third, multiplier effects would ripple through the economy due to decreased sales by agriculturally dependent businesses (farm input suppliers, food manufacturing, transportation, retail grocery, and food service) and tourism.

- Fourth, the government could bear significant costs, including eradication and containment costs, and compensation to producers for destroyed animals.

Depending on the erosion of consumer confidence and export sales, market prices of the affected commodities may drop. This would affect producers whose herds or crops were not directly infected, making the event national in scale even if the disease itself were contained to a small region.

For food types or product lines that are not contaminated, however, demand may become stronger, and market prices could rise for those products. Such goods may include substitutes for the food that was the target of the attack (e.g., chicken instead of beef), or product that can be certified not to come from regions affected by the attack (e.g., beef from another region of the country or imported beef). When Canada announced the discovery of mad cow disease [or bovine spongiform encephalopathy (BSE)] in May 2003, farm-level prices of beef in Canada dropped by nearly half, whereas beef prices in the United States remained very strong at record or near record levels.[4]

Consumer confidence in government may also be tested depending on the scale of the eradication effort and means of destroying animals or crops. The need to slaughter perhaps hundreds of thousands of cattle (or tens of millions of poultry) could generate public criticism if depopulation methods are considered inhumane or the destruction of carcasses is questioned environmentally. Dealing with these concerns can add to the cost for both government and industry.

Depending on the disease and means of transmission, the potential for economic damage depends on a number of factors, such as the disease agent, location of the attack, rate of transmission, geographical dispersion, how long it remains undetected, availability of countermeasures or quarantines, and incident response plans. Potential costs are difficult to estimate and can vary widely based on compounding assumptions.

The ability of farm commodity programs to compensate for losses due to agroterrorism is limited. Government income support programs subsidize about 25 agricultural commodities (such as corn, wheat, soybeans, rice, and cotton). These supported commodities represent about one-third of gross farm sales. The list of commodities that normally do not receive direct support includes meats, poultry, fruits, vegetables, nuts, hay, and nursery products. These nonsupported commodities account for about two-thirds of gross farm sales.

The food products more vulnerable to attack (meats, fruits, and vegetables) do not have existing federal farm income support programs, nor are there income support programs beyond the farm gate for food processors or retailers. Thus, any federal assistance to producers or processors stemming from an agroterrorist attack would likely come in the form of *ad hoc* disaster assistance. Making disaster payments to producers who do not normally receive government payments is technically more difficult than supplementing regular program payments due to drought or flood.

Federal Recognition of Agroterrorism Threats

Agriculture and food production generally have received relatively less attention, or sometimes were overlooked, in counterterrorism and homeland security. After what many observers claim to be a slow start after September 11, 2001, agriculture now is garnering more attention in the expanding field of terrorism studies and policies (Figure 16.3).

Congress has held hearings on agroterrorism and, while addressing terrorism more broadly, has implemented laws and appropriations with provisions important to agriculture. The Government Accountability Office has studied aspects of food safety, border inspections, and physical security with respect to agroterrorism. The executive branch has responded by implementing the new laws, issuing several Presidential directives, and creating terrorism and agroterrorism task forces.

In its report, the 9/11 Commission (National Commission on Terrorist Attacks upon the United States) does not make any direct references to agroterrorism or terrorist attacks on the food supply. However, agriculture obviously would be affected, along with other sectors of the economy, by some of the commission's recommendations regarding coordination of intelligence, information sharing, and first responders.

Congressional Hearings and Laws

On November 19, 2003, the Senate Committee on Governmental Affairs held a hearing titled, "Agroterrorism: The Threat to America's Breadbasket," including witnesses from the Administration, state governments, and a private think tank.

This was the first congressional hearing devoted entirely to agroterrorism since October 27, 1999. At that time, the Subcommittee on Emerging Threats of the Senate Committee on Armed Services held a hearing titled, "Agricultural Biological Weapons Threat to the United States." During the four years between

Figure 16.3 Agriculture is garnering more attention in policy discussions and terrorism studies.

these hearings, a few individual panelists at more general hearings on food safety, homeland security, or terrorism discussed agroterrorism in reference to other topics.

Bioterrorism Preparedness Act

The Public Health Security and Bioterrorism Preparedness and Response Act (P.L.107-188, June 12, 2002) contained several provisions important to agriculture. These provisions accomplish the following:

- Expand Food and Drug Administration (FDA) authority over food manufacturing and imports (particularly in Sections 303–307).
- Tighten control of biological agents and toxins ("select agents" as discussed in sections 211–213, the "Agricultural Bioterrorism Protection Act of 2002") through rules issued by the Animal and Plant Health Inspection Service (APHIS) and the Centers for Disease Control and Prevention (CDC).
- Authorize expanded agricultural security activities and security upgrades at USDA facilities (Sections 331–335).
- Address criminal penalties for terrorism against enterprises raising animals (section 336) and violation of the select agent rules (Section 231).

New FDA Rules on Food Processors and Importers

The Bioterrorism Preparedness Act responded to long-standing concerns about whether the FDA in the Department of Health and Human Services (HHS) had the authority to assure food safety. FDA was instructed to implement new rules for

- Registration of food processors
- Prior notice of food imports
- Administrative detention of imports
- Record keeping

Registration of Food Processors

The act required FDA to establish a one-time registration system for any domestic or foreign facility that manufactures, processes, packs, and handles food. All food facilities supplying food for the United States were required to register with the FDA by December 12, 2003.

Registering involved providing information about the food products (brand names and general food categories), facility addresses, and contact information. Restaurants, certain retail stores, farms, nonprofit food and feeding establishments, fishing vessels, and trucks and other motor carriers were exempt from registration requirements. However, many farms had a difficult time determining whether they needed to register based on the amount of handling or processing they performed.

Registration documents are protected from public disclosure under the Freedom of Information Act. The registry provides, for the first time, a complete list of companies subject to FDA authority, and will enhance the agency's capability to trace contaminated food. Critics argued that registration created a record keeping burden without proof that facilities will be able to respond in an emergency.

Prior Notice of Imports

As of December 12, 2003, importers are required to give advance notice to FDA before importing food. Electronic notice must be provided by the importer within a specified period before arrival at the border (within two hours by road, four hours by air or rail, and eight hours by water). With prior notice, FDA can assess whether a shipment meets criteria that can trigger an inspection. If notice is not given, the food will be refused entry and held at the port or in secure storage. Some critics are concerned that the administrative cost of compliance may raise the price of food. Others have argued that perishable imports are subject to increased spoilage if delays arise, or that certain perishables (especially from Mexico) are not harvested or loaded onto trucks before the two-hour notification period. However, implementation of the new system generally has not caused delays and most shippers have been accommodated.

To facilitate compliance, FDA and the Department of Homeland Security (DHS) Bureau of Customs and Border Protection (CBP) integrated their information systems to allow food importers to provide the required information using CBP's existing system for imports. In December 2003, the two agencies agreed to allow CBP officers to inspect imported foods on FDA's behalf, particularly at ports where FDA has no inspectors.

Security for Biological Agents and Toxins

In December 2002, the USDA APHIS issued regulations to reduce the threat that certain biological agents and toxins could be used in domestic or international terrorism. APHIS determined that the "select agents" on the list have the potential to pose a severe threat to agricultural production or food products.

The select agent regulations (9 CFR 121 for animals, 7 CFR 331 for plants) establish the requirements for possession, use, and transfer of the listed pathogens. The rules affect many research institutions, including federal, state, university, and private laboratories, as well as firms that transport such materials. The laboratories have had to assess security vulnerabilities and upgrade physical security, often without additional financial resources. Some have been concerned that certain research programs may be discontinued or avoided because of regulatory difficulties in handling the select agents.

Extensive registration and background checks of both facilities and personnel were conducted. However, because of delays at the Federal Bureau Investigation (FBI) in processing security clearance paperwork, provisional registrations were issued to laboratories that had submitted paperwork by established deadlines.

Homeland Security Act

The main purpose of the Homeland Security Act of 2002 (P.L. 107-296, November 25, 2002) was to create the DHS, primarily by transferring parts or all of many agencies throughout the federal government into the new cabinet-level department. In doing so, the law made two major changes to the facilities and functions of the Department of Agriculture. The Homeland Security Act transferred

- Agricultural border inspections from APHIS to DHS.
- Possession of the Plum Island Animal Disease Center in New York from USDA to DHS.

Agricultural Border Inspections

Section 421 of the Homeland Security Act authorized the transfer of up to 3200 APHIS border inspection personnel to DHS. As of March 1, 2003, approximately 2680 APHIS inspectors became employees of DHS in the Bureau of Customs and Border Inspection (CBP). Because of its scientific expertise, USDA retains a significant presence in border inspection.

Historically, the APHIS Agricultural Quarantine Inspection (AQI) program was considered the most significant and prominent of agricultural and food inspections at the border. Because of this prominence, AQI was one of the many programs selected for inclusion when DHS was created. Some drafts of the bill creating the new department would have transferred all of APHIS (including, for example, animal welfare and disease eradication) to DHS. Concerns from many farm interest groups about the impact this might have on diagnosis and treatment of natural plant and animal diseases prompted a legislative compromise that transferred only the border inspection function and left other activities under USDA.

DHS–CBP personnel now inspect international conveyances and the baggage of passengers for plant, animal, and related products that could harbor pests or disease organisms. They also inspect ship and air cargo, rail and truck freight, and package mail from foreign countries.

Although the border inspection functions were transferred to DHS, the USDA retains a significant presence in border activities. APHIS employees who were not transferred continue to preclear certain commodities, inspect all plant propagative materials, and check animals in quarantine. APHIS personnel continue to set agricultural inspection policies to be carried out by DHS border inspectors, and negotiate memoranda of understanding to assure that necessary inspections are conducted. APHIS manages the data collected during the inspections process, and monitors smuggling and trade compliance. USDA is also statutorily charged in section 421 (e)(2)(A) of the act to "supervise" the training of CBP inspectors in consultation with DHS.

This separation of duties is designed to allow for consolidated border inspections for intelligence and security goals, but preserve USDA's expertise and historical mission to set agricultural import policies.

Adding Agricultural Specialists

Under the CBP cross-training initiative in 2003 (also known as "one face at the border"), most CBP inspectors are trained to perform inspections in all three areas of customs, immigration, and agriculture. However, due to criticism from USDA, inspection unions, and the agricultural industry, DHS created another class of inspectors called agricultural specialists. Agricultural specialists will staff, primarily, secondary inspection stations. These specialists will include former APHIS inspectors who decided not to convert to CBP generalist inspectors and new agricultural specialist trainees.

Before DHS was created, APHIS trained its inspectors in a nine-week course that had science prerequisites. The initial DHS cross-training program announced in 2003 had only 12–16 hours for agriculture in a 71-day course covering customs, immigration, and agriculture. With the creation of the agricultural specialist position, DHS created a 43-day training program for agricultural specialists.

Although DHS is training new agricultural specialists, the future size of the agricultural specialist corps is not certain, given the eventual attrition of former APHIS inspectors. Also, details are not available as to how these inspectors will be deployed and how many ports of entry will be staffed with agricultural specialists (compared with the APHIS deployment prior to DHS). Without agricultural specialists, primary agricultural inspections—the first line of defense for agricultural security—may be conducted by cross-trained inspectors with limited agricultural training.

Executive Branch Actions

Shortly after September 11, 2001, USDA created a Homeland Security Staff in the Office of the Secretary to develop a department-wide plan to coordinate agroterrorism preparedness plans among all USDA agencies and offices. Efforts have been focused on three areas: food supply and agricultural production, USDA facilities, and USDA staff and emergency preparedness. The Homeland Security Staff also has become the department's liaison with Congress, the DHS, and other governmental agencies on terrorism issues.

The White House's National Security Council Weapons of Mass Destruction (WMD) preparedness group, formed by Presidential Decision Directive 62 (PDD-62) in 1998, included agriculture, especially in terms of combating terrorism. Many observers note that, as a latecomer to the national security table, USDA has been invariably overshadowed by other agencies.

Homeland Security Presidential Directive 7

In terms of protecting critical infrastructure, agriculture was added to the list in December 2003 by Homeland Security Presidential Directive 7 (HSPD-7), "Critical Infrastructure Identification, Prioritization, and Protection." This directive replaces the 1998 Presidential Decision Directive 63 (PDD-63) that omitted agriculture and food. Both of these critical infrastructure directives designate the physical systems that are vulnerable to terrorist attack and are essential for the minimal operation of the economy and the government.

These directives instruct agencies to develop plans to prepare for and counter the terrorist threat. HSPD-7 mentions the following industries: agriculture and food; banking and finance; transportation (air, sea, and land, including mass transit, rail, and pipelines); energy (electricity, oil, and gas); telecommunications; public health; emergency services; drinking water; and water treatment.

Homeland Security Presidential Directive 9

More significant recognition came on January 30, 2004, when the White House released Homeland Security Presidential Directive 9 (HSPD-9), "Defense of United States Agriculture and Food." This directive establishes a national policy to protect against terrorist attacks on agriculture and food systems.

HSPD-9 generally instructs the Secretaries of Homeland Security (DHS), Agriculture (USDA), and HHS, the Administrator of the Environmental Protection Agency (EPA), the Attorney General, and the Director of Central Intelligence to coordinate their efforts to prepare for, protect against, respond to, and recover from an agroterrorist attack. In some cases, one department is assigned primary responsibility, particularly when the intelligence community is involved. In other cases, only USDA, HHS, and/or EPA are involved regarding industry or scientific expertise.

The directive instructs agencies to develop awareness and warning systems to monitor plant and animal diseases, food quality, and public health through an integrated diagnostic system. Animal and commodity tracking systems are included, as is gathering and analyzing international intelligence. Vulnerability assessments throughout the sector help prioritize mitigation strategies at critical stages of production or processing, including inspection of imported agricultural products.

Response and recovery plans are to be coordinated across the federal, state, and local levels. A National Veterinary Stockpiles (NVS) of vaccine, antiviral, and therapeutic products is to be developed for deployment within 24 hours of an attack. A National Plant Disease Recovery System (NPDRS) is to develop disease and pest-resistant varieties within one growing season of an attack in order to resume production of certain crops. The Secretary of Agriculture is to make recommendations for risk management tools to encourage self-protection for agriculture and food enterprises vulnerable to losses from terrorism.

HSPD-9 encourages USDA and HHS to promote higher education programs that specifically address the protection of animal, plant, and public health. It suggests capacity-building grants for universities, and internships, fellowships, and postgraduate opportunities. HSPD-9 also formally incorporates USDA and agriculture into the ongoing DHS research program of university-based "centers of excellence."

As a presidential directive, HSPD-9 addresses the internal management of the executive branch and does not create enforceable laws. Moreover, it is subject to change without Congressional consent. Although Congress has oversight authority of federal agencies and may ask questions about implementation of the directive, a public law outlining an agroterrorism preparedness plan would establish the statutory parameters for such a plan, and, as a practical matter, might result in enhanced oversight by specifically identifying executive branch entities responsible for carrying out particular components of such a plan.

In implementing HSPD-9, the USDA Homeland Security Staff and other agencies are drawing upon HSPD-5 (regarding the national response plan) and HSPD-8 (regarding preparedness). Implementing many of the HSPD-9 directives depends on the executive branch having sufficient appropriations for those activities.

Possible Pathogens in an Agroterrorist Attack

Of the hundreds of animal and plant pathogens and pests available to an agroterrorist, it is likely that less than two dozen represent significant economic threats. Determinants of this level of threat are the agent's contagiousness and potential for rapid spread, and its international status as a "reportable" pest or disease (i.e., subject to international quarantine) under rules of the World Organization for Animal Health [also commonly known as the Office International des Epizooties (OIE)].

A widely accepted view among scientists is that livestock herds are much more susceptible to agroterrorism than crop plants. Much of this has to do with the success of efforts to systematically eliminate animals diseases from U.S. herds, which leaves current herds either unvaccinated or relatively unmonitored for such diseases by farmers and some local veterinarians. Once infected, livestock can often act as the vector for continuing to transmit the disease, facilitating an outbreak's spread, especially when live animals are transported. Certain animal diseases may be more attractive to terrorists because they can be zoonotic, or transmissible to humans (source: *Small*

Scale Terrorist Attacks Using Chemical and Biological Agents: An Assessment Framework and Preliminary Comparisons, by Dana Shea and Frank Gottron).

In contrast, a number of plant pathogens continue to exist in small areas of the United States and continue to infect limited areas of plants each year, making outbreaks and control efforts more routine. Moreover, plant pathogens are generally more technically difficult to manipulate. Some plant pathogens may require particular environmental conditions of humidity, temperature, or wind to take hold or spread. Other plant diseases may take a longer time than an animal disease to become established or achieve destruction on the scale that a terrorist may desire.

Animal Pathogens

The Agricultural Bioterrorism Protection Act of 2002 (Subtitle B of P.L. 107-188, the Public Health Security and Bioterrorism Preparedness and Response Act) created the current, official list of animal pathogens that are of greatest concern for agroterrorism. The list is specified in the select agent rules implemented by USDA APHIS and the CDC. The act requires that these lists be reviewed at least every two years.

The select agent list for animal pathogens draws heavily from the enduring and highly respected OIE lists of high-concern pathogens. The select agent list is composed of an APHIS-only list (of concern to animals) and an overlap list of agents selected both by APHIS and CDC (of concern to both animals and humans) (source: United States Animal Health Association's "Gray Book," http:www.bt.cdc.gov/agent/agentlist-category.asp).

OIE List

Before the Agricultural Bioterrorism Protection Act, the commonly accepted animal diseases of concern were all of the OIE's "List A" diseases and some of the "List B" diseases. In 2004, the OIE replaced its Lists A and B with a single list that is more compatible with the Sanitary and Phytosanitary Agreement of the World Trade Organization. The new OIE list in Table 16.1 classifies diseases equally; giving each the same degree of importance in international trade. Many of these OIE-listed diseases are included in the select agent list (source: *Terrestrial Animal Health Code*, 13th Edition, May 2004).

The OIE's List A diseases were transmissible animal diseases that had the potential for very serious and rapid spread, irrespective of national borders. List A diseases had serious socioeconomic or public health consequences and were of major importance in international trade. List B diseases were transmissible diseases considered to be of socioeconomic or public health importance within countries and significant in international trade. In creating the new list, OIE reviewed its criteria for including a disease, and the disease or epidemiological events that require member countries to file reports.

Table 16.1 Single OIE Disease List

Animal Diseases and Agents/ Toxins Listed Exclusively by		*Overlap Diseases and Agents/Toxins Listed by Both*	
APHIS 9 CFR 121.3(d)	*OIE Class*	*APHIS and CDC 9 CFR 121.3(b)*	*OIE Class*
African horse sickness	E	Anthrax (*Bacillus snthracis*)	M
African swine fever	S	Botulinum neurotoxins	
Akabane		Botulinum neurotoxin-producing species of *Clostridium*	
Avian influenza (highly pathogenic)	A		
Bluetongue (exotic)	M	Brucellosis of cattle (*Brucella abortus*)	B
Bovine spongiform encephalopathy	B	Brucellosis of sheep (*Brucella melitensis*)	C
Camel pox		Brucellosis of swine (*Brucella suis*)	S
Classical swine fever	S	Glanders (*Burkholderia mallei*)	E
Contagious caprine pleuropneumonia	C	Melioidosis (*Burkholderia pseudomallei*)	
Contagious bovine pleuropneumonia	B	Botulism (*Clostridium botulinum*)	
Foot-and-mouth disease (FMD)	M	*Clostridium perfringens* epsilon toxin	
Goat pox	C	(Valley fever) *Coccidioides immitis*	
Heartwater (Cowdria ruminantium)	M	Q fever (*Coxiella burnetii*)	M
Japanese encaphalitis	E	Eastern equine encephalitis	E
Lumpy skin disease	M	Tularemia (*Francisella tularensis*)	L
Malignant catarrhal fever	B	Hendra virus (of horses)	
Menangle virus		Nipah virus (of pigs)	

(*continued*)

Table 16.1 (*Continued*) Single OIE Disease List

Animal Diseases and Agents/ Toxins Listed Exclusively by		*Overlap Diseases and Agents/Toxins Listed by Both*	
APHIS 9 CFR 121.3(d)	*OIE Class*	*APHIS and CDC 9 CFR 121.3(b)*	*OIE Class*
Newcastle disease (exotic)	A	Rift Valley fever	M
Peste des petits ruminants	C	Shigatoxin	
Rinderpest	B	Staphylococcal enterotoxins	
Sheep pox	C	T-2 toxin	
Swine vesicular disease	S	Venezuelan equine encephalitis	E
Vesicular stomatitis	M		

Select Agents List

The regulations establishing the select agent list for animals (9 CFR 121.3) set forth the requirements for possession, use, and transfer of these biological agents or toxins. They are intended to ensure safe handling and for security to protect the agents from use in domestic or international terrorism. APHIS and CDC determined that the biological agents and toxins on the list have the potential to pose a severe threat to agricultural production or food products.

The 23 animal diseases listed exclusively by APHIS in 9 CFR 121.3(d)—the left column of the previous table—include 20 of the OIE-listed diseases and three other disease agents (Akabane, Camel pox, and Menangle) considered to be emerging animal health risks for terrorism. The much larger OIE list includes other diseases that are not listed as "select agents." However, the select agent list was created to account for the additional risks perceived to be posed by terrorism.

The 21 diseases and overlap agents/toxins included by both APHIS and CDC in 9 CFR 121.3(b)—the right column of the previous table—pose a risk to both human and animal health. In June 2002, CDC convened an interagency working group to review the list of select agents and develop recommendations regarding possible changes.

The overlap list includes 10 OIE-listed diseases, including Anthrax, Brucellosis of cattle, Brucellosis of sheep, Brucellosis of swine, Glanders, Rift Valley fever, Q fever, Eastern equine encephalitis, Tularemia, and Venezuelan equine encephalitis.

Agent Analysis

It is important to note that the select agent list designates and regulates pathogens, not diseases. Thus, the overlap list between APHIS and CDC is somewhat more comprehensive than a disease-only list, particularly because certain pathogens may not cause a disease, *per se*, but may cause symptoms such as food poisoning or central nervous systems responses.

Some of the pathogens in the select agent list receive more attention than others in discussions about agroterrorism. One reason is that the select agent list was designed to regulate access to and handling of high-consequence pathogens, not the diseases directly.

For example, the causative agent of BSE (or "mad cow disease") is considered dangerous enough to be a select agent, even though mad cow disease is less likely to be a terrorist's choice than other diseases. With BSE, infection is not certain, symptoms take years to manifest, and the disease may not be detected—all making credit for an attack more doubtful.

On the other hand, foot and mouth disease (FMD) is probably the most frequently mentioned disease when agroterrorism is discussed, because of its ease of use, ability to spread rapidly, and potential for great economic damage. In testimony before the Senate Governmental Affairs Committee on November 19, 2003, Dr. Thomas McGinn of the North Carolina Department of Agriculture described a simulation of an FMD attack by a terrorist at a single location. Only after the fifth day of the attack would the disease be detected, by which time it may have spread to 23 states. By the eighth day, 23 million animals may need to be destroyed in 29 states.

Widespread animal diseases such as brucellosis, influenza, or tuberculosis receive relatively less attention than FMD, hog cholera, or Newcastle disease. However, emerging diseases such as Nipah virus, Hendra virus, and the H5N1 strain of avian influenza (zoonotic diseases that have infected people, mostly in Asia) can be lethal since vaccines are elusive or have not yet been developed.

Plant Pathogens

The Agricultural Bioterrorism Protection Act of 2002 (Subtitle B of P.L. 107-188) also instructed APHIS and CDC to create the current official list of potential plant pathogens. The federal government lists biological agents and toxins for plants in 7 CFR 331.3 (Table 16.2). The act requires that these lists be reviewed at least every two years, and revised as necessary.

Prior to the act, there was not a commonly recognized list of the most dangerous plant pathogens, although several diseases were usually mentioned and are now included in the APHIS select agent list.

The list of nine biological agents and toxins in 7 CFR 331.3 was compiled by the Plant Protection and Quarantine (PPQ) program in APHIS, in consultation with

Table 16.2 Plant Diseases and Causing Agents from 7 CFR 331.1

Plant Diseases Caused by	*Select Agents Listed in 7 CFR 331.3*
Citrus greening	*Liberibacter africanus, L. asiaticus*
Philippine downy mildew (of corn)	*Peronosclerospora philippinensis*
Soybean rust	*Phakopsora pachyrhizi*
Plum pox (of stone fruits)	*Plum pox potyvirus*
Bacterial wilt, brown rot (of potato)	*Ralstonia solanacearum,* race 3, biovar 2
Brown strip downey mildew (of corn)	*Sclerophthora rayssiae* var. *zeae*
Potato wart or potato canker	*Synchytrium endobioticum*
Bacterial leaf streak (of rice)	*Xanthomonas oryzae* pv. *Oryzicola*
Citrus variegated chlorosis	*Xylella fastidiosa*

USDA's Agricultural Research Service; Forest Service; Cooperative State Research, Education, and Extension Service; and the American Phytopathological Society. The listed agents and toxins are viruses, bacteria, or fungi that can pose a severe threat to a number of important crops, including potatoes, rice, soybeans, corn, citrus, and stone fruit. Because the pathogens can cause widespread crop losses and economic damage, they could potentially be used by terrorists.

Other plant pathogens not included in the select agent list possibly could be used against certain crops or geographic regions. Examples include Karnal bunt and citrus canker, which both currently exist in the United States in regions quarantined or under surveillance by the USDA. As with other agents, the effectiveness of an attack to spread such a disease may be dependent on environmental conditions and difficult to achieve.

Countering the Threat

The goal of the U.S. animal and plant health safeguarding system is to prevent the introduction and establishment of exotic pests and diseases, to mitigate their effects when present, and to eradicate them when feasible. In the past, introductions of pests and pathogens were presumed to be unintentional and occurred through natural migration across borders or accidental movement by international commerce (passengers, conveyance, or cargo). However, a system designed for accidental or natural outbreaks is not sufficient for defending against intentional attack.

Consequently, the U.S. system is being upgraded to address the reality of agroterrorism.

The National Research Council outlines a three-pronged strategy for countering the threat of agroterrorism (source: National Research Council, 2003, pp. 41–59):

- Deterrence and prevention
- Detection and response
- Recovery and management

Even though no foreign terrorist attacks on crops or livestock have occurred in the United States, government agencies and private businesses have not taken the threat lightly. Because of the importance of brand names in marketing, many agribusinesses have prepared response plans or added security measures to protect their product line, looking at threats ranging from the source of their inputs to their retail distribution network. Since the terrorist attacks of 2001, biosecurity is an increasing priority among food manufacturers, merchandisers, retailers, and commercial farmers nationwide.

Deterrence and Prevention

Primary prevention and deterrence interventions for foreign pests and diseases include international treaties and standards (such as the International Plant Protection Convention, and those of the OIE/World Organization for Animal Health), bilateral and multilateral cooperative efforts, offshore activities in host countries, port-of-entry inspections, quarantine, treatment, and postimport tracking of plants, animals and their products.

Every link in the agricultural production chain is susceptible to attack with a biological weapon. Traditionally the first defense against a foreign animal or plant disease has been to try to keep it out of the country. Agricultural inspectors at preclearance inspections and at the U.S. borders are the first line of defense (source: National Research Council 2004). Smuggling interdiction efforts can act as deterrents before biological agents reach their target.

DHS and USDA already conduct such inspection and quarantine practices, but continued oversight is necessary to determine which preparedness activities and threats require more attention. Offshore activities include preclearance inspection by APHIS of U.S. imports before products leave their port of origin. APHIS has personnel in at least 27 host countries. Although many of these inspections programs were built to target unintentional threats, they are being augmented with personnel and technology to look for intentional threats.

Various U.S. intelligence and law enforcement agencies collect information about biological weapons that could be used against U.S. agriculture. Building and maintaining a climate of information sharing between USDA, DHS, and the intelligence community is necessary, especially so that agriculture is not overlooked compared to other infrastructure and human targets.

Once inside the United States, many parts of the food production chain may be susceptible to attack with a biological weapon. For example, terrorists may have unmonitored access to geographically remote crop fields and livestock feedlots. Diseases may infect herds more rapidly in modern concentrated confinement livestock operations than in open pastures. An undetected disease may spread rapidly because livestock are transported more frequently and over greater distances between farms, and to processing plants. Processing plants and shipping containers need to be secured and/or tracked to prevent tampering.

An important line of defense is biosecurity, or the use of preventive security measures. On the farm, biosecurity is the use of farm management practices that both protect animals and crops from the introduction of infectious agents and contain a disease to prevent its rapid spread within a herd or to other farms. Biosecurity practices include structural enclosures to limit outside exposure to people and wild animals, and the cleaning and disinfection of people, clothing, vehicles, equipment, and supplies entering the farm.

Most farm specialists agree that livestock farmers are increasingly aware of the importance of biosecurity measures, particularly since the FMD outbreaks in European cattle and the avian flu and exotic Newcastle infections in U.S. poultry. More farm operators are requiring visitors to wear boot covers to guard against bringing in disease. Regardless of the reason for following biosecurity measures (terrorism or accidents), these precautions help prepare farms against agroterrorism.

Detection and Response

Biological attacks on crops and livestock may not be immediately apparent. Therefore, existing frameworks for detecting, identifying, reporting, tracking, and managing natural and accidental disease outbreaks are being applied to combating agroterrorism. Appropriate responses are being developed based on specific pathogens, targets, and other circumstances that may surround an attack.

DHS and USDA have responded with a more detailed and coordinated plan to secure the food supply, particularly with the announcement of HSPD-9. The departments are cooperating on research funding, detection technology, surveillance, partnerships with private industry, and state and local response coordination.

Within private industry, the Food and Agriculture Information Sharing and Analysis Center (ISAC) shares information with government intelligence bureaus through the DHS. The Food and Agriculture ISAC includes more than 40 of the primary trade associations representing food and agriculture. Such ISAC centers exist in several industries and are one of the primary partnerships between government and industry for counterterrorism cooperation. By combining information among members in the same industry, security problems or attacks may become apparent more quickly than observations within individual companies. In the event of a terrorist incident, the ISAC would facilitate communication within the industry and coordinate response efforts with government officials. The Food and Agriculture ISAC was created in February 2002 and is administered by the Food Marketing Institute. In 2003, three sub-ISACs were created to cover more specific threats and information sharing for (source: Food Marketing Institute, "Food and Agriculture ISAC (Information Sharing and Analysis Center), at http://www.fmi.org/isac/.)

- Agriculture
- Food manufacturing and processing
- Retail

In addition to the ISAC, DHS recently created the Food and Agriculture Sector Coordinating Council, which will oversee food security and incident management. The Council includes seven subcouncils: plant producers, animal producers, manufacturers/processors, restaurants/food service, retail, warehousing, and agricultural production inputs (source: Food Chemical News, "Food Industry Creates New Homeland Security liaison groups," July 12, 2004).

The exact methods for control and eradication operations are difficult to predict. Past experience and simulations have shown that day-to-day decisions would be made using "decision trees" that include factors such as the geographical spread, rates of infestation, available personnel, public sentiment, and industry cooperation. Response procedures are outlined in the APHIS PPQ *Emergency Programs Manual* and the APHIS Veterinary Services *Federal Emergency Response Plan for an Outbreak of Foot-and-Mouth Disease or Other Highly Contagious Diseases.*

In an outbreak, damage is proportional to the time it takes to first detect the disease. If a foreign disease is introduced, responsibility for recognizing initial symptoms rests with farmers, producers, veterinarians, plant pathologists, and entomologists. Cooperative Extension Service agents at state universities are receiving additional training on recognizing the likely symptoms of an agroterrorism attack.

Effective detection depends on a heightened sense of awareness, and on the ability to rapidly determine the level of threat (e.g., developing and deploying rapid disease diagnostic tools). Lessons from disease outbreaks, including the recent FMD outbreaks in Europe and avian flu in Asia and the United States, show that the speed of detection, diagnosis, and control, spell the difference between an isolated incident and an economic and public health disaster.

However, in recent years, the number of veterinarians with experience to recognize many foreign animal diseases has declined. This is because the United States has been successful in eradicating many animal diseases. Also, the number of veterinarians available across the country with large animal experience and within APHIS has declined. In light of this trend, APHIS has initiated efforts to increase training for foreign animal diseases and create registries of veterinarians with appropriate experience.

Most of the initial response to the diagnosis of a foreign animal disease is at the state and local level. If an outbreak spreads across state lines or if state and local efforts are unable to control the outbreak, federal involvement quickly follows. Numerous simulation exercises have been conducted by federal, state and local authorities to test the response and coordination efforts of an agroterrorism attack. Examples of such simulations include the Silent Prairie exercise in Washington on February 11, 2003), the Silent Farmland exercise in North Carolina (August 5, 2003), and Exercise High Stakes in Kansas (June 18, 2003).

The last line of defense, and the costliest, is the isolation, control, and eradication of an epidemic. The more geographically widespread a disease outbreak, the costlier and more drastic the control measures become. Officials gained valuable experience from recent agricultural disease outbreaks such as avian influenza in the United States, Canada, and Asia; FMD in the U.K.; and citrus canker in Florida. Each one of these epidemics has required the depopulation and destruction of livestock and crops in quarantine areas, indemnity payments to farmers, and immediate suspension of trade.

Of all lines of defense, mass eradication is the most politically sensitive and difficult. Actions taken in each of these outbreaks have met with varying degrees of resistance from groups opposed to mass slaughter of animals, citizens concerned about environmental impacts of destroying carcasses, or from farmers who fear the loss of their livelihood. During the 2001 outbreak of FMD in the United Kingdom, the public was clearly opposed to the large piles of burning carcasses. The disposal of millions of chicken carcasses in British Columbia, Canada, during 2004 also caused a significant public debate. Thus, scientific alternatives are needed for mass slaughter and carcass disposal. Citrus canker eradication efforts in Florida's residential neighborhoods illustrate how science-based measures have been challenged and delayed in the courts, or how farmers may be reluctant to voluntarily test crops or livestock.

Laboratories and Research

Since September 11, 2001, the United States has expanded its agricultural laboratory and diagnostic infrastructure, and created networks to share information and process samples. So far, 19 universities and institutions have been tapped for the USDA-funded National Plant Diagnostic Network and its sister group, the National Animal Health Laboratory Network. A main goal of each is to improve the diagnostic and detection system in the event of a deliberate or accidental disease outbreak.

The effectiveness of these networks will require coordinated outreach, observers say, and cooperative extension services will take on new prominence in their role of providing information about diseases such as soybean rust to farmers and others who have regular contact with farms.

Within the USDA, several agencies have upgraded their facilities to respond better to the threat of agroterrorism by expanding laboratory capacity and adding physical security. These programs include the ARS research on foreign animal diseases at the Plum Island Animal Disease Center in New York (the physical facility is now managed and operated by DHS) and the ARS Southeast Poultry Research Laboratory in Athens, Georgia.

Also at USDA, three major laboratories are consolidating operations in a new BSL-3 facility in Ames, Iowa. These include the ARS National Animal Disease Center, the APHIS National Veterinary Services Laboratories (NVSL), and the APHIS Center for Veterinary Biologics. The complex will be USDA's largest animal health center for research, diagnosis, and product evaluation. The NVSL is especially visible because it makes the final determination of most animal diseases when samples are submitted for testing.

USDA also cooperates with other federal agencies on counterterrorism research and preparedness, including the ARS and APHIS partnership with the U.S. Army Medical Research Institute for Infectious Diseases at Fort Dietrick, Maryland. The Fort Dietrick site offers USDA access to additional high-level biosecurity laboratories. In recent years, USDA has conducted research on soybean rust at Fort Dietrick.

In April 2004, the DHS Science and Technology Directorate announced the department's first university research grants for agriculture as part of its "centers for excellence" program. The University of Minnesota and Texas A&M will share $33 million over three years. Texas A&M's new Center for Foreign Animal and Zoonotic Disease Research will study high consequence animal diseases. The University of Minnesota's new Center for Post-Harvest Food Protection and Defense will establish best practices for the management of and response to food contamination events. Texas A&M is partnering with four universities and will receive $18 million; Minnesota is partnering with 10 universities and will receive $15 million.

Federal Authorities

When a foreign animal disease is discovered, whether accidentally or intentionally introduced, the Secretary of Agriculture has broad authority to eradicate it or prevent it from entering the country. The use of these authorities is fairly common, as shown recently by the import restrictions imposed during the 2004 outbreak of avian influenza in Asia. Federal quarantines and restrictions on interstate movement within the United States are also common for certain pest and disease outbreaks, such as for sudden oak death in California and citrus canker in Florida. In addition to federal authorities, most states have similar authorities, at least for quarantine and import restrictions.

For example, if an animal disease outbreak is found in the United States, the Secretary of Agriculture is authorized, among other things, to

- Stop imports of animals and animal products into the United States from suspected countries (7 U.S.C. 8303).
- Stop animal exports (7 U.S.C. 8304) and interstate transport of diseased or suspected animals (7 U.S.C. 8305).
- Seize, quarantine, and dispose of infected livestock to prevent dissemination of the disease (7 U.S.C. 8306).
- Compensate owners for the fair market value of animals destroyed by the Secretary's orders (7 U.S.C. 8306(d)).
- Transfer the necessary funding from USDA's Commodity Credit Corporation to cover costs of eradication, quarantine, and compensation programs (7 U.S.C. 8316).

Similar authorities cover plant pests and diseases (7 U.S.C. 7701-7772).

Recovery Management

Several activities such as confinement and eradication start in the response phase but continue throughout the management and recovery phase. Long-term economic recovery includes resuming the husbandry of animals and plants in the affected areas, introducing new genetic traits that may be necessary in response to the pest or disease, rebuilding confidence in domestic markets, and regaining international market share.

Confidence in food markets, by both domestic and international customers, depends on continuing surveillance after the threat is controlled or eradicated. Communication and education programs would need to inform growers directly affected by the outbreak, and inform consumers about the source and safety of their food. The social sciences and public health institutions play a complementary role to the agricultural sciences in responding to and recovering from agroterrorism.

If eradication of the pest or disease is not possible, an endemic infestation would result in a lower equilibrium level of production or quality. Resources would be devoted to acquiring plant varieties with resistance characteristics and breeds of animals more suitable to the new environment. This is the goal of the NPDRS mentioned in HSPD-9 and being initiated by APHIS.

Summary

In the wake of the events of September 11, 2001, it must be understood that the terrorist threat exists in the nation at all times, and it is certainly possible that some form of agroterrorism, perhaps in conjunction with biological or chemical threats, could happen and therefore, preparation is necessary. This is especially true in the realm of agroterrorism, where such an incident, even one that is in reality relatively minor, could have severe effects on consumer confidence, the supply-and-demand economy, and the various associated businesses that would be affected by some form of terrorism-caused outbreak related to an American farm. Agroterrorism will not carry the shock value that bombings and hijackings do, and the effect on human life may not be as severe, but significant economic issues can arise that affect many facets of the population beyond the farmer. These issues could expand all the way out to U.S. import and export markets, to the federal government itself, which could incur significant costs to contain and eradicate the threat, as well as potentially compensating farmers for destroyed animals. Such a destruction of animals en masse because of the discovery of a terrorist threat could also raise environmental and other health issues that must be addressed, using more time and resources at all levels.

After September 11, 2001, there was a significant increase in the attention paid to a variety of terrorist threats, both large and small. But for various reasons, there was less attention paid to agroterrorism. This has been addressed in the past couple of years, through acts such as HSPD-7, which added agriculture to the list of critical infrastructure that must be protected. HSPD-9 took this a step further by establishing a national policy to protect against terrorist attacks on agriculture and food systems. In addition, the President's annual budget request to Congress now includes a cross-cutting budget analysis of homeland security issues, and from USDA, six agencies and three offices receive or have requested funding related to homeland security. Such funding is categorized based on six mission areas (functions), as defined in the National Strategy for Homeland Security.

There have only been a limited number of occurrences of agroterrorism on U.S. soil. Maladies that could potentially strike in large scales against herds of cattle and other types of animals have been thought to be dealt with through vaccination and other programs to educate farmers on their potential dangers. However, scientists now believe that livestock herds are much more susceptible to agroterrorism than crops, because current herds are either not vaccinated against threats or are relatively unmonitored against such threats, because they may have been thought to be eradicated previously. Certain animal diseases may be more attractive to terrorists because they can be transmissible to humans.

The Agricultural Bioterrorism Protection Act of 2002 created the current, official list of animal pathogens that are of greatest concern for agroterrorism. The act requires that these lists be reviewed at least every two years. In addition, there is overlap between the CDC and APHIS because some pathogens on the list may not cause a disease, but may cause symptoms such as food poisoning or responses in the central nervous system. One pathogen, FMD, is mentioned often when agroterrorism is discussed, because of its ease of use, ability to spread quickly, and potential for tremendous economic damage. There is also a similar list of plant pathogens, as required by the Agricultural Bioterrorism Protection Act of 2002. The goal of the U.S. animal and plant health safeguarding system is to prevent the introduction and establishment of exotic plants and diseases to mitigate their effects and eradicate them where necessary/possible. Part of this effort requires coordination and cooperation between federal agencies to not only safeguard domestic products and resources, but also those that may be imported in from foreign countries. Through inspection of cargo and the requirement of importers to report specific types of cargo that could fall into an agroterrorism issue, agencies have the power to fight the entry of foreign diseases or agents. However, should a foreign animal disease be discovered, whether accidentally or intentionally introduced, the Secretary of Agriculture has broad authority to eradicate or prevent it from entering this country. The use of these authorities is fairly common. Federal quarantines and restrictions on interstate movement within the United States are also common for certain pest and disease outbreaks.

The federal government has, through the involvement of many agencies and offices, taken steps to prevent agroterrorism wherever possible, and to respond to an incident should an outbreak occur.

References

1. Henry S. Parker, Agricultural Bioterrorism: A Federal Strategy to Meet the Threat, McNair Paper 65, National Defense University, March 2002.
2. Monterey Institute of International Studies.
3. Peter Chalk, "Hitting America's Soft Underbelly: The Potential Threat of Deliberate Biological Attacks Against U.S. Agriculture and Food Industry," Rand National Defense Research Institute, January 2004.
4. Charles Hanrahan and Geoffrey Beckier, *Mad Cow Disease and U.S. Beef Trade*, CRS Report RS21709, August 4, 2004.

Chapter 17

Pandemic Preparedness

Douglas Himberger

Contents

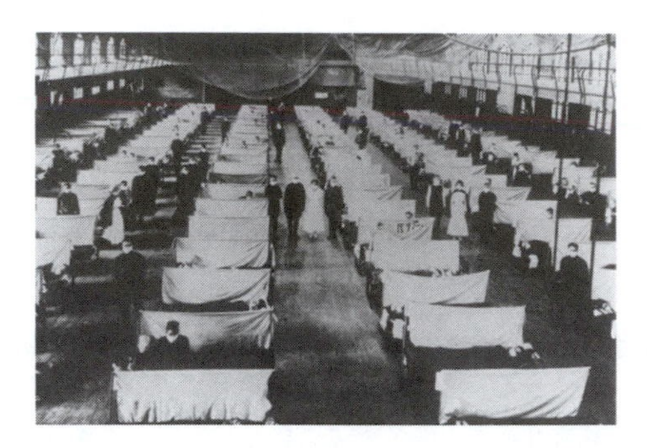

Figure 17.1 Dealing with pandemics has often been a seemingly insurmountable problem, but preparedness planning might change this outlook. (Office of the Public Health Service Historian.)

Pandemic preparedness is a topic of periodic and great concern, as are pandemics themselves. Often, the public's awareness of these events is raised to a heightened level, although the same is not necessarily true for the public's overall preparedness. Enormous sums have been spent on such preparedness—literally billions of dollars in the United States alone between 2004 and 2010. However, many would say that our preparedness for these infectious disease events has changed little from a decade earlier. This state of affairs is attributed to a combination of several effects: the nature of pandemics as compared with other crises; the unique preparedness requirements of pandemics; and the somewhat challenging and burdensome pandemic preparedness planning activities. This chapter discusses each of these effects, along with activities during and after a pandemic, and the way ahead in pandemic preparedness planning (Figure 17.1).

The Nature of Pandemics

Health Concerns of Pandemics

A pandemic is defined as "... an epidemic (a sudden outbreak) that becomes very widespread and affects a whole region, a continent, or the world."[1] The *Collaborative International Dictionary of English* describes a pandemic as an "everywhere epidemic." Pandemics are often thought of in terms of notable historical events, such as the first (of as many as seven, to date[2]) Asiatic Cholera Pandemic of 1817; the "Spanish Flu" or "Great Pandemic" of 1918; and the Severe Acute Respiratory Syndrome (SARS) Pandemic of 2003 (probably more appropriately deemed an "epidemic" because of its relatively limited geographic scope). Figure 17.2 (adapted from data provided by Phil Hoad[3]) depicts notable pandemics over the ages [this figure includes the recent 2005 to 2009 "Avian Influenza" (type A/H5N1 variant) and 2009 to 2010 "Swine Influenza" (type A/H1N1 variant) pandemics].

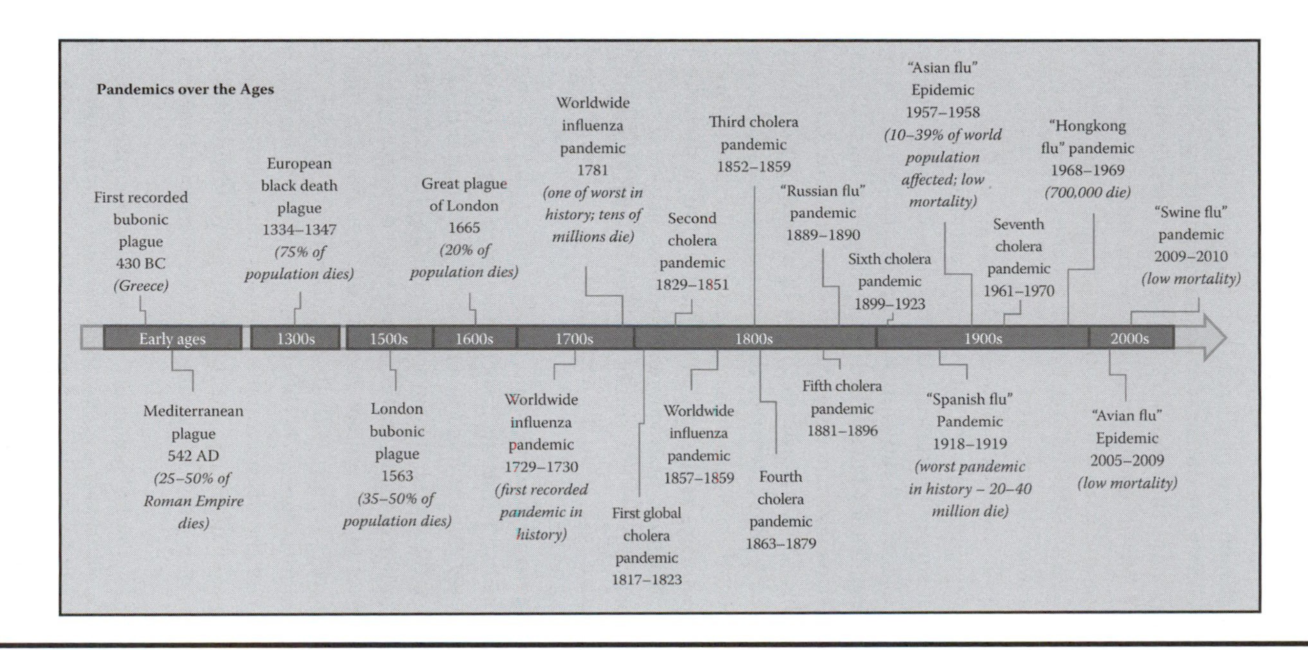

Figure 17.2 Pandemics have occurred for centuries and show no signs of lessening either their frequency or their impact. (Adapted from Hoad, Phil, 2003, "Pandemics Timeline," http://www.guardian.co.uk/society/2003/apr/02/health.lifeandhealth?INTCMP=SRCH.)

Pandemic crises are recurring on average three times per century, and preparing for "the next one" should be continuous. Dr. David Nabarro, former Senior United Nations System Coordinator for Avian and Human Influenza and a senior expert at the World Health Organization (WHO), said in 2005 that he believed a major pandemic would erupt soon, and predicted that it might kill roughly 5 million to 150 million people.[4] Clearly, the threat is real, and preparedness is imperative.

In each major pandemic, the most devastating impact was in the area of health care. The scores of people sickened and killed by the infectious diseases affected families, communities, and even entire countries or global regions. The diseases placed "sudden and intense demands on health systems,"[5] and overarching community structures also were often broken.

The illness and death caused by these pandemics resulted in societal disruption. The "psyche" of populations was so severely degraded that the outlook of those citizens was not simply depressed—the pandemics colored the very future of their societies. As Figure 17.2 shows in stark terms, the toll can be devastating. Continuing cholera pandemics have killed staggering numbers of people worldwide (although no reliable estimates are available, particularly for the early cholera pandemics, roughly 33,000 deaths occurred in a *single day* during the second global pandemic of 1829–1851[3]), whereas the worst single pandemic, the "Spanish Flu" Pandemic* of 1918–1919 killed an estimated 20 million to 40 million people globally (with some estimates as high as 50 million to 100 million deaths), and as many as 675,000 in the United States alone.[6] Numbers like these take a heavy toll on affected populations not only in terms of treating the sick and dying but also in dealing with mass burials in a culturally acceptable way and tending to the associated mental health impact.

In fact, in absolute terms of sickness and death (or morbidity and mortality), the recent "Avian Flu" potential pandemic associated with the type A/H5N1 virus was relatively minor: only 549 cases resulted in a relatively high mortality of 320 deaths.[7] However, the effect that these cases had on society worldwide was chilling: the "worried well" continued to tap valuable health resources, and the fear generated by the disease spanned many countries and continents, disproportionally affecting behavior. Similarly, compare these effects with those caused by the 2009 to 2010 "Swine Flu" (correctly referred to as type A/H1N1 influenza virus) pandemic: in the final analysis, the virulence of the disease might or might not be less than some of the more profound historical pandemics, but the resultant worldwide fear and anxiety affected many aspects of society. For example, Vice President Biden at one point

* The pandemic was referred to as the "Spanish Flu" pandemic, although the disease is believed to have possibly started at a military base in Kansas (note that other research indicates that it might have begun in the Far East or Austria) and been carried by World War I U.S. troops to other parts of the world. Because Spain was a neutral country during the conflict and had no wartime censorship in place, it was one of the few news outlets reporting on the pandemic (particularly when it moved from France to Spain)—hence, the term "Spanish Flu" might have been coined as a result. The variant of the influenza virus dominant in the Spanish Flu pandemic was H1N1.

said that he would advise his family not to travel on subways or airlines, and many citizens followed suit.

The most troubling characteristic of these viruses may be their unpredictability. Public health academic Philip Alcabes, author of "Dread," says that instead of looking to physicians to predict epidemics, "... we should leave the job of seeing the future to the mystics, prophets, and fortunetellers." This level of uncertainty breeds anxiety.

Nonetheless, there are certainties in pandemics: pandemics will recur, and they will have great impact on society. In a prescient statement in 2005, a Congressional Budget Office (CBO) report stated, when referring to the threat of the H5N1 Avian Flu virus, "... [the H5N1 virus] could evolve in a way that rendered it harmless, and a pandemic could arise from an entirely different virus subtype."* Although the H5N1 Avian Flu virus cannot yet be deemed "harmless," certainly the type A/H1N1 Swine Flu virus became a public pandemic concern.

Community Continuity Concerns of Pandemics

The impact of pandemics goes beyond the health implications discussed. The entire socioeconomic system is deeply affected, with effects being felt well beyond the health community. "[Pandemics] expose existing weaknesses in these systems and, in addition to their morbidity and mortality, can disrupt economic activity and development."[5] Nearly all facets of a community, or even a nation, can be affected gravely, from mundane daily tasks to broad strategic operations.

These operations are disrupted on several levels. The people required for any level of community activities—governance, education, healthcare, transportation, food distribution—are the parts of the infrastructure that are affected profoundly. Not only are there fewer people to conduct these vital activities, but the people with appropriate skills are also affected in unpredictable ways. Although most experts forecast high levels of absenteeism, the specific absentees themselves are not predictable. Although as much as 40% of a population may be unable to function because of their own or a family member's sickness or death,[8] there is no way to predict exactly which individuals would be affected. Critical functions might be affected to even a greater degree, and overall continuity of operations could be compromised severely.

Communities need a plan for pandemic preparedness, but how do they know when these plans must be put into practice? Several tools, or yardsticks, exist for measuring the pandemic threat. Two of the best known are WHO's pandemic alert

* Congressional Budget Office, "A Potential Influenza Pandemic: Possible Macroeconomic Effects and Policy Issues," December 8, 2005, revised July 27, 2006, www.cbo.gov/ftpdocs/69xx/doc6946/12-08-BirdFlu.pdf. This report draws heavily on multiple sources, including the World Health Organization (WHO), *Avian Influenza: Assessing the Pandemic Threat* (Geneva: WHO, January 2005); Howell, Pugh, *Pandemic, The Cost of Avian Influenza, Contingencies* (September/October 2005), pp. 22–27; and Garrett, Laurie, *The Next Pandemic, Foreign Affairs* (July/August 2005), pp. 3–23.

level tool,[9] and the U.S. Federal Government's [Department of Health and Human Safety (HHS)] response stage tool.[10] Figure 17.3 depicts these tools.

For each WHO phase or HHS stage, there are "triggers" that lead either organization to elevate the levels. These same triggers inform communities and populations about changes in the pandemic's nature. As Figure 17.3 illustrates, WHO and HHS monitor these triggers, but lower level entities, including state and regional public health organizations, local community health authorities, and even businesses, also should monitor the triggers. The more aware these organizations are, the more prepared they will be to implement appropriate preparedness actions.

<table>
<tr><th colspan="2">WHO Phases</th><th colspan="2">Federal Government Response Stages</th></tr>
<tr><td colspan="4">Interpandemic period</td></tr>
<tr><td>1</td><td>No new influenza virus subtypes have been detected in humans. An influenza virus subtype that has caused human infection may be present in animals. If present in animals, the risk of human disease is considered to be low.</td><td rowspan="2">0</td><td rowspan="2">New domestic animal outbreak in at-risk country</td></tr>
<tr><td>2</td><td>No new influenza virus subtypes have been detected in humans. However, a circulating animal influenza virus subtype poses a substantial risk of human disease.</td></tr>
<tr><td colspan="4">Pandemic alert period</td></tr>
<tr><td rowspan="2">3</td><td rowspan="2">Human infection(s) with a new subtype, but no human-to-human spread, or at most rare instances of spread to a close contact.</td><td>0</td><td>New domestic animal outbreak in at-risk country</td></tr>
<tr><td>1</td><td>Suspected human outbreak overseas</td></tr>
<tr><td>4</td><td>Small cluster(s) with limited human-to-human transmission but spread is highly localized, suggesting that the virus is not well adapted to humans.</td><td rowspan="2">2</td><td rowspan="2">Confirmed human outbreak overseas</td></tr>
<tr><td>5</td><td>Larger cluster(s) but human-to-human spread still localized, suggesting that the virus is becoming increasingly better adapted to humans, but may not yet be fully transmissible (substantial pandemic risk).</td></tr>
<tr><td colspan="4">Pandemic period</td></tr>
<tr><td rowspan="4">6</td><td rowspan="4">Pandemic phase: increased and sustained transmission in general population.</td><td>3</td><td>Widespread human outbreaks in multiple locations overseas</td></tr>
<tr><td>4</td><td>First human case in North America</td></tr>
<tr><td>5</td><td>Spread throughout United States</td></tr>
<tr><td>6</td><td>Recovery and preparation for subsequent waves</td></tr>
</table>

Figure 17.3 WHO and federal pandemic response levels. (From U.S. Department of Health and Human Services, www.PandemicFlu.gov, 2009.)

The WHO phases measure the *transmissibility* of a virus, not the *severity* of a resultant pandemic. This is a key difference because many pandemic plans are based on severity of the illness. Although the WHO might measure a pandemic as being at a high level (e.g., levels "5" or "6"), the WHO is indicating that triggers show the virus to be transmissible at the levels indicated in the WHO phase chart (e.g., "widespread human infection" for levels 5 and 6), not that resulting illnesses are necessarily severe (although that might also be the case). The WHO recently made clarifications to their phase structure,* but the phases themselves still relate to the transmissibility of the virus.

With respect to severity, at this writing, the WHO has not yet developed a system for measuring this characteristic of a pandemic. "Severity" refers to not only the degree of virulence of a virus (e.g., number of severe illnesses and deaths, contagiousness of the virus, age distribution of cases, prevalence of chronic health problems and malnutrition of the population, viral mutations, number of waves of illness, and quality of health services) but also the overall socioeconomic impact of the outbreak.[11]

Although the WHO may be developing such a severity index, HHS already has one in place. Based on hurricane classifications, this system is shown in Figure 17.4.[12] The HHS Pandemic Severity Index is part of the agency's guidance on community interventions for combating a pandemic. The index is based on case fatality rates (CFR), with a CFR of 2% or greater signaling the most severe pandemic (Category 5). The earlier described outbreaks of 1957 and 1968 would be rated as Category 2 events, with CFRs between 0.1% and 0.5%.[12] Note that there is potential for public confusion when using this index in combination with the WHO pandemic phases, which may have led to WHO's delay in presenting a similar index. HHS has presented numerous cross-references between their index and WHO's and has included characteristics and interventions for each severity level.

The WHO and HHS tools depend on faithful and timely reporting of the relevant triggers. During the 2009 and 2010 Swine Flu pandemic, it was reassuring to see that no apparent cover-up existed on the part of any nation with respect to reporting the illness (as some had believed might have been the case for reporting the 2002 and 2003 SARS event). However, there was concern not only that some time had passed—perhaps three weeks or more[13]—between the first cases in Mexico and the first WHO warnings, but also that some time had passed before the Centers for Disease Control and Prevention (CDC) identified the novel variant

* The World Health Organization (WHO), http://www.who.int/csr/disease/avian_influenza/phase/en/index.html. Of note, "In the 2009 revision of the phase descriptions, WHO has retained the use of a six-phase approach for easy incorporation of new recommendations and approaches into existing national preparedness and response plans. The grouping and description of pandemic phases have been revised to make them easier to understand, more precise, and based on observable phenomena. Phases 1–3 correlate with preparedness, including capacity development and response planning activities, while Phases 4–6 clearly signal the need for response and mitigation efforts. Furthermore, periods after the first pandemic wave are elaborated to facilitate postpandemic recovery activities."

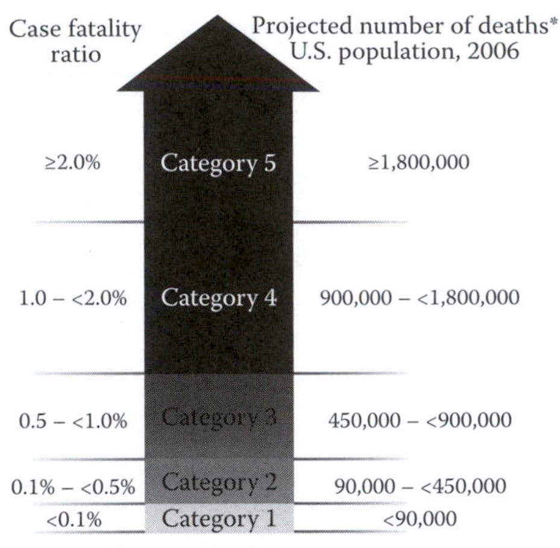

Figure 17.4 Pandemic severity index. (From U.S. Department of Health and Human Services, "Community Strategy for Pandemic Influenza Mitigation," February 2007.)

of the virus. Going forward, these delays must be eliminated so that our pandemic preparedness plans can be implemented when they can have the greatest impact.

Psychosocial Concerns of Pandemics

As stated earlier, the effects of a pandemic go well beyond health concerns. The cyclical nature of pandemics, as well as the rapid transmission of the diseases, generates fear. The "worried well" further overwhelm an already besieged health care infrastructure. Individuals whose family members or friends have been sickened or have died suffer from mental health issues that could disable even the strongest among us. Other practical issues result; in 1918, during the Spanish Flu pandemic, Philadelphia suffered such losses of adult citizens that the Bureau of Child Hygiene determined that it was unable to care for the large number of orphans.[14] Even worse, the city was unable to cope with the numbers of dead. If a pandemic of similar virulence were to occur today, there could be "1.7 million deaths in the United States and 180 million to 360 million deaths globally,"[15] clearly overloading the health care system, as well as the funereal communities, and taking these essential services beyond the breaking point.

One does not need to look to horrific pandemic data to sense the fragility of a community's health care infrastructure. In 2006, a two-week heat wave in

California caused as many as 141 deaths, leading "coroners ... [to deal with] the large jump in the number of bodies that were stuffed, some piled on top of others, into the freezers at the Fresno County morgue."[16]

Another consideration for planning for pandemics is the very nature of human behavior. It is well known that such behavior is nonlinear, and therefore might seem unpredictable. However, if sufficient information is known about a community's culture, historical actions, and motivations, behavior can be predicted with some fidelity. Unfortunately, a deep understanding of all these factors is seldom possible, particularly after a pandemic has started. Therefore, the more a community knows about itself and its underlying structure, and the more that knowledge is factored into preparedness planning, the more effective that planning will be.

When developing a plan that fits a community, it is important to consider all aspects of that population. There are five major aspects to consider: (1) culture of the group; (2) policies, strategies, and plans in place; (3) economics, management, and budgeting of the entities involved; (4) governance and operations; and (5) technology implemented to deal with a crisis (e.g., information databases, communication networks).[17]

Planners must thoroughly understand all five aspects. For instance, if the culture of a group is to be prepared, but the policies for preparedness are not in place, or the technologies to enable the group to be thoroughly prepared are not implemented, preparedness might not be possible. Similarly, developing plans for such preparedness must take into account all these elements, or the plans will not work.

Economic Impacts of Pandemics

As described in the section "Psychosocial Concerns of Pandemics," the effects of pandemics go well beyond health, and specifically impact the economic wellbeing of a community. The economic toll that pandemics can impose is tremendous and far-reaching.

The economic cost occurs at both ends of a pandemic. Costs to prepare the population for pandemics can be enormous, as can costs to respond to and recover from the health crisis itself. For example, within days of the first outbreaks of the Swine Flu in early 2009, the World Bank alone responded to the pandemic with "fast disbursing funds" ($25 million for drugs and supplies, and $180 million for epidemiologic, regulatory, institutional, and operational activities[18]), and many other agencies and institutions followed course. These amounts are likely only a portion of the total amount of funding applied; where those funds were found and what other programs were impacted is unknown at this writing, but costs associated with a pandemic were and will be high.

As another example, in November 2005, the Bush administration requested $7.1 billion in emergency funding for pandemic preparedness[19] (although less funding was ultimately made available to planners and responders, the majority of this funding request was allocated). The activities supported included development of plans, stockpiling of antivirals, production of vaccines, and similar key actions. Long-term infrastructure would ultimately be strengthened, including source control and surveillance, vaccine research and development, antiviral drug research and development, and health care system readiness.[19]

The last of these is of particular importance. When it comes to pandemics, our health care infrastructure is fragile and insufficient. During a severe pandemic, there would be far more demand on the U.S. health care system than the system could accommodate (as many as 5 million to 10 million sick individuals,[20] exceeding the roughly 970,000 staffed hospital beds and 100,000 ventilators,[19] with 75% of those in use at any given time under normal, nonpandemic situations). Consequently, building a stronger health care infrastructure before a pandemic would be key. Even a mild pandemic such as the 2009 and 2010 Swine Flu pandemic (considered mild in severity, although transmissible at a relatively high level) demonstrated to communities that the stress on the health infrastructure could be enormous. A sizable percentage (about 25%) of the federal stockpile of antiviral drugs was distributed quickly to states and localities; however, it was clear that local heath departments would have difficulty distributing them rapidly to the population as a result of budget cuts and layoffs at local levels. Treating a sick population would have been difficult, or even impossible. Supporting the recovery of those who had been ill would add another level of stress to the health care system.

The indirect costs (e.g., decreased supply from the shrinking workforce, and a dramatic decline in demand for goods and services because people would avoid shopping malls, restaurants, and other public places) during and after a pandemic can be even more staggering. The economic impact of recent pandemics, even relatively minor ones such as the SARS pandemic of 2003, has been enormous. By some estimates, the indirect costs attributed to disruption of activity in business, civil, and governmental domains related to the SARS event were as much as $30 billion to $100 billion (note that the CBO and others continue to debate these figures). In 2005, the World Bank estimated that a 2% loss of global gross domestic product (GDP) would result from a pandemic of similar severity to 1918, which translates into $800 billion in losses for a year.[21] Recent estimates raise this expected impact to nearly 5% of global GDP, or about $3 trillion.[22] This impact is "greater than in recent recessions and roughly the same size as the average postwar recession."[19]

Unique Preparedness Requirements of Pandemics

It is clear that we must prepare for pandemics. But what is the goal of such preparedness? This issue continues to be debated, but the CBO (drawing on WHO and others) has identified the following goals for the federal government:[19]

- Support for the efforts of governments of other countries and international organizations to contain [virus] strains and control their evolution to diseases that are transmitted easily from person to person.
- Building stockpiles of vaccines, improving antiviral drugs, and putting in place new technologies that allow effective vaccines to be produced more rapidly and in large quantities.
- Improving the capacity of the health system to care for many people in all parts of the country who are sick simultaneously.

Plans at other levels (states, regions, localities, and individual organizations and entities) have parallel goals appropriate for their own needs, but the last objective typically is the driving goal for planners.

Pandemics at Hand—Pandemic Influenzas: Avian and Swine

A pandemic of great recent concern is *avian flu*. This flu infects birds, such as domestic poultry (e.g., chickens) and wild fowl (e.g., ducks). "Avian flu ... is caused by influenza viruses that occur naturally among wild birds. Low pathogenic [avian flu] is common in birds and causes few problems [to humans]. Highly pathogenic H5N1 is deadly to domestic fowl, can be transmitted from birds to humans, and is deadly to humans. There is virtually no human immunity, and human vaccine availability is very limited."[23] In fact, "viruses of the H5 subtype are not known to have ever circulated among the human population, which means that there would be little immunity to it."[19] Therefore, it is this highly pathogenic version of avian flu that is a risk to communities and has the potential for pandemic impact.

Similarly, *swine flu* is caused by influenza viruses that occur in swine. However, although the immunity and human vaccine availability is like that of avian flu, the 2009 and 2010 Swine Flu pandemic appeared to be a variant that was a mutation (or "reassortment") of avian flu, swine flu, and human flu.* The resultant virus appeared to be transmitted with some efficiency from human to human, leading the WHO to raise the pandemic alert phase to 5 in April 2009. This phase carried with it the fact that the virus was "characterized by human-to-human spread ... into at least two countries in one WHO region. Although most countries [would] not be affected at this stage, the declaration of Phase 5 [was] a strong signal that a pandemic [was] imminent and that the time to finalize the organization, communication, and implementation of the planned mitigation measures [was] short."[24] A

* "According to virologists at the U.S. Centers for Disease Control and Prevention (CDC), the influenza A subtype H1N1 isolated from ... patients in April [2009] was a genetic reassortment of four different influenza virus strains, including human influenza gene segments, swine influenza from North America and Eurasia, and avian gene segments from North America, never before reported among swine or human isolates from anywhere in the world." Food and Agricultural Organization [FAO] (2009). "The human influenza due to a novel subtype H1N1." www.fao.org/ag/againfo/programmes/en/empres/AH1N1/Background.html.

significant number of cases had been reported worldwide, with numbers rising and at levels that many considered alarming. ("As of 27 December 2009, worldwide, more than 208 countries and overseas territories or communities have reported laboratory confirmed cases of pandemic influenza H1N1 2009, including at least 12,220 deaths."*) The WHO subsequently raised the outbreak to a full pandemic level, Phase 6, indicating a global pandemic was underway.

Note that every community already suffers significant impact from *seasonal flu*, a common respiratory illness transmitted from person to person. Although many people have some immunity to this illness, and vaccines are available annually, each year in the United States alone an average of 36,000 deaths occur as a result of seasonal influenza. These deaths, along with more than 200,000 U.S. hospitalizations and more than $10 billion in U.S. economic cost,[25] accompany more than one-quarter to one-half million deaths worldwide annually.[26] This toll should lead to a heightened awareness of hygiene principals—principals that would be effective for not only a pandemic but also seasonal flu. Unfortunately, some say our culture has adopted a *laissez-faire* approach to this disease and to measures effective in preventing its transmission.

There are encouraging developments, however. Recently, public service announcements (PSAs) and other media communications have been emphasizing the importance of hygiene for limiting transmission of infectious disease (e.g., suggesting to citizens that they wash their hands regularly, taking the washing time necessary to sing the tune "Happy Birthday" twice). Although it has been suggested that the public had been suffering from "information fatigue" as a result of extensive coverage of the Avian Flu pandemic, and that subsequently media reports had waned, recent coverage had again risen, particularly as a result of the Swine Flu pandemic. This coverage can be helpful in preparing the public for pandemics, but only if the messages are clear and easily understood.

Numerous interesting new tools and techniques related to pandemics are being introduced—not only to the health care community and emergency preparedness professionals, but also to the public. For example, keying on Internet searches, Google's Flu Trends looks at the "relative popularity of a slew of flu-related search terms to determine where in the U.S. flu outbreaks may be occurring." "What's exciting about Flu Trends is that it lets anybody—epidemiologists, health officials, moms with sick children—learn about the current flu activity level in their own

* U.S. Centers for Disease Control and Prevention (CDC), "Pandemic (H1N1) 2009—update 81," http://www.who.int/csr/don/2009_12_30/en/index.html. As of May 1, 2009, 331 cases were reported in 11 countries, with 10 resultant deaths. By May 10, 2009, only nine days later, the reported cases had risen dramatically, to 4379, with 49 deaths. (World Health Organization [WHO], http://www.who.int/csr/don/2009_05_10/en/index.html.)

state based on data that's coming in this week."* Given the popularity of Google web tools, including Google Earth (now being used to track avian flu outbreaks and mutations worldwide), the public will have techniques never before available to aid them in their awareness and sensitivity to preparedness activities. Pandemic preparedness plans must include these tools, and planners must use the tools in exercises.

Persistence and Pervasiveness of Pandemics

Pandemics are different from other crises. Events such as floods, tornadoes, hurricanes, and wildfires are typically limited to relatively small areas. By definition, pandemics are crises that affect wide areas, often covering entire countries, continents, or even the globe.

This geographic coverage greatly complicates preparedness planning. For most emergency scenarios, planners can depend on other resources and aid from areas not affected by the crisis. Firefighters often travel to serious fires; first responders are often brought in to disaster scenes such as floods or hurricanes; and food and other resources are typically flown to stricken areas. Planners for pandemics must assume that potentially the pandemic will similarly hit *all* other areas, and people in other areas will be unable to assist. Pandemic plans must build on locally available staff, food, water, power, transportation, and other resources. The plans must call on the affected areas to weather the pandemic organically. As is sometimes said, the citizens must be "their own first responders."

Temporal Requirements of Pandemic Preparedness

Pandemic planners must not only assume that little or no outside help will exist but also accommodate another unique aspect of pandemics: typically, pandemics are not bounded by a short, easily defined timeframe. Rather, they often occur in "waves" (each with a several-week duration) separated by some time period. These waves can be individually devastating; collectively, they can sap a population's energy and resources as response, recovery, and planning phases begin to overlap one another.

These waves can be separated by many years, as has been the case for the global cholera pandemics (e.g., seven of which have occurred from 1817 to the present).

---

* Landau, Elizabeth (2008), "Google tool uses search terms to detect flu outbreaks," http://edition.cnn.com/2008/HEALTH/conditions/11/11/google.flu.trends/index.html. As further stated in the article, "The Centers for Disease Control and Prevention collaborated with Google on the project, helping validate and refine the model, and has provided flu tracking data over a five-year period." Furthermore, "In the 2007 and 2008 flu season, Google accurately estimated current flu levels one to two weeks faster than published CDC reports in each of the nine U.S. surveillance regions, Google said in a statement." Google's Flu Trends can be found at www.google.org/flutrends.

Pandemics separated by such long periods are more manageable in some ways. Preparedness for a successive wave can build on lessons learned from a previous wave, with sufficient time elapsing for replenishment of vital medical and other resources. However, if pandemic waves are separated by shorter periods (measured in weeks or months), the collective impact can be demoralizing and overwhelming. For example, the Spanish Flu pandemic of 1918 and 1919 attacked in three waves: the first occurring in the spring/summer of 1918; the second, in the fall of 1918; and the third and final, in the spring of 1919. Future influenza pandemics are expected to follow a similar pattern, causing great concern to planners and responders alike.

Pandemic Preparedness Planning

The concept of planning for a pandemic is not only a notion of practical or intuitive need but also one of policy. The highest level federal planning guidance addressing all crisis preparedness was the Department of Homeland Security's (DHS's) National Response Plan (NRP) (issued in December 2004), which was superseded by the National Response Framework (NRF) (issued on March 22, 2008). The NRP, and subsequently the NRF, is an "all-discipline, all-hazards plan intended to establish a single, comprehensive framework for managing domestic incidents."[27] The NRF describes federal support to be implemented through activation of 15 emergency support functions (ESFs), including several relevant to pandemic planning: Mass Care, Emergency Assistance, Housing, and Human Services (ESF #6); Public Health and Medical Services (ESF #8); Public Safety and Security (ESF #13); and Long-Term Community Recovery (ESF #14). Each of the 15 ESFs, including those relevant to pandemics, includes details about crisis planning, implementation, training, and exercising.[28]

Furthermore, the federal government has developed specific detailed guidance with respect to pandemics—the National Strategy for Pandemic Influenza[25]—motivated largely by the threat of the type A/H5N1 virus ("Avian Flu"). Although the National Strategy was developed in response to the Avian Flu, it addresses the preparation for and response to any pandemic, and it is fully consistent with higher level policy and planning documents such as the NRP and the subsequent NRF. The National Strategy "guides our preparedness and response to an influenza pandemic, with the intent of (1) stopping, slowing, or otherwise limiting the spread of a pandemic to the United States; (2) limiting the domestic spread of a pandemic, and mitigating disease suffering and death; and (3) sustaining infrastructure and mitigating impact to the economy and the functioning of society."[25] There are three "pillars" in the National Strategy: (1) preparedness and communication; (2) surveillance and detection; and (3) response and containment. Each has considerable detailed guidance for communities and the general population in business, civil, and governmental situations.

Following the National Strategy, the National Strategy for Pandemic Influenza Implementation Plan was released.[29] This plan presents specific actions for more than 300 federal departments and agencies (e.g., Department of Defense, HHS, Veteran's Administration), and lower level implementation plans by these agencies have followed.

This section addresses the process for developing pandemic preparedness plans, and focuses on essential elements of not only the plans themselves but also the implementation of those plans.

Developing a Pandemic Preparedness Plan

The National Strategy includes key overarching goals (described previously). (The National Strategy goals are elaborated in a U.S. Homeland Security Council publication.[30]) Any lower level pandemic preparedness plan must be in consonance with these high-level goals. Such lower level plans may well be developed carefully, and are almost certainly well intentioned, but many are "not legally or logistically feasible." Furthermore, it has been said that "lessons from simulations had not been drawn on to revise plans."[22] In short, the plans must be realistic, appropriate to the entity or entities being addressed, and exercised thoroughly. These exercises must be carried out repeatedly so that participants will become trained and plans can be updated, based on exercise results. Situations for an affected population may change, and the plan must be agile enough to accommodate these changes, but these can be appreciated only in the context of realistic exercises. All too often, a preparedness plan (for pandemics or for other crises) becomes "shelf-ware"—developed with care, but then relegated to a shelf until a crisis hits. At this point, it is too late to update the plan or exercise the potentially affected population.

Numerous tools are in the pandemic preparedness plan toolbox, including the following: (1) vaccines, (2) antiviral medications, (3) infection control measures, and (4) community mitigation measures (Figure 17.5). Each is complex and could potentially have a broad effect on a pandemic; however, none is likely to completely address the severe impact of a pandemic. Most likely, all these tools (and perhaps others) will be needed. Lisa Koonin, senior advisor for the Influenza Coordination Unit and lead for Pandemic Medical Countermeasures at the CDC, likens this issue to swiss cheese. A single mitigation tool is like a single slice of cheese—full of holes. However, when we lay one slice over another, and then another, and then another, we find that the holes may be partially—or even completely—filled. Similarly, these preparedness toolbox techniques are additive; as many as possible must be used to achieve the maximum mitigating effect.

Vaccines are a unique and troublesome element of a pandemic preparedness plan. Typically, vaccines can be quite effective, providing a measure of immunity from the virus to an individual. They are in widespread use for other viruses, including seasonal influenza, where they have been shown to be effective. Typically,

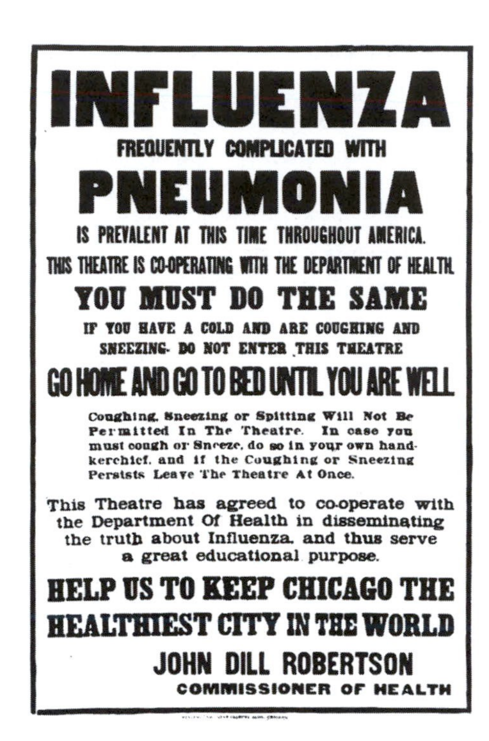

Figure 17.5 Infectious disease transmissibility mitigation techniques are long-standing and likely will be employed again as part of overall pandemic plans and their implementation. (Office of the Public Health Service Historian.)

pandemic preparedness plans specifically address the availability, distribution, and use of vaccines.

However, a vaccine must be matched precisely to the virus in circulation. The laborious process for developing this "match" typically takes months.* This time lag is crucial; if a highly transmissible virus were to cause a pandemic, many thousands (even millions) of people could be infected, many of whom could die before a vaccine were to become available. Furthermore, the time lag is not the only issue concerning vaccines. It is possible or even likely that the vaccines, once developed, would not be available in sufficient quantities to be distributed to all affected populations. Therefore, prioritization of recipients will be required (e.g., health care workers and emergency management responders could be first on such

* There is hope for great improvement in this area. New vaccine development techniques promise dramatic decreases in time required to develop the vaccine. For instance, the National Institute of Standards and Technology and the University of Queensland in Australia have "successfully demonstrated ... findings that could reduce the time it takes to produce a vaccine from months to weeks...." National Institute of Standards and Technology, *NIST Tech Beat*, December 8, 2008).

a list). Practical and ethical issues are certain to arise. For example, should those in failing health (unrelated to the influenza) be administered the vaccine rather than those who are younger and more vital? Should "key" staff be administered a vaccine before those deemed less critical for operations? Government and health care officials would be expected to resolve these endless issues. Other issues also would need to be considered, including topics such as use and efficacy of prepandemic vaccines (ones that might be available before a pandemic but not perfect matches to the circulating virus).

Antivirals do not function like vaccines. Antivirals do not provide immunity, but they may make the illness less severe or shorten the course of the illness. They are not specific to a virus in circulation (although considerable differences exist in effectiveness and the like for each variation of antiviral), and large stockpiles (e.g., tens of millions of regimens in the United States alone) already have been developed and stored. Antivirals have demonstrated some effectiveness in prophylactic use, but they are not intended to immunize a recipient to a virus.

However, antivirals also have shortcomings. Because they can be used prophylactically, there is fear that they will be overused or cause the virus strain to develop resistance. Resistance, whether a result of overuse or not, can be significant. For the 2008 and 2009 seasonal flu, virtually all flu cases have shown resistance to Oseltamivir,* a popular antiviral. Furthermore, although large quantities of antivirals are stockpiled, there still might be a need to prioritize who gets them and when.

Another tool is infection control. Many control measures are commonplace and intuitive, but others are less so. Encouraging standard hygiene processes (e.g., hand washing, covering coughs and sneezes) has a strong impact on limiting the transmission of an influenza virus. Other personal protective equipment (PPE) such as surgical masks and other respiratory devices (e.g., N95 masks, respirators) has been shown to be effective, although such equipment can be burdensome, expensive, possibly misused, and effective only in certain situations. The key issue for all these techniques is compliance; without constant and vigilant use, these tools are ineffective. As discussed earlier, training is a key element in developing compliance; the more communities are trained, the more likely that they will comply with these infection control approaches.

Finally, community mitigation techniques are an important part of pandemic preparedness planning. A key mitigation approach is social distancing, requiring that people maintain a minimum distance between one another to "reduce the duration and/or intimacy of social contacts and thereby limit the transmission of

---

* *L.A. Times*, "Tamiflu No Longer Works for Dominant Flu Strain," February 4, 2009, http://articles.latimes.com/2009/feb/07/science/sci-flu7. Although this news report states specifically that only the H1N1 virus is showing this resistance (thus avoiding such resistance from the H5N1 Avian Flu variant), this caused further concern over the H1N1 Influenza A (Swine Flu) variant that posed a global threat in 2009.

influenza,"[30] either through shifts in operation times and/or locations,* or individual behavior modification. The CDC and OPM published guidance about such behavior modification for the 2009 and 2010 Swine Flu crisis: maintain a distance of 6 feet from each other unless PPE masks are used. Other guidance for mitigation included using flexible work schedules to reduce face-to-face interactions.[31] Other approaches could be used: minimizing mass gatherings (e.g., schools, churches, malls); practicing isolation (for people who are sick or tending to the sick); or even quarantining (a method not typically expected to be used extensively in a modern pandemic, but might be used in severe circumstances). These tools are available to planners and responders, and pandemic plans will almost certainly include several of these techniques.

Many excellent examples of pandemic preparedness plans have been developed, most of which are available online at state emergency management sites (an example of such a site is the Virginia Emergency Management website, http://www.vaemergency.com/library/plans/index.cfm), as well as others specific to businesses and civil entities. Each plan typically includes guidance regarding assumptions, scenario descriptions, roles and responsibilities, a concept of operations, incident management actions, and maintenance of the plan itself.

Of note, the WHO released a checklist for pandemic influenza preparedness planning,[32] based on their guiding document, "Pandemic Influenza Preparedness and Response."[33] This guidance draws on "extensive practical experience ... gained from responding to outbreaks of highly pathogenic avian influenza A (H5N1) virus infection in poultry and humans, and from conducting pandemic preparedness and response exercises in many countries. There is greater understanding that pandemic preparedness requires the involvement of not only the health sector, but the whole of society."[33] Furthermore, "WHO decided ... to update its guidance to enable countries to be better prepared for the next pandemic."[33]

Training for and Exercising Pandemic Preparedness

Training is critical to the success of pandemic plans. It is often said that the military is successful largely because they "train like they fight, and fight like they train." Pandemic plans may call for significant changes in behavior (e.g., social distancing described previously); these changes will come about only with frequent training and exercising, particularly if behaviors are cultural in nature. For example, take the common practice of shaking of hands when meeting. If not practiced often, *not* shaking hands might be perceived as rude rather than a way to lessen the transmission of infectious diseases.

---

* These shifts can notably include "telework." OPM announced a new government-wide telework policy in April 2009, which combined components of the Telework Improvements Act (H.R. 1722) and the Telework Enhancement Act (S. 707).[31]

Training and exercising of preparedness plans is becoming more commonplace, even for pandemic plans. Setting the tone for such exercises is the series of federal exercises, Top Officials (TOPOFF), and the subsequent National Exercise Program (NEP).* These exercises address emergencies such as the release of biological agents, radiological dispersal devices, and chemical agents. TOPOFF was a "congressionally mandated, national terrorism exercise that was designed to identify vulnerabilities in the nation's domestic incident management capability by exercising the plans, policies, procedures, systems, and facilities of federal, state, and local response organizations against a series of integrated terrorist threats and acts in separate regions of the country." Top officials were engaged in the decision-making processes they would face in a real-world disaster, from public health concerns to communications challenges. "The purpose of the open exercise design was to enhance the learning and preparedness value of the exercise through a "building-block" approach, and to enable participants to develop and strengthen relationships in the national response community. Participants at all levels stated that this approach has been of enormous value to their domestic preparedness strategies."[34]

In particular, lessons learned from the exercises have helped pandemic planners design exercises at federal, regional, state, or local levels for pandemic scenarios: PANEX 07, a Federal Emergency Management Agency–hosted, joint federal–state exercise; and FBIIC/FSSCC Pandemic Flu Exercise of 2007, a Department of the Treasury–hosted exercise involving the banking and financial services sectors. The latter example has clear goals: "enhance the understanding of systemic risks ... [on] sector[s]; provide an opportunity for firms to test their pandemic plans; and examine how the effect of a pandemic flu on other critical infrastructures will impact ... sector[s]."[35] These should be the goals for any exercise intended to validate a pandemic preparedness plan.

Specific tools and approaches that are components of pandemic plans must also be trained and exercised. Often, compliance with these techniques, once used, is key to their effectiveness. As an example, although social distancing appears to be an effective mitigation technique, it is not easy nor foolproof. A recent study[36] showed that the process can be described, a plan for implementation developed, and a study group tested—but that the results depend on overall compliance of the subjects, so clearly diligent training and exercising is critical to this mitigation approach.

Other guidance for developing exercises intended to validate pandemic plans also is available (e.g., WHO's "Exercise Development Guide for Validating Influenza Pandemic Preparedness Plans"[37]). The leadership and advice that these documents

* "With the establishment of the NEP and the Post-Katrina Emergency Management Reform Act (PKEMRA), requirements were set forth to conduct all-hazards exercises annually. Accordingly, the congressionally mandated TOPOFF series, which focused on terrorist attacks, has evolved into the NLE series, which takes an all-hazards approach, focusing on preparation for catastrophic crises ranging from terrorism to natural disasters." (U.S. Department of Homeland Security (DHS). Frequently Asked Questions (FAQ) National Level Exercise 2011 (NLE 11), http://www.ready.gov/nle2011/_downloads/NLE2011_FAQ.pdf.)

provide will increase the effectiveness of our pandemic plans, but only if they are used to thoroughly and frequently exercise the plans themselves, and involve those who will be affected by a pandemic.

Dynamically Replanning for Pandemic Preparedness

Any plan must be revisited regularly and adjusted for changes in the threat, environment, and participants. Therefore, plan(s) must be reviewed and exercised not only periodically but also when circumstances necessitate review, such as a change in guiding policy and a change in the nature of a virus.

During and after a Pandemic

Even with diligent prepandemic planning, some issues need to be addressed when a population is in the midst of a pandemic. Pandemics may come in waves, and successive planning, taking into account lessons learned from a previous wave, can drastically reduce the socioeconomic impacts of subsequent waves. This section presents some issues that might be faced.

Responding to Pandemic Infection

One question that will be particularly relevant to any population is whether individuals are infected and who these people are. There is considerable risk in overreacting to this question (ranging from social "shunning" to possibly illegal quarantining); therefore, screening, detection, and response will be necessary, and all should be considered and included in a holistic pandemic plan. Of interest is a set of new techniques for this screening and testing. For example, in early 2009, CDC developed a rapid diagnostic test kit to detect the type A/H1N1 virus and distributed these kits to all 50 U.S. states, the District of Columbia, Puerto Rico, and internationally. Although of lower sensitivity than viral culture or other similar traditional tests, these kits and others like them will increase rapid testing capacity and are likely to result in a more accurate, rapid picture of the impact of this disease.

Similarly, also in early 2009, the U.S. Food and Drug Administration (FDA) cleared a new, more rapid test for the detection of type A/H5N1.* "This test is an important tool for helping quickly identify emerging influenza A/H5N1 infections and reducing exposure to large populations," said Daniel G. Schultz, director of the FDA's Center for Devices and Radiological Health. "The clearance of this test

* "The test, called AVantage A/H5N1 Flu Test, detects influenza A/H5N1 in throat or nose swabs collected from patients who have flu-like symptoms. The test identifies in less than 40 minutes a specific protein (NS1) that indicates the presence of the influenza A/H5N1 virus subtype. Previous tests that the FDA cleared to detect this influenza A virus subtype can take three to four hours to produce results." (U.S. Food and Drug Administration. FDA News release, April 7, 2009.)

represents a major step toward protecting the public from the threat of pandemic flu."

Communicating during a Pandemic

As with any emergency situation, effective, transparent, and complete communication is of utmost importance. This section describes the emerging tools and techniques relevant for pandemic events.

Social networking tools are emerging as a powerful means of communication in a broad set of scenarios. Not only are they used routinely for interpersonal or social interactions, but also they are becoming dominant for these purposes in many arenas. In fact, "... social networks and blogs are now the 4th most popular online activity ... [and are] ahead of personal email. Member communities are visited by 67 percent of the global online population, [and the] time spent is growing at three times the overall internet rate, accounting for almost 10 percent of all internet time."[38] Beyond the personal applications, business social networks also are growing. One of the fast-growing business networks, LinkedIn, is enjoying a surge of activity; "LinkedIn membership is up to 85 million [as of December 2010] ... every second that ticks by, LinkedIn gets a new user."[39]

It is in this context of broad community participation that these networks are becoming effective tools for emergency communication. This is certainly true as it applies to pandemics. As examples, many such tools have set up specific sections for pandemic discussion and communication, including LinkedIn, Facebook, and Twitter. These sites are in addition to the numerous, more traditional websites devoted to these topics; excellent government sites also exist (e.g., http://www.pandemicflu.gov/; http://www.cdc.gov/flu/Pandemic/), as do many outstanding private or civic sites (e.g., h1n1alliance.org/; cidrap.umn.edu/). Add to these a wealth of other communication venues, such as PSAs, special public audio/video conferences and symposia,* and the paths to communicate the pandemic message (in the preparedness phase and in the response and recovery phases) are many and diverse (Figure 17.6).

One particular concern regarding pandemic communications is the notion of alarming the public unnecessarily. Certainly, public awareness is important, but avoiding information fatigue (or even being perceived as "crying 'swine'" as one report quoted[40]) is equally key. Communication must be continually measured in terms of its intent, its process, and its effectiveness.

* A powerful example of this media communication is the Public Broadcast System (PBS) video, *Predicting Pandemics—How Do We Fight Both the Swine Flu Pandemic and Our Fear of It?*, of May 8, 2009.

Figure 17.6 Public health messages are not a new technique for addressing pandemics, but one that will need to continue. (U.S. Department of Health and Human Services.)

Recovering after a Pandemic

Pandemics have a powerful effect on a population. Nearly every measure of a community or nation can be affected. For example, the average life span of U.S. citizens dropped nearly 12 years in the year following the Spanish Flu pandemic of 1918 and 1919. The way citizens behave, and the way they see themselves and each other, fundamentally changes when a pandemic devastates their way of life. Recovery, therefore, is neither an easy nor a quick process.

Specific guidance addresses recovery from pandemics. For example, DHS has prepared a document, "Pandemic Influenza Preparedness, Response, and Recovery Guide for Critical Infrastructure and Key Resources."[41] This document focuses on businesses (stating as its purpose to "stimulate the U.S. private sector to act now"), but it specifically includes consideration of recovery activities following a pandemic. "Continuity of Operations–Essential (COP-E) is the central concept in the guidance, wherein COP-E is an extension and refinement of current business contingency and continuity of operations planning that fully exploits existing efforts and integrates ... the suite of business disaster plans. The COP-E process assumes severe pandemic-specific impacts to enhance and complement existing business continuity plans."[41] Relevant to recovery, COP-E incorporates an approach for "survival" under distinct COP-E scenarios, and it enhances business continuity planning to address other catastrophic disasters. This holistic approach to continuity of operations is at the heart of recovery for a nation, region, or community.

Summary

Pandemic preparedness is an imperative for all communities, as rampant infectious diseases can have broad and sometimes devastating consequences. Although the public's awareness of potential pandemic events may be raised through media coverage, the same may not be true for overall preparedness. Pandemics are unpredictable, but this much is known: they will happen again, and they will more than likely have great impact on society. The impact goes beyond the health implications; the entire socioeconomic system of a population will be affected.

Rapid transmission of the disease generates enormous fear in the population. As discussed, certainly those who are ill stress the community's ability to respond; the "worried well," driven by that fear, overwhelm the already besieged health care infrastructure.

When developing a tailored pandemic preparedness plan for a community or large population group, several aspects must be considered: culture; policies, strategies, and plans; economics, management, and budgeting; governance and operations; and technology.

Planning for a pandemic must be consistent with existing culture and policy. The U.S. Government has several goals for preparing for pandemics, and all lower

Figure 17.7 Pandemics have been a part of our lives for centuries, and will be again in the future, but will we be prepared? (Office of the Public Health Service Historian.)

level plans should match these objectives: stopping, slowing, or otherwise limiting the spread of a pandemic; mitigating disease suffering and death; sustaining the infrastructure; and mitigating impact to the economy and functioning of society (Figure 17.7).

Training participants and exercising pandemic preparedness plans is critical to the ultimate success of implementing those plans. In the balance is the protection of citizens but ultimately the recovery of society.

We have seen that pandemic preparedness is not only a government construct but also a process that must involve nongovernmental entities (nonprofits, faith-based organizations, and businesses) to reach a level of preparedness that enables resilience to be woven into the process of mitigating effects and accelerating recovery.

Preparedness is neither an easy nor a quick process. Pandemics may pose one of the more difficult scenarios for planners, but also may be one of the most important—another pandemic is most certainly on its way.

References

1. MedicineNet.com. http://www.medterms.com/script/main/art.asp?articlekey=4751.
2. Nevondo, T. S., and T. E. Cloete. 2001. "The Global Cholera Pandemic." http://scienceinafrica.co.za/2001/september/cholera.htm.
3. Hoad, Phil. 2003. "Pandemics Timeline." http://www.guardian.co.uk/society/2003/apr/02/health.lifeandhealth?INTCMP=SRCH.
4. Nabarro, David. 2005. "CIDRAP News, 29Sep05." http://www.cidrap.umn.edu/cidrap/content/influenza/avianflu/news/sep2905avian2.html.
5. World Health Organization (WHO). "Communicable Disease Surveillance and Response." http://www.wpro.who.int/southpacific/sites/ccd/csr/.
6. U.S. Department of Health and Human Services (HHS). "The Pandemic." The Great Pandemic—the United States 1918–1919. http://1918.pandemicflu.gov/the_pandemic/index.htm.
7. World Health Organization (WHO). "Cumulative Number of Confirmed Human Cases of Avian Influenza A/(H5N1) Reported to WHO, 11 April 2011." http://www.who.int/csr/disease/avian_influenza/country/cases_table_2011_04_11/en/.
8. U.S. Department of Health and Human Services (HHS). "Pandemic Planning Assumptions." http://www.flu.gov/professional/pandplan.html.
9. World Health Organization (WHO). "Current WHO phase of pandemic alert for avian influenza H5N1." http://www.who.int/csr/disease/avian_influenza/phase/en/index.html.
10. U.S. Department of Health and Human Services (HHS). "HHS Pandemic Influenza Implementation Plan." http://www.hhs.gov/pandemicflu/implementationplan/intro.htm.
11. CIDRAP News. "WHO: H1N1 Flu More Contagious than Seasonal Virus." http://www.cidrap.umn.edu/cidrap/content/influenza/swineflu/news/may1109severity.html.
12. U.S. Department of Health and Human Services (HHS). "Community Strategy for Pandemic Influenza Mitigation," February 2007. http://www.flu.gov/professional/community/commitigation.html.

13. *New York Times*, "A Spotty Response to the Flu Threat." http://www.nytimes.com/2009/05/02/opinion/02sat1.html?scp=1&sq=A%20Spotty%20Response%20to%20the%20Flu%20Threat&st=cse.
14. U.S. Department of Health and Human Services (HHS). "The Great Pandemic—the United States 1918–1919." http://1918.pandemicflu.gov/the_pandemic/01.htm.
15. Osterholm, Michael T. 2005. "Preparing for the Next Pandemic," *New England Journal of Medicine* 352, 1839–1842.
16. Maitre, Michelle. 2006. "Heat Linked to More Than 130 Deaths." *Oakland (CA) Tribune*, July 29, 2006.
17. Sulek, D., R. Cowell, and M. Delurey. 2008. Mission Integration—A Whole of Government Strategy for a New Century. Booz Allen Hamilton white paper, December 2008.
18. The World Bank. "Influenza A (H1N1): Questions and Answers." http://web.worldbank.org/WBSITE/EXTERNAL/NEWS/0,contentMDK:22160388~pagePK:64257043~piPK:437376~theSitePK:4607,00.html.
19. Congressional Budget Office. "A Potential Influenza Pandemic: Possible Macroeconomic Effects and Policy Issues," December 8, 2005, revised July 27, 2006. http://www.cbo.gov/ftpdocs/69xx/doc6946/12-08-BirdFlu.pdf.
20. Lister, Sarah. 2005. "Pandemic Influenza: Domestic Preparedness Efforts," CRS Report for Congress RL33145 (Congressional Research Service, November 10, 2005), pp. 10.
21. Brahmbhatt, Milan. 2005. "Avian and Human Pandemic Influenza—Economic and Social Impacts." http://web.worldbank.org/WBSITE/EXTERNAL/NEWS/0,contentMDK:20715087~menuPK:3325365~pagePK:34370~piPK:42770~theSitePK:4607,00.html.
22. Worsnip, Patrick. 2008. "Bird Flu Pushed Back, Pandemic Threat Remains—UN." http://uk.reuters.com/article/healthNews/idUKTRE49K8BY20081021.
23. U.S. Department of Health and Human Services (HHS). "Pandemic Terms Defined." http://www.pandemicflu.gov/popup.html.
24. CARE. "WHO updates the definitions of pandemic phases (but does not raise the alert level)." http://avianflunetwork.blogspot.com/2009/04/who-updates-definitions-of-pandemic.html.
25. U.S. Homeland Security Council. The National Strategy for Pandemic Influenza, November 1, 2005.
26. World Health Organization (WHO), Influenza (Seasonal)—Fact Sheet No. 211. http://www.who.int/mediacentre/factsheets/fs211/en/index.html.
27. U.S. Department of Homeland Security. National Response Plan, December 2004.
28. U.S. Department of Homeland Security. National Response Framework, March 22, 2008.
29. U.S. Homeland Security Council. The National Strategy for Pandemic Influenza Implementation Plan, May 3, 2006. www.whitehouse.gov/homeland/nspi_implementation.pdf.
30. U.S. Homeland Security Council. "National Strategy for Pandemic Influenza Implementation Plan One Year Summary," July 2007. http://www.flu.gov/professional/federal/pandemic-influenza-oneyear.pdf.
31. U.S. Office of Personnel Management (OPM). Memorandum for Heads of Executive Departments and Agencies—Advice to Federal Employees and Agencies on Preventing the Spread of the Current Flu and Maintaining Readiness to Use HR Flexibilities if Necessary, April 26, 2009.

32. World Health Organization (WHO). "WHO Checklist for Influenza Pandemic Preparedness Planning." http://whqlibdoc.who.int/hq/2005/WHO_CDS_CSR_GIP_2005.4.pdf.
33. World Health Organization (WHO). "Pandemic Influenza Preparedness and Response." http://www.who.int/csr/disease/influenza/pipguidance2009/en/.
34. U.S. Department Of Homeland Security (DHS). Top Officials (Topoff) Exercise Series: Topoff 2—After Action Summary Report, December 19, 2003. www.Dhs.Gov/Xlibrary/Assets/T2_Report_Final_Public.Doc-2003-12-19.
35. FBIIC FSSCC SIFMA and U.S. Department of the Treasury. The FBIIC/FSSCC Pandemic Flu Exercise of 2007 After Action Report. http://www.fspanfluexercise.com/Pandemic%20Flu%20AAR.pdf.
36. Magoon, M. A., D. E. Himberger, and J. M. Bishop. 2009. "Social Distancing and Hygiene as an Influenza Pandemic Mitigation Strategy: Employee Compliance and Performance." Organizational Behavior Management Network (OBMN) Leadership Conference, Cocoa Beach FL, February 20, 2009.
37. World Health Organization (WHO). "Interim Report," February 2006. http://www.wpro.who.int/NR/rdonlyres/DA340E3E-D27E-47A6-9833-452E7AAC9ED5/0/EDTedDRAFT1ExerciseDevelopmentGuide.pdf.
38. The Nielson Company. "Global Faces and Networked Places—A Nielsen report on Social Networking's New Global Footprint," March 2009. http://blog.nielsen.com/nielsenwire/wp-content/uploads/2009/03/nielsen_globalfaces_mar09.pdf.
39. Slutsky, Irina. "Why LinkedIn Is the Social Network That Will Never Die." http://adage.com/article/digital/linkedin-social-network-die/147475/.
40. Associated Press. 2009. "Swine Flu Warnings Totally Overblown, Some Say," May 7, 2009. http://www.msnbc.msn.com/id/30627377/.
41. U.S. Department of Homeland Security (DHS). "Pandemic Influenza Preparedness, Response, and Recovery Guide for Critical Infrastructure and Key Resources." http://www.flu.gov/professional/pdf/cikrpandemicinfluenzaguide.pdf.

Chapter 18

Special Events

Patrick J. Jessee

Contents

Municipalities and jurisdictions all have different resources. The resources of New York, New York, and Hot Springs, Arkansas, are not necessarily the same. Although some common resources such as law enforcement, fire, EMS, and public works exist, the assets they have are distinctly different. Despite having these clearly different levels of equipment, experience, and training, they have the same common goals of protecting and providing for the civilian population. This includes normal operations throughout the year as well as special gatherings and events.

Special event planning is a unique challenge for emergency managers and public safety professionals. Emergency managers and public safety professionals typically plan for potential events or catastrophic responses. For them, many of the planning phases are for events that may occur, rather than an actual event that will occur. Special events present a known specific event that creates an extra stress on the resources of a municipality beyond normal expectations. Planning for these special events in advance can help alleviate the burden on municipal resources that would occur during a special event.

Should an incident occur during a special event, a lack of planning would create an unpleasant and hazardous environment for those attending the event. Therefore, it is critical for emergency managers and public safety professionals to analyze potential issues associated with special events to help create a secure venue for these special events. Fiscally responsible emergency managers and public safety professionals may consider the benefits of the following analysis to help plan and prepare for special events, which are applicable to all municipalities—from densely populated urban areas to small and sparsely populated villages throughout the country.

Special Event Types

When thinking about special events, the first thought that comes to mind are large-scale festivals such as the Olympics or 4th of July celebrations. These special events require significant planning by emergency managers and public safety professionals. An acceptable definition of a *special event* should be determined. Many different types of events may qualify as a special event, but what constitutes a special event is partially determined by the emergency managers and public safety professionals.

An event may be determined as a special event for a number of reasons: who is in attendance, what the event represents (e.g., a protest rally or march), the significance of the event for the city or community, etc.

Special events may occur in a certain area because of yearly celebrations or festivals, such as the 4th of July celebrations or county fairs. They may also occur because of championship games for sporting events. Another possibility is dignitary visits to an area (e.g., a congressman or foreign ambassador). All of these events fulfill the broad definition of special events, yet they all require significantly different planning considerations. Emergency managers and public safety professionals can use the same basic considerations for each of these events and then augment the planning process as needed to fulfill the objectives.

Certain events that occur throughout the United States qualify as National Security Special Events (NSSEs) and are deemed of particular importance. These large-scale incidents of national significance bring a large amount of media attention, political importance, and high crowd densities, and present a significant threat for homeland security and counterterrorism experts. When an event is designated as an NSSE, the Secret Service, Federal Bureau of Investigation (FBI), and Federal Emergency Management Agency (FEMA) take lead roles in managing the event.

Some examples of previously designated NSSEs are Super Bowl games, political national conventions (both Democratic and Republican), and even the Academy Awards. This section will not discuss the NSSEs because of the federal level involvement in planning for NSSEs and the resources that the federal government brings to these events. Additional open source information on NSSEs can be found online at the DHS or FEMA websites.

Large-scale events require considerable attention and planning by emergency managers and public safety professionals. These events, such as holiday celebrations, may be planned out up to a year in advance (events such as the Olympics may be planned many years in advance). These events may require extra security, fire and rescue, EMS, or first aid tents. Public works may need to contribute with extra sanitation vehicles or roadblocks, to name a few. These agencies all contribute to making a full-scale operation that operates in the background of a public event.

In some special events, emergency managers and public safety professionals only have days to plan as they are last-minute events added into schedules. In order to accommodate this wide array of events and timing, emergency managers and public safety professionals must be flexible and resourceful. The most important factor, however, is to be informed of the event and to share open-source information with all stakeholders. This will allow the rapid synthesis of a plan as all parties involved to efficiently contribute to the development of the response plan.

Smaller municipalities may face more challenges in dealing with a special event because they have less resources or personnel. This should not hinder the emergency managers and public safety professionals. Instead, they should think outside the box and use creative problem solving to develop sound and economically feasible solutions. To a smaller municipality, county fairs, art shows, or even parades could be events that emergency managers and public safety professionals wish to have plans for.

Why are special events of particular importance for emergency managers and public safety professionals? These events represent a potential for a mass-casualty event in case of either natural (floods, tornados, fires) or man-made disasters (riots, shootings, terrorism). The sheer number of people gathered in a venue presents a target-rich environment for violent terrorist activities that frequently occur in the presence of media practitioners.

Additionally, many events have limited access or egress from the venue, which creates a bottleneck for both incoming and outgoing crowds. This type of situation can potentially lead to a high number of injuries in case an incident occurs in that location or as a result of trampling as people make a mass exodus because of the incident.

Particular attention should be paid to access points. Establishing multiple access points helps to minimize bottlenecks as people enter the venue. However, it has the drawback of requiring more personnel and equipment to man these posts. Having multiple entrance points also helps to facilitate the exit of people during a potential threat or disaster. Most people, when presented with a threat, will leave by the way in which they entered. A consideration to account for this would be to have many points of egress established throughout the venue that can be opened as needed. Ultimately, the decision to establish multiple points arises from the event planners.

Emergency managers and public safety professionals should give special attention to the layout of the special event venue. A thorough knowledge of the building or topography of the area will help emergency managers and public safety professionals create secondary access/egress points should an event happen. EMS personnel will be able to understand the quickest routes to gain access and remove patients from the area. Law enforcement may be able to understand where potential weaknesses in the boundaries exist. Fire departments will have a knowledge of water resources and building construction. Hazmat will be able to identify inherent chemicals/hazards present within the venue and monitor continuously throughout for unknowns.

Special events also typically capture the attention of those who are not even in attendance. Many of these places have media present also. Local or national news crews may be on-site to report on the events of the festival or gathering. Print media may be present, mingling with the visitors. Television media may have cameras or a stage established to report from. Of course, the ever-present Internet presence is there as people update their status on social networks such as Facebook, MySpace, or Twitter. A constant stream of reports on the atmosphere of the gathering is inevitable. The atmosphere as well as the public service presence, whether overt or low-profile (intermingled with the crowd), are constantly updated through persistent open-source updates.

This constant presence of media reporting can provide feedback to personnel operating at the scene should it be monitored through the Internet. This constant media presence creates a singular focus for the world should an event, natural or man-made, occur. Any injuries that occur from these incidents turns into an instantaneous news story, one that captivates the audience, especially during the first few hours of an event as it unfolds. This is one primary reason to plan for these events—to quickly respond and rectify any incidents as they occur.

Intelligence Sharing

It is implicitly understood that intelligence sharing must occur for numerous agencies to work together under one Joint Operations Center (JOC). The decision to plan for a special event generally comes from an elected official. This person may choose what response is appropriate or he may delegate it to his subordinates for the planning sessions. At other times, significant threat information from law enforcement or even a meteorologist may cause the formation of a special event response plan. Regardless of the reason, the information that directed the decision should be shared as much as possible with all parties involved in the planning stages.

Not all information can be shared readily between agencies. On occasion, law enforcement may have credible information regarding a particular threat or have an understanding of protestors who may become unruly. It is the law enforcement agencies responsibility to choose how to handle this information with an understanding that other agencies need to have some degree of justification for the resources that are to be used in the special event. All agencies understand that some of the information is For Official Use Only or Classified. Emergency managers and public safety professionals must be flexible in these considerations when planning for special events.

Numerous technologies can be combined in the sharing of intelligence that is open source, which may help planners share information. Online geographical resources, such as Google Earth, exist to help visualize the operations area. Other software, such as CAMEO (Computer Aided Management of Emergency Operations), allow hazmat TIER 2 reports (quantities of hazardous chemicals onsite), streets, hydrants, gas mains, and other layered information to be presented on top of a map for planners to use. It is ultimately the planners's decisions as to which technologies to use.

Personnel

In planning for special events, emergency managers and public safety professionals have three levels of "players" that are needed for these events. The emergency managers and public safety professionals need to have planners—those who specialize in moving around resources and accommodating the needs of the emergency managers and public safety professionals. They also need to have managers—those who can manage the operations conducted during the special event. Finally, they need operators—operators are the end product of all the planning and are conducting the "leg work" of the operations. Without the coordination of these three groups, no special event operations can be accomplished.

When establishing a special event responder selection process, a determination needs to occur to establish how to staff. Most special events do not require a full-time complement of personnel. The following issues need to be considered when establishing a special events team:

- Selection of qualified personnel
- Straight pay or overtime rates for those participating
- Compensatory time off
- Contract regulations for unionized special events responders

The selection of the planners, managers, and operators also needs to consider operational security (OPSEC) dependent on the sensitivity of the event. All individuals should be able to pass a standard background check conducted by normal law enforcement agencies as needed. They should understand the importance of not discussing the details of the event with family members and friends, even in a casual setting. Posting event information on their activities via blogs, Twitter, or Facebook may unintentionally compromise the mission of the special events planning.

OPSEC may be more of a concern for higher profile events. Should an NSSE be established, the lead agencies may dictate special needs for the OPSEC. During the creation and recruitment stage of a special events response team, these lead federal agencies should be contacted to determine if they have additional requirements for any NSSE OPSEC.

Planners

Planners are the chiefs, commissioners, and department heads who represent the authority of the department. Once it has been decided that a special event is going to need the support of emergency managers and public safety professionals, the planners should begin establishing what resources are needed and how they can be brought to the event without taxing existing services. A joint meeting with all involved in the special event should be conducted as soon as possible to begin bringing the resources needed for the special event. This will help establish what resources will be needed and who can provide them for the event.

It is important that all stakeholders are present in this meeting so an open discussion of operational objectives can be discussed, and which groups may contribute to fulfilling these objectives. By doing this, the groups can hear what others are bringing and help reduce any unintentional redundancy. If possible, these meetings should be conducted in person rather than through teleconferencing or Internet communications. This helps create an open dialogue between people and organizations and keeps the planners' attention on the task at hand.

Having open communications with the other planners during these sessions is critical. By having an open dialogue with the other planners, they will be able to obtain information and true operational status by discussing what is needed and what can be brought to the special event. If possible, informal meetings with potential stakeholders for special events should be conducted. These meetings may function as a means to hear what other agencies are doing and what resources they have. These simple "meet and greets" may function to help build open communications between the planners. This degree of familiarity will help the planning sessions for future special events as they occur.

Managers

Managers are the field supervisors who will supervise operations during the special event. These may be fire battalion chiefs, police sergeants or lieutenants, EMS shift supervisors, or foreman of public works. Managers should possess specialized skills or have shown an ability to function well with the public and with other organizations. Unfortunately, the nature of special events planning requires additional training and personnel qualifications that not all managers possess. A 30-year member who has been promoted up internally may not be as qualified as a 20-year employee who has an interest in the field of special events planning or proven competencies with technologies. In order to ensure a smooth operation, it is necessary for a selection process to be conducted by the planners in order to find who has the knowledge, skills, and abilities to accomplish the objectives and be highly efficient during the special event.

Consider possibly using managers for special events who have previously been operators at special events. These managers will understand the dynamics and challenges of working within a special event. Those skills, combined with an understanding of the proper management of a group, would produce a well-qualified manager for a special event. It is understandable, however, that this is not always feasible, especially when using groups at a special event is new. Discretion should then be used in selecting qualified managers from the existing roster.

The managers for the events can function in a variety of positions. Depending on the number or personnel resources that are being used, multiple managers may be needed to conduct the special events operations. The effective span of control should be considered when determining the number of managers. In order to have an effective span of control, no more than seven units should be assigned to an individual manager. The optimal target should be four or five per manager. If strike teams, task forces, or small operations groups are being used for a special event, each group may qualify as a single entity for reporting responsibilities. These suggestions are consistent with the National Incident Management System (NIMS).

Managers may also function over responsibilities other than simply managing personnel resources. They may function as logistic officers handling equipment, food, and perishables during a sustained event. Managers may function as communications officers, dispatching the special events resources to areas that need it. They may also function as liaisons with other organizations or as the Public Information Officer for the event. Obviously, not every manager will have these abilities, so the most qualified person should be placed in these roles.

Managers should have a working knowledge of the organization that they arise from so that they can bring their considerable expertise and experience to help produce a good response for the event. Each manager should operate within their specialty and not cross over into other specialties for the response agencies. Obviously, you do not want a fire supervisor making law enforcement decisions. Ultimately, the responsibilities of managers are flexible with the needs of the event and it is up to the planners to determine what their objectives are.

It is very probable that the managers will be the individuals who are operating within a JOC or a unified Command Post (CP) should either be established during the event. This is dependent on the planners' directions during the planning phases for the special event. Should the managers operate out of the JOC or CP, they must have certain assets at their disposals.

Any person operating out of a JOC/CP should have access to all media outlets (television, radio, Internet). Sometimes these are the quickest ways to find out about something happening within the special event venue. The JOC/CP should have access to maps showing the physical layout of the area. A gridded map would be a much better tool than a simple aerial photograph. A grid layout helps to define boundaries of the operational area and produces a faster response than giving operators directions based on landmarks or approximate locations. This would allow the operators a quicker way to determine where they are to respond to an incident or issue demanding their attention.

The JOC/CP should also have access to an Incident Action Plan (IAP). An IAP is a single document that summarizes all pertinent preplanned information for an event. IAPs will be covered in greater detail in the documents section. This document allows a quick reference for all involved with information on the special event planning. IAPs are just the start of the documentation needed for special events.

The JOC/CP should also have access to communications to the planners, managers, and operators of the special event. Communication with the managers allows status reports as needed and adds a fluid dynamic to the special events planning so that it can adapt based on real-time data. Following a strict chain of command is important when operating with a JOC/CP. This keeps all players in the information loop. However, there are times when the JOC/CP may need direct information from the operators. This may occur because of a critical event about to occur or new information that is time sensitive. Giving the JOC/CP the capability to reach directly to the operators is good planning.

Operators

Operators are the end product of the tasks to be completed. The operators at a special event fulfill the tasks assigned to them and are ultimately the public representation of the managers and planners, who are determining the scope, objectives, and logistics, that are supporting the operators. The operators should have a clear understanding of what the group's objectives are and what tasks are beyond the scope of their capability. These objectives must be clearly delineated at predeployment meetings.

Selection of operators should begin with examination of the objectives of the group. These operators should be experienced in their respective field so that they may bring a degree of professionalism and ability to the special event. Years of experience should not be the only qualifier for selection for special event operators. The most seasoned operators may be selected for special events or newer members who have special skill sets that can aid in the mission objectives of the special event response. Planners should keep these considerations in mind as they are selecting their operators for an event.

Operators should have the ability to quickly adapt to the situation and overcome obstacles as they occur. To this degree, operators should have a degree of latitude in order to complete tasks but keep their group on target of the mission responsibilities. This is not always an easy task and when staffing small groups, selection of operators should be considered who can facilitate this task.

Special event operators should fulfill the minimum qualifications needed to conduct the operations determined by the various organizations they represent. They should also be able to interact well socially, communicate clearly with others, and have the level of physical fitness required to complete the tasks. Some special event venues are only within a building, whereas some require extensive walking through a number of city blocks. When creating a selection pool for these types of events, the planners and managers should keep these additional considerations in mind and inform potential candidates for special events operations of these requirements. Using letters of reference or interviews may help establish individuals who have the personality types that can fit well into these groups.

Training

Before any type of special event program becomes operational, it is critical for established training protocols to be conducted. While special events utilize basic skills learned during candidacy or new-employee hiring, it also uses skills developed outside of this initial training. The training needed for special events is dependent on the agency but there are several common training objectives that may be covered by all agencies.

Foundational training should include training in NIMS courses. Many of the initial NIMS courses (ICS 100, 200, 700, 800) can be found online through FEMA's online training website (training.fema.gov) as well as numerous other job-specific courses. Some of these brief courses are already required in police and fire departments so that municipalities may apply for homeland security grants. These basic courses provide information on NIMS and the federal preparedness guidelines in place throughout the country. The planners, before establishing special events groups, may wish to use these quick online courses as basis for consideration into the groups capable of responding to special events. They may also wish to require additional online courses that are subject-dependent as prerequisites into the special event response teams. Other more advanced classes in the ICS series help round out this initial training with instruction on the numerous NIMS documents and IAP creation. ICS 300 and 400 are courses taught by various groups. The DHS Training Consortium does teach them on-site at no cost to the attendees at the Center for Domestic Preparedness in Anniston, Alabama.

One type of training to consider before establishing a group is a training program designed to prepare an individual for special events responses. This could be a once-a-year program that brings a presentation from each group involved with special events responses. A brief lecture or PowerPoint session by members of each group (fire, law enforcement, FBI, public health, public works, etc.) should be prepared and presented to those who are in consideration for the groups. This training could be a one-time training requirement for entry into the groups, or it could also be an annual refresher. This training could be limited to just one day in length or it could be longer. It is dependent on what the planners want to cover in the topics and how long they wish the training to be to establish sufficient qualifications.

Establishing this special event training as an annual refresher has several benefits. First, it allows opportunities for new members to be included into the group on a regular basis. This helps bring more qualified people into the pool of members who may work special events. Second, it provides a mechanism to remove people from the list should they no longer be interested. A yearly refresher also allows all levels of people to meet from various agencies, which helps establish the familiarity needed prior to actually working a special event.

Outside training may be required by planners or members wishing to join special events responder groups. Members may be required to attend specialized communications classes, incident command classes, or other classes deemed important for their specialization. DHS hosts numerous courses throughout the United States that cover many of the important homeland security/weapons of mass destruction topics. These courses, under the banner of the Training Consortium, are mostly free for attendees and provide good, consistent information on many topics that special events responders may need. Each state also allows for other specialized training on incident command, hazmat, fire, EMS, tactical law enforcement, or technical rescue. The authority having jurisdiction (AHJ) needs to make the decision of which courses it will recognize as acceptable training courses.

Utilizing these courses as the basis for special events responders helps to separate the special events responders from the normal operational responders and also has the dual benefit of producing individuals who have a variety of skill sets that can be utilized by the planners, managers, and operators. Those working in special events should be chosen from the "rank and file" of the groups they represent as individuals who have skill sets that make them particularly desirable to support these functions.

The more skill sets that each operator has makes that individual more desirable as a special events responder. A law enforcement agent who is also a bomb technician, or a firefighter who is also a paramedic, helps to increase the number of support functions to their operations group while reducing the number of people in the group. This helps to establish a manageable span of control for the operations groups. It also creates special events team members who have a much more comprehensive understanding of the goals and objectives that are to be met by the responders.

Planning

Before any planning sessions, a memorandum of understanding (MOU) should be established. The MOUs should contain information from each agency as to the resources that they bring to the special events planning, the duration of the support of the program, and the signatures of the person in charge of each represented organization. The MOUs create a legally binding document between all stakeholders and the limitations of their involvement. All organizations that may be stakeholders in these events may fall under the Emergency Support Functions (ESFs) found within the National Response Framework. ESFs include such positions to be addressed as firefighting, EMS, public works, as well as nearly a dozen other roles. The functions represented in special events planning is entirely dependent on the special event type and the planners' decisions for usage.

The special events responses can be of numerous types. An AHJ may choose to have a visual presence of police, fire, and public works personnel. They may want to establish uniformed officers throughout the venue, a first aid tent staffed by EMS workers, or public works equipment prestaged. The nature of the venue helps dictate how much of a visible presence is needed.

A large visual presence can create an atmosphere of unease, if for example a large law enforcement contingency is present, or there could be a public relations benefit in having a first aid tent or fire apparatus for the attendees to see. Larger venues such as festivals may effectively allow the placement of equipment or first aid tents that can be used to help handle the event. Other smaller venues may, however, not be able to support this type of presence.

Some events, such as sporting events or dignitary events, may dictate a more low-profile presence. The presence of equipment (such as first aid equipment or meters) effectively makes the ability to be covert impractical, yet the lack of uniforms may allow them to be more conspicuous. This low-profile appearance, which is incident-dependent (e.g., suits and ties for a dignitary event, sweatshirts and winter coats for an outdoor sporting event) may make movement throughout the venue more fluid and allow rapid access to areas for treatment, investigation, or intervention. This can help decrease the reflex time to handle a situation and to begin mitigation of an incident. It can also help eliminate any excess response if the incident is unfounded or of a smaller magnitude that can be handled by small teams, which does not require a full-scale response. Ultimately, having an appropriate response helps make the event more successful and the attendees less fearful.

Volunteers

Volunteer organizations exist as a potential manpower pool to help augment the response to a special event. Volunteers who receive minimal training can function to fulfill some of the low-skill positions that exist. Volunteers can be used to help manage access or egress from an area. They can be a network of ears and eyes for special event planners that can relay pertinent information back to a liaison with the special event managers. Medically trained volunteers can staff first aid shelters.

Although volunteer organizations may not be able to bring in highly skilled, highly trained operators to an event, they definitely bring a vast amount of resources and sheer numbers in volunteers to support the special events plan. They should not be discounted and should be considered depending on the venue and the event response objectives.

Communications

When planning for special events, communication equipment and abilities is critical. A clear communications plan should be established and tested before the dates of the special events including testing equipment in the special event venue. This allows the planners, managers, and operators to discover where dead zones are and what limitations the equipment has. It may be necessary to build in redundancies for communications by using other technologies.

When selecting the equipment for the special event, the appearance of the operators within the special event venue should be considered. These operators may be directed to function in a more low-profile status than a uniformed response group. For this purpose, smaller, more compact radios with unobtrusive earpieces or throat microphones can be considered. However, if the group is to function in an overt fashion such as fast response fire or EMS response teams, normal radios may be appropriate. The JOC/CP may utilize portable radios or, if possible, use more of a base station–style radio if operating out of a mobile command post or if space is available in the JOC/CP. The use of repeaters may help boost the signals for radios and should be considered.

The frequency channels used for communications should be clearly defined within the IAP for all operators, managers, the JOC/CP, and municipal dispatch. The level of security for the radio encoding should be determined by the emergency managers and public safety professionals The frequencies do not need to be known by all surrounding regular agencies. Should communications need to occur between internal venue response and external responding resources, the JOC/CP may initiate this request. If bandwidth is available, a dedicated special event frequency can be utilized by those operating within the special event venue with the local common frequencies being available to the special event planners, managers, and operators.

At times, normal radio traffic may neither be available nor appropriate. Consider the ambient sound surrounding an operator that is functioning at a rally or concert as examples. When information is passed during loud ambient times, the message may be lost because of the sounds. Using email or SMS (text) messages may function as an alternate source of communicating should the primary option fail. Communication plans may include alternate radio frequencies or cellular phone numbers to use should the primary lines of communications fail. With the prevalence of Smart phones in communities, email may be accessible as a form of communication but is not as reliable as direct radio or cellular phone communications. Email may be considered for establishing information before an event or having a digital version of the IAP accessible by the operators. If any technologies are to be used by the special events teams (radio, cellular phones, computers), they should all be provided by the agencies and not utilize personnel equipment.

Common terminology should be utilized as much as possible, and use of slang or inaccurate description of events or locations should be discouraged. "Ten codes" have been replaced by common language in most NIMS planning. It helps to deliver a clear, succinct message summarizing the events that are occurring or what the operations groups are to accomplish. Using these communication checkmarks helps to create a professional atmosphere for all involved.

Paperwork

IAPs are the cornerstone document upon which operations at a special event should be based. IAPs are a singular document that collect basic information on the details of the event. During the planning sessions, the sections of an IAP should be completed by the appropriate responsible agencies. This information should then be sent back to a single group to compile and produce a single workable document.

After an initial draft is prepared for an IAP, the planners should carefully review the document and consider any changes. The nature of some IAPs cause them to be open documents until nearly the beginning of the event. Some of the information included on the IAPs that may be late additions are as follows: changing itineraries of VIPs for the special event, weather conditions, personnel assigned to various roles, as well as other issues as dictated by the event.

Some information is a bit static that is contained within an IAP. Information such as street closings, radio frequencies, access/egress points, mass casualty collection points, or prestaged resources such as first-aid tents may be planned well in advance. As is clearly evident, some of this information comprises last-minute additions, whereas others are in place significantly in advance of the actual special event.

The IAP should also contain emergency plans for both special event operators and outside personnel responding to the event. This includes preplanned staging areas or rally points should groups be split or become threatened because of a change in the venue status. These may occur if peaceful protests turn into a more moblike behavior. It also may help if inclement weather appears and disrupts the attendees such as if they were outside and a rush occurs for people to get out of the weather. Having this information readily available, instead of informing after the event, helps responders to move into a response/recovery phase much quicker for all involved.

Other paperwork may be needed beyond the IAP for an event. An IAP only functions as a summary of information. Numerous other documents may be prepared before an event. Attending a FEMA incident command course that contains ICS 100, 200, 300, and 400 may be of particular benefit for managers and planners. Additionally, as resources are available, operators should be minimally trained in ICS 100 and 200, but preferably trained in ICS 300 and 400 as well. This helps all involved understand the documents that they are seeing.

Benefits

Planning for a special event is a challenge for even the most seasoned emergency manager or public safety professional. It requires a degree of flexibility while still maintaining a clear view of the objective goals of the response purpose. The development and establishment of many special events at local, state, and federal levels have brought numerous ideas and tactics that have been proven time and time again.

In developing these plans, planners should reach out to those in other agencies that have had experiences, both successes and failures, to see the lessons learned from the events. This can help streamline future planning for special events. This will produce an overall better-orchestrated response that can protect those who are in the special event venue.

The public relations for the agencies and organizations involved with a special event response plan can be beneficial for all involved. The public views a combined group including various agencies that are working well together. This creates a very positive public perception of the organizations involved. It shows government funds being actively used, instead of passively being consumed by sitting in the firehouse or squad car. It helps convey a message of concern for everyone's enjoyment of these special events.

Using special events planning as a foundation for long-term deployment exercises is also an added benefit. The small-scale special events that occur on a regular basis may be used to create foundations of cooperation between agencies. This cooperation may then be beneficial as the experience and lessons learned from the events are applied during a disaster or large-scale event. The overarching success of special events will then help establish a more robust and dependable emergency response community within the municipality that has trained and experienced personnel to help protect the public.

Chapter 19

Mass Care, Sheltering, and Human Services

Michael Steinle

Contents

Introduction

Communities across the country face a variety of natural and man-made hazards that may necessitate mass care and sheltering of evacuees in the aftermath of an emergency or catastrophic incident. But more than just putting a roof over disaster victims' heads and providing food and water, mass care involves a whole host of vital services as a method of providing an effective on-ramp to full recovery.

The National Response Framework (NRF), Emergency Support Function (ESF)-6, Mass Care, Emergency Assistance, Housing, and Human Services, defines broad services to support the immediate needs of disaster victims displaced from their homes and communities (Figure 19.1).

The broad services defined for ESF-6 are as follows:

- **Mass Care** includes sheltering, feeding operations, emergency first aid, bulk distribution of emergency items, and collecting and providing information on victims to family members.
- **Emergency Assistance** involves services required by individuals, families, and communities
  - Evacuation support including registration and tracking of evacuees
  - Family reunification
  - Provision of aid and services to those with function needs
  - Evacuation, sheltering, and other emergency services for household pets and services animals

Figure 19.1 Emergency Support Function-6 (ESF-6) functions.

  - Support to specialized and medical shelters and nonconventional shelter management
  - Coordination of donated goods and services
  - Coordination of voluntary agency (VOLAG) assistance
- **Housing** includes options such as rental assistance, repair, loan assistance, replacement, factory-built housing, semipermanent and permanent construction, referrals, identification and provision of accessible housing, and access to other sources of housing assistance.
- **Human Services** includes implementation of disaster assistance programs to help disaster victims recover nonhousing losses, replace destroyed personal property, and assistance in obtaining disaster loans, food stamps, crisis counseling, disaster unemployment, disaster legal services, and support and services for those with functional needs.

These services are complex and, even during a relatively small-scale event, local communities may find it difficult to provide all services at the level required. The purposes of this chapter include

- To provide an overview of the federal support structure for ESF-6 services.
- To introduce the Federal Emergency Management Agency's (FEMA) "Guidance on Planning for Integration of Functional Needs Support Services in General Population Shelters" (hereinafter referred to as the FNSS Guidance), released in late 2010.
- To provide a conceptual framework to support local implementation of comprehensive ESF-6 services.

The FNSS Guidance contains significant changes to traditional methods of implementing shelter plans. Based on post-Hurricane Katrina legal actions, the Post-Katrina Emergency Management Reform Act (PKEMRA), and other regulations and legal findings, the FNSS Guidance specifies that the distinction between General Population Shelters and Special Needs Shelters is no longer appropriate. In essence, all shelters must be able to accommodate the general population and those with functional needs. Medical shelters or alternate care sites will accommodate those who require acute medical care in the event that hospitals cannot accommodate all patients. These concepts are discussed in depth later in this chapter.

Federal Support Structure for ESF-6 Functions

As set forth in the NRF ESF-6 Annex,[1] federal agencies are assigned and equipped to support various mass care functions when local and state resources are exhausted. In addition to resources, the federal response assets and organization in the ESF-6 Annex provide a model for assimilating local and state plans and resources to support the complex functions required under ESF-6.

This section describes legal authorities, the concept of operations, and federal resources available to support ESF-6 operations. A recent report submitted to the President and Congress in October 2010 by the National Commission on Children and Disasters also provides new guidance that is likely to change certain ESF-6 functions to foster a more conducive environment to allow children to recover from catastrophic incidents. The topic of children and disasters is discussed in Chapter 20.

Legal Authorities

The legal authorities providing the context, resources, and structure for federal agencies to support ESF-6 operations are summarized below.[2]

- **Robert T. Stafford Disaster Relief and Emergency Assistance Act**—This act authorizes the President to issue a major disaster declaration to speed a wide range of federal aid to states determined to be overwhelmed by catastrophic incidents including financing through the Disaster Relief Fund (DRF) and administered by the Department of Homeland Security (DHS). This act also authorizes temporary housing, grants for immediate needs of families and individuals, repair of public infrastructure, emergency communications systems, and other forms of assistance.
- **Homeland Security Act of 2002**—This act established the DHS and consolidated the operations of 22 existing federal government agencies including FEMA and other operations, which support domestic defense measures and emergency response.
- **Homeland Security Presidential Directive 5**—This directive established the National Incident Management System (NIMS), which is designed to cover the prevention, preparation, response, and recovery from terrorist attacks, major disasters, and other emergencies. NIMS sets forth procedures to foster effective collaboration among all levels of government including consistent use of the Incident Command System.
- **Post-Katrina Emergency Management Reform Act of 2006**—Among many other issues, this comprehensive act clarified authorities, roles, and responsibilities to
  - Prepare for, respond to, and recover from disasters
  - Enhance emergency communications
  - Provide assistance to disaster-affected areas and populations
  - Support regional preparedness and cooperation
  - Improve timely delivery of goods and services in disaster incidents
  - Change contracting practices to enhance preparedness and strengthen accountability
  - Apply specific expertise (Disability Coordinator, Small State Advocate, and Modeling and Analysis) to disaster planning, response, and recovery activities

- **Pets Evacuation and Transportation Standards Act of 2006**—This act requires states seeking FEMA assistance to accommodate pets and service animals in emergency planning initiatives including evacuation planning.
- **Public Health Service Act**—This act authorizes the Department of Health and Human Services (HHS) to establish emergency services under the U.S. Public Health Service. As amended, this act also provides other constructs pertinent to ESF-6 operations such as the Health Insurance Portability and Accountability Act of 1996.
- **Social Security Act of 1935**—This act authorizes the Social Security Administration (SSA) to provide Social Security Disability, Social Security Retirement, Social Security Survivors, Special Veterans, and Supplemental Security Income benefits and provides a structure to ensure continuity of service to beneficiaries.
- **Americans with Disabilities Act of 1990**—This act is a broad civil rights law that prohibits *discrimination* based on *disability*. As it relates to ESF-6, this act is one of the guiding principles upon which the recently released FNSS Guidance was developed.

Concept of Operations

During an event requiring mass care, sheltering, and human services, initial response operations should focus on life/safety issues and the immediate needs of victims. Evacuation from hazard zones, feeding and watering, and basic shelter from the elements are first-priority issues in the immediate aftermath of an incident. Initially, resources to support ESF-6 operations may be scarce, which makes prioritizing these activities important. Also, during the initial operational periods, close coordination between ESF-6 and ESF-8, Public Health and Medical Services, is critical as many victims will require first aid and medical treatment before evacuation.

As the emergency progresses, recovery efforts can be initiated that include housing considerations and other human services such as crisis counseling, employment assistance, legal assistance, and provision of other federal and state benefits.

As the lead federal agency for ESF-6, FEMA coordinates federal response and recovery operations in close coordination with local, tribal, and state governments, volunteer organizations, and the private sector. As ESF-6 response operations escalate, resources are requested progressively through the chain from the local level through federal agencies as depicted in Figure 19.2.

To support expedited resource requests, it is recommended that, to the extent possible, local and state agencies consider developing prescripted mutual aid requests, Emergency Management Assistance Compact requests, and Federal Action Request Forms. Prescripting these requests based on reasonably understood scenarios can facilitate timely submission of requests during a time of need.

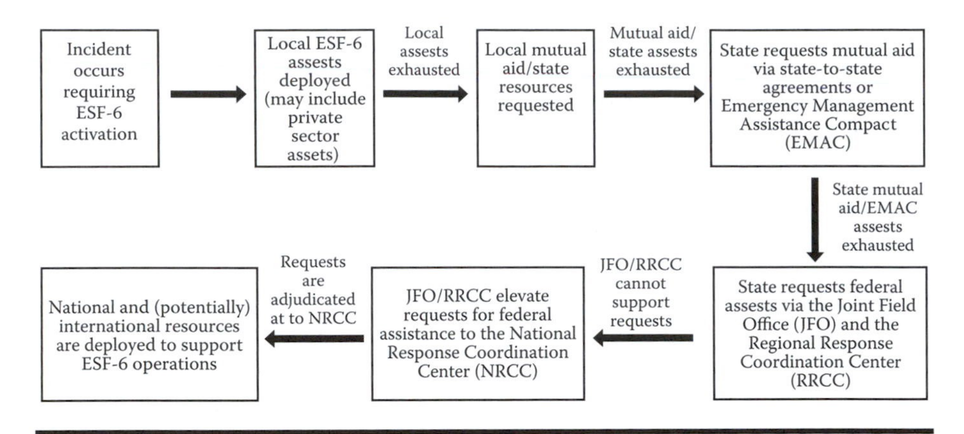

Figure 19.2 Escalating resource request flow.

Federal Resources Supporting ESF-6

Federal agencies are assigned to support ESF-6 in various functional areas. Likewise, most local, tribal, and state governments have assigned various functions to appropriate agencies, volunteer organizations, nonprofits, and, in some cases, private sector partners. Although federal resources to support ESF-6 may change periodically based on a variety of factors, resources are described below that may be available to support local, tribal, and state governments during a crisis.

Mass Care Services

Mass care services are initiated at the local level with support as needed from tribal and state governments. Working through a state-designated agency directed by the Governor of that state, FEMA coordinates to provide federal resources necessary to support various mass care services as defined below.

- **Shelter Services.** Emergency shelter may include use of designated shelter sites in existing structures within the affected area as well as additional sites designated by local government. Shelter sites that are accessible to individuals with disabilities should be selected.
- **Feeding.** Feeding can be accommodated through any combination of fixed sites, mobile feeding units, and bulk distribution sites. During large-scale catastrophic events, shelf-stable meals may be an option to accommodate short-term sustenance needs. As resources and facilities become available, feeding operations can shift to a prepared meal format.
- **Bulk Distribution.** To accommodate logistical challenges associated with large-scale incidents, state and federal agencies supporting mass care services may institute bulk distribution operations to distribute emergency relief

items to meet urgent needs. Under these circumstances, federal agencies will coordinate with local, tribal, and state agencies as well as volunteer and private-sector organizations to identify bulk distribution sites to accommodate expedited and scaled distribution of goods. Assets that may be distributed in bulk include food, water, and other commodities to support mass care operations.

- **Emergency First Aid.** In concert with ESF-8, Public Health and Medical Services, local, tribal, and state agencies should coordinate emergency first aid at mass care facilities and designated sites. In the initial phases of a large-scale event, such aid may occur at make-shift field stations. As situational awareness and resource needs are identified, federal resources can be requested including activation of the National Disaster Medical System, the Public Health Service Commissioned Corps, and other HHS assets.
- **Disaster Welfare Information.** As experienced during the aftermath of Hurricane Katrina, family members can be separated for long periods without a coordinated tracking system. Local, tribal, and state agencies should institute a system to collect Disaster Welfare Information, which includes communicating with family members outside of an affected area and information regarding individuals residing within the affected area. Family tracking can also support reunification of family members within the affected area.

The services described above are supported at the federal level by FEMA as the lead agency with support from the United States Department of Agriculture (USDA) Food and Nutrition Service (FNS). Federal agencies will coordinate with local, tribal, state governments, and volunteer organizations to distribute food and food supplies when capabilities do not keep up with demand. Federal support may include support to private-sector feeding operations, securing food commodities, developing feeding plans, and obtaining warehouse space. In bulk distribution operations, support by federal agencies may include transportation, technical support, and other mission-critical tasks.

Emergency Assistance

To the extent feasible, emergency assistance is initiated at the local level with the initial focus on evacuation. Local ESF-6 responders should coordinate with local ESF-9, Search and Rescue, to designate and operate collection points for evacuees. Further coordination with ESF-8, Public Health and Medical Services, may be required to support evacuees with medical needs and to accommodate medical evacuation. Emergency assistance services are defined in greater detail below:

- **Mass Evacuation.** Mass evacuation planning should include use of any local bus services that may be available as well as mutual aid agreements with neighboring Emergency Medical Services providers to support medical evacuation.

To the extent feasible, local evacuation plans should identify routes out of potential hazard zones. In the event that local, tribal, and state resources are exhausted, various federal agencies may be able to support evacuation needs. For example, under the Public Assistance Program, FEMA may be able to activate ambulance services under certain conditions. Also in support of transportation resources, FEMA can request transportation support through the Department of Defense (DOD) if commercial transportation is exhausted. If aerial evacuation is necessary, FEMA may also coordinate with the Federal Aviation Administration to manage aviation traffic near the emergency impact area.

- **Facilitated Reunification.** During mass evacuation and sheltering operations, it may be necessary to support reunification of family members. The National Mass Evacuation Tracking System (NMETS) is a manual and computer-based system designed to assist local, tribal, and state jurisdictions in tracking transportation-assisted evacuees. In lieu of NMETS, a formalized system of tracking should be implemented to support facilitated reunifications of family members.
- **Household Pets and Service Animals.** As stated in the PETS Act (2006), a household pet is a domesticated animal, such as a dog, cat, bird, rabbit, rodent, or turtle, that is traditionally kept in the home for pleasure rather than for commercial purposes, can travel in commercial carriers, and be housed in temporary facilities. Household pets do not include reptiles (except turtles), amphibians, fish, insects, arachnids, farm animals (including horses, and animals kept for racing purposes). Service animals are defined as any guide dog, signal dog, or other animal individually trained to provide assistance to an individual with a disability including, but not limited to, guiding individuals with impaired vision, alerting individuals with impaired hearing to intruders or sounds, providing minimal protection or rescue work, pulling a wheelchair, or fetching dropped items. Although assignments for pets and service animals may vary at the local, tribal, and state levels, the NRF assigns care of pets and service animals to ESF-6 with ESF-8, Public Health and Medical Services, and ESF-11, Agriculture and Natural Resources, providing support.

The overall goal is to establish the operational framework to foster pet health and safety and to coordinate effective use of public and private partnerships for the care and wellbeing of animals classified as either household pets or service animals. The USDA Animal and Plant Health Inspection Service can provide technical support regarding the safety and wellbeing of household pets. The USDA FNS can provide food and nutritional assistance to support animal care. HHS Veterinary Medical Services can support animal care through provision of qualified veterinary medical personnel and through environmental health services such as disease and vector control.

- **Sheltering.** One of the most complex tasks assigned to ESF-6 is to shelter those displaced during a disaster. Each disaster is different and may involve sheltering a surge of evacuees from an impacted area in a different community and/or finding shelter within an impacted area. As discussed in FNSS Overview, the FNSS Guidance requires accommodations for those with functional need in general population shelters. To the extent feasible, local, tribal, and state agencies with ESF-6 responsibilities should identify shelters that comply with American's with Disabilities Act (ADA) guidelines and provide resources set forth in the FNSS Guidance. Federal agencies will provide assistance, resources, and technical assistance in support of local, tribal, and state governments, VOLAGs, and host states when conventional and nonconventional congregate care systems and shelter-in-place activities require additional resources. The National Shelter System (NSS), a Web-based database, provides information for shelters posted to the NSS during response to disasters and emergencies. Other nonconventional sheltering may include:
  - Hotels, motels, and other single-room facilities
  - Temporary facilities such as tents, prefab module facilities, trains, and ships
  - Specialized shelters and functional and medical support shelters (through coordination with ESF-8 and the affected or host state)
  - Support for other specialized congregate care areas that may include respite centers, rescue areas, and decontamination processing centers

Initially, during a catastrophic incident, disaster survivors may form loose groupings of individual shelters or group tents erected in the affected area and near their homes. These temporary shelters will provide minimum shelter but the affected population will rely on support for food, water, first aid, and information. As situational awareness is gained, local, tribal, state, and federal coordination can ensue to allow identification of suitable congregate care facilities, transportation, and other resource needs to support sheltering. The American Red Cross (ARC), nongovernmental organizations (NGOs), and faith-based organizations provide critical support during initial and ongoing sheltering operations. Preplanning with these organizations locally is crucial to facilitate timely and effective response to sheltering needs.

FEMA supports sheltering via activation of the Individual Assistance–Technical Assistance Contract for mass care services. This contract allows FEMA to support a variety of sheltering missions. FEMA also facilitates mass care resource requests from NGOs via FEMA Voluntary Agency Liaisons (VALs). The ARC provides personnel to staff FEMA regional offices in support of ESF-6 mass care activities and can deploy specially trained liaisons to work at designated locations to support mass care activities. The Salvation Army can also deploy liaisons to work at designated locations to support mass care activities and can provide subject-matter expertise on regulations, policy, and relevant mass care issues. Other national organizations, such as the Corporation for National and Community Service and the National Voluntary Organizations Active in Disaster (NVOAD), provide personnel to support a variety of mass care and sheltering operations.

- **Support to Unaffiliated Volunteers and Unsolicited Donations.** Use of unaffiliated volunteers and unsolicited donations during the immediate aftermath of a disaster can present some daunting challenges. There are risks associated with deploying spontaneous volunteers, particular those in professional positions such as doctors and nurses. Likewise, lack of chain of custody of unsolicited donations presents risks especially as it relates to medications and perishable products. Local, tribal, and state officials should establish proactive mechanisms for managing spontaneous volunteers and unsolicited donations. FEMA also has resources such as the Donations Management Unit to support management of volunteers and donated goods. Procedures, processes, and activities to support spontaneous volunteers and unsolicited donations are defined in the Volunteer and Donations Management Support Annex.
- **Voluntary Agency Coordination.** As all emergencies start and end at the local level, proactive planning with local volunteer organizations, NGOs, faith-based organizations, and the private sector can provide valuable and validated resources to support citizens during a time of need. FEMA VALs can support coordination with VOLAGs affiliated with NVOAD, other NGOs, and private-sector entities to support ESF-6 operations.

The services described above are supported at the federal level by FEMA, HHS, USDA, ARC, DOD, and other support agencies to support expedited emergency service and to establish a long-term recovery strategy to address the unmet needs of all individuals and families.

Housing

Once the immediate mass care and emergency assistance services are established and an initial roadmap to recovery begins to form, support for housing can begin. Although local resources are critical to support housing needs, substantial federal resources, managed by FEMA and other federal agencies, are available as identified in the National Disaster Housing Strategy, which defines the full scope of options for disaster housing assistance:

- **Temporary Roof Repair**—Expedited repairs to damaged roofs on private homes allows residents to return to and remain in their own homes while performing permanent repairs.
- **Repair Program**—Financial assistance is available to homeowners for repair of their primary residence, utilities, and residential infrastructure.

- **Replacement Program**—Financial assistance is also available to victims to replace their destroyed primary residence.
- **Existing Housing Resources**—A centralized location for available housing resources from the private sector and other federal agencies [Department of Housing and Urban Development (HUD), Department of Veterans Affairs (VA), and USDA properties] provides a clearinghouse to support victim placement.
- **Rental Assistance**—Financial assistance is available to individuals and families for rental of temporary accommodations.
- **Noncongregate Facilities**—Facilities may be available that provide private or semiprivate accommodations, but are not considered temporary housing such as cruise ships, tent cities, military installations, school dorm facilities, or modified nursing homes.
- **Transportation to Other Locations**—Assistance may be available to relocate individuals and families outside of the disaster area where short- or long-term housing resources are available. Transportation services may include return to the predisaster location.
- **Permanent Construction**—Direct assistance may be available to victims and families for permanent or semipermanent housing construction.
- **Direct Financial Housing**—Payments may be made directly to landlords on behalf of disaster victims.
- **Hotel/Motel Program**—During transition periods, temporary accommodations for individuals and families from congregate shelters or other temporary environments may be available for those unable to return to their predisaster dwelling.
- **Direct Housing Operations**—Temporary units, usually factory-built, may be available. This option is utilized only when other housing resources are not available. Units will be appropriate to the community needs and include accessible units.

In addition to FEMA, other federal agencies described below provide housing support services:

- **Small Business Administration (SBA) Disaster Loan Program.** SBA provides low-interest, long-term disaster loan assistance for qualified homeowners and renters, nonagricultural businesses of all sizes, and nonprofit organizations to fund the repair and replacement of disaster-damaged property. Loans may also be used for relocation, mitigation, refinancing of existing liens, code-required upgrades, and one-year insurance premiums.
- **Department of Housing and Urban Development.** HUD provides access to and information on available habitable housing units, including housing units accessible to individuals with disabilities, owned, or in HUD possession, within or adjacent to the incident area for use as temporary housing.

This program provides enforcement of the Fair Housing Act and compliance with other civil rights statutes.

- **USDA—Rural Development (RD).** USDA RD provides information on USDA-financed, currently available, habitable housing units that are not under lease or under agreement of sale. This program assists eligible recipients to meet emergency housing assistance needs resulting from presidentially declared emergencies or major disasters.
- **Veterans Administration.** VA provides available facilities suitable for mass shelter and assists veterans affected by disasters to help them avoid defaulting on existing home mortgages and/or foreclosure on their homes. VA also provides assistance to veterans with disabilities to retrofit homes with necessary accessibility measures (e.g., wheelchair ramp). VA also maintains plans to make available housing assets available to survivors in catastrophic disasters.

Human Services

Human services generally fall into three categories: (1) recovery or replacement of nonhousing losses and personal property; (2) financial assistance such as disaster loans and unemployment, and food stamps; and (3) specialized services such as crisis counseling, disaster legal services, and support and services for those with functional needs. Local and state volunteer organizations, NGOs, and faith-based organizations may be able to support these needs. Federal agencies and programs to support human services are identified below.

- Recovery/Replacement Services
  - **Department of the Treasury—Alcohol and Tobacco Tax and Trade Bureau (TTB).** TTB provides federal alcohol and tobacco excise tax refunds to businesses that have lost assets in a disaster.
  - **Department of the Treasury, Internal Revenue Service (IRS).** The IRS provides tax counseling and assistance to taxpayers whose property has been damaged or lost in a federally declared disaster area.
  - **Department of the Treasury, Bureau of the Public Debt.** The Bureau assists disaster victims by expediting replacement or redemption of U.S. Savings Bonds and may waive the minimum holding period for Series EE and I Savings Bonds presented to authorized paying agents for redemption.
  - **Victims of Crime Assistance, Department of Justice (DOJ).** The DOJ supports crime victim compensation in incidents resulting from terrorism or acts of criminal violence, as appropriate.
  - **Veterans Assistance Program (VAP).** The VAP provides insurance settlements, adjustments to home mortgages, and death benefits and ensures continuity of services, such as pensions, to beneficiaries.

  - **Social Security Administration.** SSA provides Social Security Disability, Social Security Retirement, Social Security Survivors, Special Veterans, and Supplemental Security Income benefits and ensures continuity of service to beneficiaries.
  - **U.S. Postal Service (USPS).** The USPS provides extended mail services to relocated populations.
- Financial Assistance Services
  - **Cora Brown Fund.** Administered by FEMA, the Cora Brown Fund is used for uninsured or under-insured disaster-related needs of individuals or families who are unable to obtain adequate assistance from other local, tribal, state, and federal programs or from VOLAGs.
  - **Other Needs Assistance.** Administered by FEMA, awards assist with medical, dental, funeral, personal property, transportation, moving and storage, and other expenses for uninsured or underinsured eligible applicants.
  - **HHS.** HHS's broad suite of services provides financial assistance in the form of expedited claims for new federal benefits. These services ensure continuity of services to beneficiaries, such as Medicaid, Temporary Assistance to Needy Families, Child Care. This program also supports states hosting relocated populations by extending existing programs and benefits or taking other actions as needed. HHS integrates ESF-6 needs with those provided under ESF-8, Public Health and Medical Services.
  - **Disaster Unemployment Assistance (DUA), Department of Labor.** Administered by the impacted state, DUA provides financial assistance to individuals whose employment or self-employment has been lost or interrupted as a direct result of a major disaster declared by the President, and who are not covered by regular unemployment insurance.
- Specialized Services
  - **Crisis Counseling and Training.** Administered by FEMA and the Substance Abuse and Mental Health Services Administration, the Crisis Counseling Assistance and Training Program provides immediate, short-term crisis counseling services to relieve grieving, stress, or mental health problems caused or aggravated by a disaster or its aftermath. Assistance provided is short term and is at no cost to the disaster victim.
  - **Disaster Case Management.** Administered by FEMA and HHS, this program provides case management services, including financial assistance, through government agencies or qualified nonprofits to eligible individuals. Case management ensures that a sequence of delivery is followed to streamline assistance, prevent duplication of benefits, and provide an efficient referral system.
  - **Disaster Legal Services, American Bar Association (ABA)/Young Lawyers Program.** The ABA Disaster Legal Services provides free disaster legal services for low-income individuals who are unable to secure legal services to meet disaster-related needs.

FNSS Overview

Traditionally, sheltering has occurred using a variety of means including general population shelters, special needs shelters, and medical shelters, none of which share universal definitions across the country. After Hurricane Katrina, many lawsuits were filed regarding the manner in which mass care and sheltering was performed in the immediate aftermath of the hurricane and for many months following the devastation in Louisiana, Mississippi, and Alabama. In 2007, the DOJ signed an amended settlement agreement[3] calling for, among other things, development of new emergency management plans, which include provisions for accommodating people with disabilities.

In April 2010 and under contract by FEMA, Baptist Child and Family Services released a document titled, "Guidance on Planning for Integration of Functional Needs Support Services in General Population Shelters." In late 2010, FEMA issued an official guidance document under the same title. In short, this document was developed to provide guidance to integrate children and adults with disabilities and functional needs into emergency shelter planning and response. This section describes the legal foundations of the FNSS Guide, application of the guide to support integrative planning, and other considerations to support holistic shelter and mass care planning.

Legal Foundation

> "Children and adults with disabilities have the same rights to services in general population shelters as other residents."
>
> **—FEMA, Guidance on Planning for Integration of Functional Needs Support Services in General Population Shelters, 2010, p. 8**

The FNSS Guide cites the Stafford Act, PKEMRA, and federal civil rights laws as mandates to integration and equal opportunity for people with disabilities in general population shelters. In addition to citing laws and legal authorities, the FNSS Guide identifies nondiscrimination concepts and examples of how these concepts apply in sheltering and mass care operations:[4]

1. Self-Determination—People with disabilities or functional needs are most knowledgeable about their own needs.
2. One Size Does Not Fit All—People with disabilities do not all require the same assistance and do not all have the same needs. Different types of disabilities affect people in different ways. Preparations should be made for people with a variety of functional needs, including people who use mobility aids, require medication or portable medical equipment, use service animals, need information in alternate formats, or rely on a caregiver.

3. Equal Opportunity—People with disabilities must have the same opportunities to benefit from emergency programs, services, and activities as people without disabilities. Emergency recovery services and programs should be designed to provide equal choices for all people. This includes choices relating to short-term housing or other short- and long-term disaster support services.
4. Inclusion—People with disabilities have the right to participate in and receive benefits of emergency programs, services, and activities provided by governments, private businesses, and nonprofit organizations. Inclusion of people with various types of disabilities in planning, training, and evaluation of programs and services will ensure that all people are given appropriate consideration during emergencies.
5. Integration—Emergency programs, services, and activities typically must be provided in an integrated setting. Providing services such as sheltering, information intake for disaster services, and short-term housing in integrated settings keeps people connected to their support system and caregivers.
6. Physical Access—Emergency programs, services, and activities must be provided at locations that all people can access, including people with disabilities. People with disabilities should be able to enter and use emergency facilities and access the programs, services, and activities that are provided. Facilities typically required to be accessible include: parking, drop-off areas, entrances and exits, security screening areas, toilet rooms, bathing facilities, sleeping areas, dining facilities, areas where medical care or human services are provided, and paths of travel to and from and between these areas.
7. Equal Access—People with disabilities must be able to access and benefit from emergency programs, services, and activities equal to the general population. Equal access applies to emergency preparedness, notification of emergencies, evacuation, transportation, communication, shelter, distribution of supplies, food, first aid, medical care, housing, and application for and distribution of benefits.
8. Effective Communication—People with disabilities must be given information that is comparable in content and detail to that given to the general public. It must also be accessible, understandable and timely. Auxiliary aids and services may be needed to ensure effective communication. These resources may include pen and paper; sign language interpreters through on-site or video; and interpretation aids for people who are deaf, deaf–blind, hard of hearing or have speech impairments. People who are blind, deaf–blind, have low vision, or have cognitive disabilities may need large-print information or people to assist with reading and filling out forms.
9. Program Modifications—People with disabilities must have equal access to emergency programs and services, which may entail modifications to rules, policies, practices, and procedures. Service staff may need to change the way questions are asked, provide reader assistance to complete forms, or provide assistance in a more accessible location.

10. No Charge—People with disabilities may not be charged to cover the costs of measures necessary to ensure equal access and nondiscriminatory treatment. Examples of accommodations provided without charge to the individual may include ramps; cots modified to address disability-related needs; a visual alarm; grab bars; additional storage space for medical equipment; lowered counters or shelves; Braille and raised letter signage; a sign language interpreter; a message board; assistance in completing forms or documents in Braille, large print or audio recording.

The remainder of this chapter is dedicated to identifying challenges and solutions in addressing planning and response relative to the nondiscrimination concepts identified above and the FNSS Guide in general.

Practical Considerations

Mass care and shelter planning should consider functional and access needs as well as chronic medical conditions. Chronic medical conditions prevalent in our society include cancers, diabetes, heart disease, hypertension, stroke, mental disorders, and pulmonary conditions. The primary services that enable independence for those with access, functional and medical needs include

- Reasonable modifications to policies, practices and procedures
- Durable medical equipment
- Consumable medical supplies
- Personal assistance services
- Other goods and services as needed

With these services in mind, the primary challenges to modifying existing mass care and shelter plans to accommodate effective integration of people with functional needs include: (1) estimating space requirements and layout consideration; (2) identifying sources for equipment and services; and (3) modifying plans to reflect integrative policies. These issues are addressed in detail below.

Space Requirements and Layout Considerations

Throughout general population shelters, cots and other furniture should be organized to allow routes that are accessible to people who use wheelchairs, crutches, or walkers. According to the American Red Cross, 20 square feet per person should be available for short-term sheltering and up to 40 square feet per person for sheltering longer than 72 hours. People who use wheelchairs, lift equipment, service animals, and personal assistance services may require up to 100 square feet.[5] In addition to sleeping space, shelters should have space (as much as 50% extra space) to accommodate these support services:

- Cooking and dietary needs
- Service animals
- Communications
- Bathing and toileting needs
- Quiet area
- Mental health services
- Medical and dental services
- Medication storage
- Transportation services

According to the National Organization on Disabilities,[6] 54 million people in the United States have a disability and 61% of them have not made plans to quickly and safely evacuate their homes. Between 15% and 25% of the population have disabilities in most counties in the United States. Given these data, space requirement examples are provided in Tables 19.1 and 19.2.*

As shown in the calculations, a sizable amount of space is necessary to support sheltering needs. In addition to space, shelter layout must be considered to accommodate the services discussed above. Figure 19.3 provides a dimensional depiction of accommodations necessary to support people with functional needs.[7] The figure indicates a wider, higher bed to support functional needs as well as space for a wheelchair or other equipment. The figure also indicates space for a personal care attendant and service animal accommodations.

Figure 19.4 provides a conceptual layout for a 250-bed congregate care facility. The wider aisles shown in the diagram accommodate 56 beds for people with functional needs or approximately 22% of the sheltered population. The figure also indicates placement of stations or rooms to accommodate the services required at a congregate care shelter.

Identifying Sources for Equipment and Services

People who have functional or special needs may arrive at shelters without medical equipment or medications necessary to maintain health and wellbeing. Although some jurisdictions may purchase and store quantities of equipment and medications, if financial, logistics, or other constraints prohibit such purchases, it is important to work with the disability community and disability service providers in your community to identify emergency sources for such assets. To solidify availability, it is advisable to develop provider agreements with private sector service, equipment,

---

* Calculations are based on 40 square feet per person for General Population, 100 square feet per person for those with disabilities, and a 50% size factor to accommodate support services.

Table 19.1 Example Shelter Space Calculation, County 1

Data Element	*Value*
County population	150,000
% Citizens with disabilities	15
Number of citizens with disabilities	22,500
Number of citizens without disabilities	127,500
Space requirements for citizens with disabilities (sq. ft.)	2,250,000
Space requirements for citizens w/o disabilities (sq. ft.)	5,100,000
Total space requirements for citizens (sq. ft.)	7,350,000
Total space requirements for support services (sq. ft.)	3,675,000
Total space requirements (sq. ft.)	11,025,000
Space requirements—10% of population affected (sq. ft.)	1,102,500
Space requirements—20% of population affected (sq. ft.)	2,205,000
Space requirements—30% of population affected (sq. ft.)	3,307,500
Space requirements—40% of population affected (sq. ft.)	4,410,000
Space requirements—50% of population affected (sq. ft.)	5,512,500

and medication providers to ensure expedited availability during an emergency. In addition to the actual supplies, it is important to consider and plan for logistical requirements to receive supplies and services at required locations.

Modifying Plans to Reflect Integrative Policies

If existing plans address general population sheltering, special needs sheltering, and medical sheltering individually and differently, provisions should be modified to address integration of functional and medical needs in congregate care facilities. Provisions should address those support services identified at the beginning of Space Requirements and Layout Considerations.

Dietary Needs. Plans should include provisions to provide meals and snacks to children and adults with specific dietary needs and restrictions. It is important to implement a process for promptly obtaining, documenting, and communicating specific dietary needs and restrictions to those responsible for meal and snack preparation. Processes should also include provisions for responding quickly to unanticipated dietary needs.

Table 19.2 Example Shelter Space Calculation, County 2

Data Element	*Value*
County population	80,000
% Citizens with disabilities	25
Number of citizens with disabilities	20,000
Number of citizens without disabilities	60,000
Space requirements for citizens with disabilities (sq. ft.)	2,000,000
Space requirements for citizens w/o disabilities (sq. ft.)	2,400,000
Total space requirements for citizens (sq. ft.)	4,400,000
Total space requirements for support services (sq. ft.)	2,200,000
Total space requirements (sq. ft.)	6,600,000
Space requirements—10% of population affected (sq. ft.)	660,000
Space requirements—20% of population affected (sq. ft.)	1,320,000
Space requirements—30% of population affected (sq. ft.)	1,980,000
Space requirements—40% of population affected (sq. ft.)	2,640,000
Space requirements—50% of population affected (sq. ft.)	3,300,000

Service Animals. A service animal is defined as any animal individually trained to provide assistance to a person with a disability. Service animals are not pets and should not be confused with pet sheltering. Plans should include provisions to determine if an animal is a service animal including

1. Is this a service animal required because of a disability?
2. What work or tasks has the animal been trained to perform?

Service animals provide a variety of types of assistance including guiding people who are blind or have low vision, alerting people who are deaf or hard of hearing, pulling wheelchairs, carrying or retrieving items for people with mobility disabilities, assisting people with disabilities to maintain their balance or stability, alerting people to and protecting them during medical events, and working or performing tasks for individual with psychiatric, neurologic, or intellectual disabilities.

Service animal plans should address areas where animals can be housed, exercised, and toileted. In addition, a reliable source for food and supplies (water bowls,

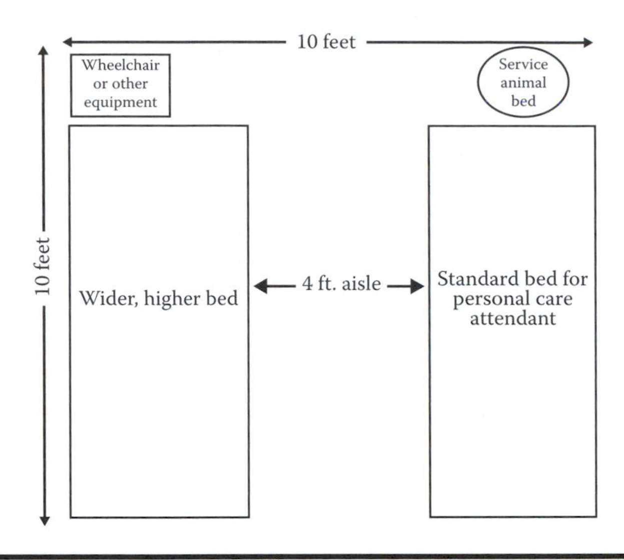

Figure 19.3 Functional needs space accommodations.

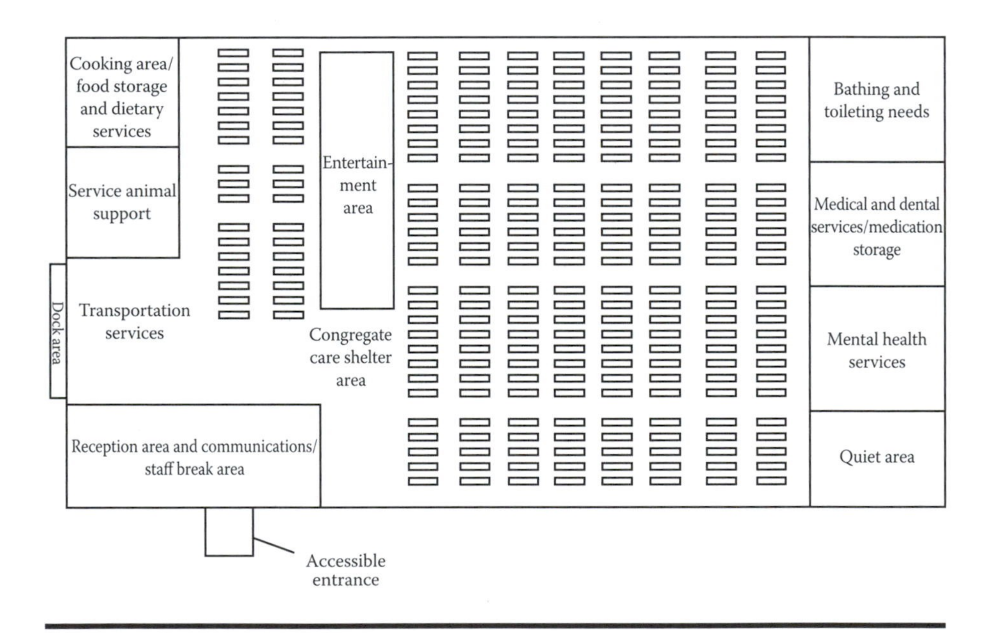

Figure 19.4 Congregate care shelter layout, 250 beds.

leashes, collars) is needed to support service animal wellbeing. It is also important to communicate information regarding service animals in alternative accessible formats to ensure that all who show up at a shelter understand processes for service animal support.

Communications. Children and adults who have access or functional needs should be given all applicable information given to the general population using methods that they understand. Such methods may include those designed to support people with vision and hearing impairments as well as those who may have language barriers. It is important to know the population in a particular jurisdiction in order to allow effective planning relative to communication barriers.

Bathing and Toileting Needs. Facilities for bathing and toileting must include accessible bathing and toileting facilities for adults as well as children. If a chosen shelter facility is lacking in accessible facilities, it should be modified at the earliest convenience or alternative facilities should be located.

Quiet Area. Plans should include a strategy to provide a quiet area within each congregate care shelter. A quiet area will allow people to relieve stress caused by the noise and crowded conditions associated with shelter operations. It is particularly important for elderly persons, people with psychiatric disabilities, parents with very young children, and those with autism to have a safe, quiet place to relax.

Mental Health Services. Congregate care shelter plans should include shelter staff with expertise regarding disabilities, functional, and access needs. Licensed mental health professional should also be available at the shelter or on call at all times. Local and state laws, rules, and regulations may also specify certain mental health requirements to support sheltering. All medical and mental health care performed at shelters should be documented.

Medical and Dental Services. Plans should include medical care that can be provided in a home setting and should be available at all shelters. It is important to preidentify medical and dental personnel and to enter contract arrangements with them to support medical services at congregate care shelters. As with mental health services, medical and dental services provided at congregate care shelters should reflect requirements of local and state laws and should be documented.

Medication Storage. Plans should include procedures for obtaining, storing, dispensing, documenting, and disposing of medications. Preidentified contract agreements are recommended to ensure timely access to critical medications. Many local public health departments have mass prophylaxis plans in place that may serve as a basis to build medication provisions for sheltering operations. Such plans generally include provisions for receipt and storage of Strategic National Stockpile assets and other locally available medical caches.

Transportation Services. Plans should include procedures to support preidentified contractual agreement with transportation providers. It is important to understand the demographics of the jurisdiction and to consider proportional needs relative to those with functional and disability needs.

During plan reviews and modifications, it is important to have thorough and substantive representation from the disability and functional needs community and from disability and functional needs service providers. Such representation will assist in developing effective plans, may provide effective and efficient options that may otherwise go unnoticed, and meets the nondiscrimination concept of self-determination in developing integrative plans.

Other Considerations

Advanced planning is crucial to the success of congregate care shelter operations. In addition to the day-to-day operational considerations discussed above, plans should include provisions for transition and recovery. Such planning should include a broad range of stakeholders and should identify staffing requirements for successful transition and recovery. Transition and recovery provisions should be in place before an event occurs and begin as soon as a shelter is opened. The primary objective of shelter operations is to provide short-term care while finding long-term solutions.

It is important to allow a reasonable amount of time and assistance to locate suitable housing and services when individuals cannot return to their homes, and every effort should be made to move residents back to the least restrictive environment. Planning considerations should include short- and long-term accessible housing, replacement of services such as personal assistant services, and ensuring that individuals have necessary medical equipment and supplies.

To facilitate transition from a shelter to the community, plans should address how local government will determine that the jurisdiction is safe to inhabit. In addition, provisions should have provisions to assure that an individual's house is safe to return to, ensure services have been restored to the area, and ensure that accessible transportation is available.

Once all citizens have been successfully transitioned out of the shelter, the facility can be restored to its previous condition and the shelter can officially close. It is important to perform an after-action review with shelter staff to capture lessons learned to improve plans, resources, and training to support future successful operations.

ESF-6 Conceptual Framework

ESF-6 is a resource- and labor-intensive function requiring many different competencies. To achieve efficient and effective operations, it is important to identify a conceptual framework that accommodates time-phased and scaled response. The conceptual framework presented herein focuses on a large-scale event, but the concepts can be scaled to meet any response need. Time-phased response refers to the periods of time progressing forward from the actual emergency as shown in Table 19.3.

Table 19.3 Time Phased Response

Response Phase	*Actions*
Phase 1—Incident Notification	Incident occurs and triggers notification of local, tribal, state, and federal officials. Situational awareness may be sporadic and fragmented.
Phase 2—Activation and Immediate Response	Local agencies activate available resources, tend to immediate response needs, assess conditions throughout their jurisdiction, develop better situational awareness, and escalate information regarding the incident and resource needs to other local, regional, and state agencies. Predefined conditions for various types of emergencies may also trigger notification of federal agencies.
Phase 3—Deployment	As situational awareness is enhanced, additional local and state resources may be deployed to facilitate properly scaled response operations. States may request mutual aid and federal assistance as required to support escalating emergencies.
Phase 4—Sustained Response	Sustained response is characterized by defined operational periods and resources necessary to support response operations. Sustained response continues until life/health issues and other critical response objectives are addressed.
Phase 5—Recovery	The recovery phase involves restoration of conditions to pre-event status, if possible. For ESF-6, recovery is characterized by placement of people in their original homes or a suitable replacement and financial and medical independence.

General Concept of Operations

Concepts for providing mass care, sheltering, and human services should consider resources available both inside and outside of a particular jurisdiction. In some rural jurisdictions, resources may not be available to support even small-scale displacement of citizens for a prolonged period. Likewise, large metropolitan areas are more likely to incur larger numbers of displaced individuals in a catastrophic event, which may require support from outside jurisdictions. Figure 19.5 describes a general concept of operations to support ESF-6 functions with consideration given to impact area support as well as support external to the impact area. This concept also provides mechanisms to support long distance relocation to other host cities. Various nodes with Figure 19.5 are described below.

- **Evacuation Assembly Sites** (EAS) are locally operated site where evacuees will be directed to receive assistance. These sites may serve as the interface between ESF-9, Search and Rescue, and ESF-6. Support should include food, water, restrooms, and any available medical support. Under this concept of operations, evacuees move from this site to a Consolidated Services Site.
- **Local Congregate Care Shelters** are facilities available within the impact zone that can accommodate integrated mass care and sheltering needs. If evacuees in a local shelter require medical support beyond existing capabilities, they may be moved to a Consolidated Services Site.
- **Consolidated Services Sites** provide a coordination point to integrate local, state, and federal resources to provide robust services for evacuees who require long-term support or who will be evacuating from their communities. As shown in Figure 19.6, these sites are intended to consolidate the following services: (1) evacuee processing including pets; (2) respite sites for food, water, personal hygiene, and short-term rest; (3) medical and mental health services including Disaster Medical Assistance Teams if warranted; and (4) emergency assistance and human services. These sites are intended to provide "one-stop-shopping" for victims and also reduce logistical burdens by consolidating complementary services in one general location. Formal evacuation tracking for individuals leaving their community is initiated at these locations. Under this concept of operations, evacuees will move from this site to a Reception Processing Center for shelter placement and other mass care and human services.
- **Emergency Respite Sites** are locations along evacuation routes between the consolidated services sites and the reception processing center that provide water and fuel.

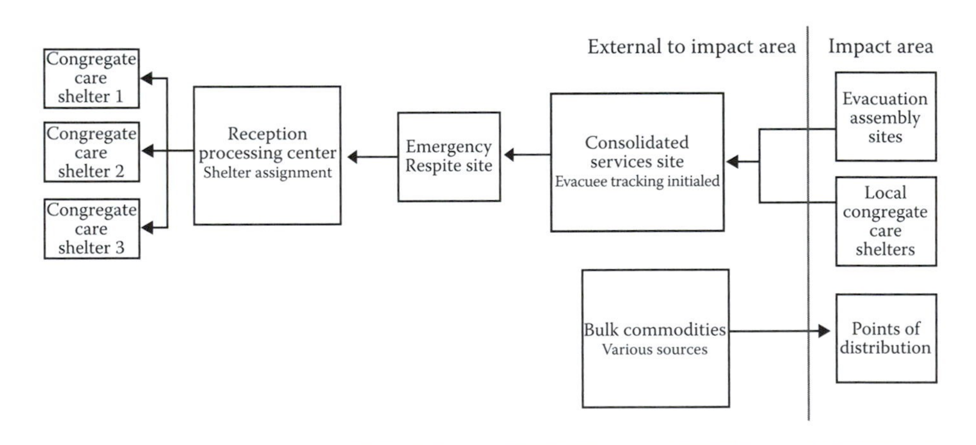

Figure 19.5 General ESF-6 concept of operations.

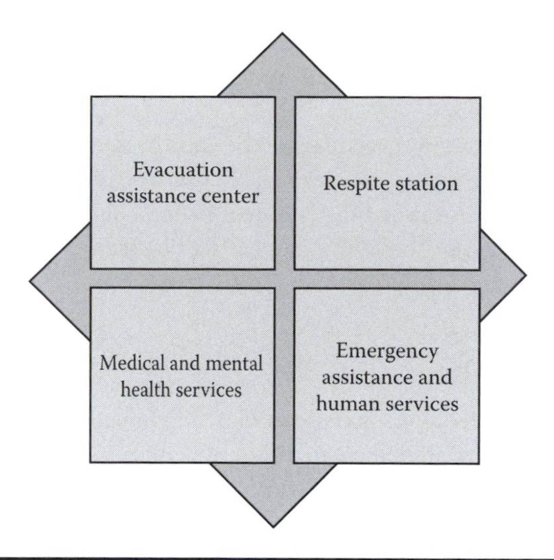

Figure 19.6 Consolidated services site.

- **Reception Processing Centers** serve as welcome centers in host cities and receive evacuees and direct them to locations where they can seek short-term shelter. These centers allow both placement in shelters as well as services to provide transportation and other resources to support placement in alternative locations such as with family members or friends. Evacuees will move from these sites to congregate care shelters or other shelter options.
- **Congregate Care Shelters** support integrative sheltering for the general population and those with functional, access, and certain chronic medical needs. They should be equipped so that individuals with access and functional needs can seek temporary lodging, food, hydration, and short-term lodging.

Evacuation

The concept of operations relative to evacuation support embarkation from an EAS operated by a local jurisdiction. If an evacuee has medical needs, he or she may be routed to a hospital or serviced via ESF-8 resources established within the impacted area. Those with no medical needs are relocated to a Consolidated Services Site. Transportation is generally accomplished by ground transportation with air assets reserved for medical emergencies.

At Consolidate Services Sites, evacuees may receive assistance from local, state, and federal agencies. At these sites, evacuees will be staged for further evacuation to Reception Processing Centers in host cities. Formal evacuee tracking will be implemented for all transportation assisted evacuees, if not already accomplished, with tracking information relayed to a formalized central point. The central point serves as a clearinghouse for those looking for friends and family. Emergency Respite Sites established along routes between the Consolidated Services Sites and Reception Processing Centers allow refueling and hydration for both transportation-assisted evacuees and self-evacuees.

Upon arrival in a host city, evacuees will be transported to Reception Processing Centers operated by the local host city jurisdiction receiving evacuees. These centers will provide central locations for evacuee shelter assignments and provision of individual assistance. From this site, evacuees will be transported, if required, to a congregate care shelter.

Feeding Operations

In a large-scale event, feeding operations can be challenging. At the local level, coordination with the local chapter of the ARC, NGOs, and faith-based organizations as well as the private sector is vital to meeting feeding needs both within shelters and for those who are able to stay at home but do not have access to food. As indicated in Figure 19.7, if local and mutual aid resources are outpaced, state resources can be requested via the State ESF-6 or mass care coordinator. Likewise, if local, state, and mutual aid resources are outpaced, the state can request support from the federal government via ESF-11. USDA can engage FNS and other resources to support feeding operations.

Bulk Distribution

Bulk distribution is the nexus between ESF-7, Logistics and Resource Management, and ESF-6. Working with local jurisdictions, sites should be assessed, selected, and established, both within the affected area and outside, for bulk distribution of commodities. The relationship between ESF-7 and ESF-6 is defined in Figure 19.8.

ESF-7 is responsible for coordinating procurement and delivery of bulk commodities through mobilization centers, incident support bases, staging areas, and other necessary nodes. The goal is to provide resources as close to the impact area as possible. ESF-6 is responsible for distribution of delivered supplies to identified survivors that may include operating Points of Distribution (PODs) to support individuals as well as resupply points that serve congregate care shelters. PODs

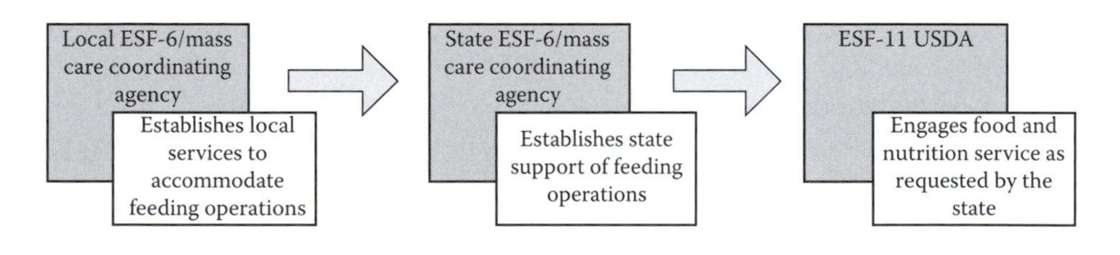

Figure 19.7 Chain of support for feeding operations.

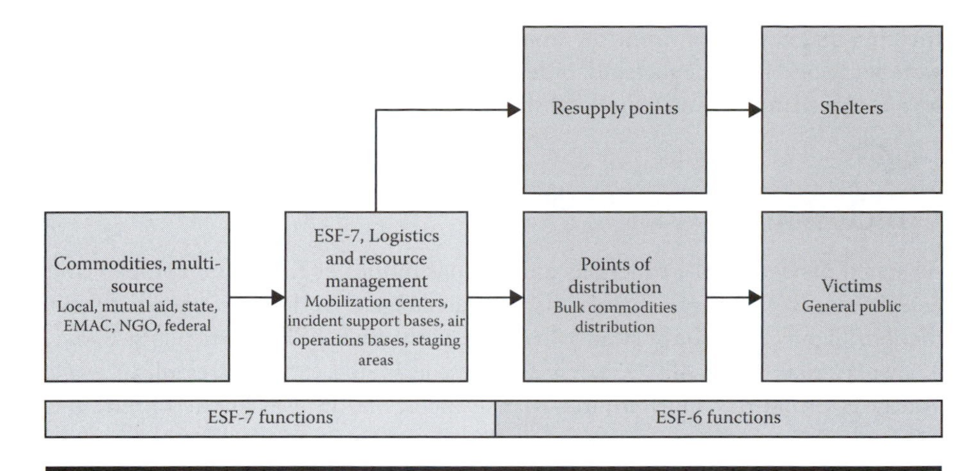

Figure 19.8 Commodity distribution flow.

should be designed to supply life-sustaining commodities to individuals including shelf-stable meals, tents, tarps, bottled water, and medications. Support for medication and medical supplies can be requested through ESF-8, Public Health and Medical Services.

Evacuation/Shelter of Household Pets and Service Animals

Animal owners have the primary responsibility for the survival and wellbeing of their animals. Although it may be logistically challenging, colocating pet shelters near congregate shelters allows owners to spend time with and to care for their pets. The general concept of operations for pets provided in Figure 19.9 is inclusive of evacuation operations and mass care. Local, state, federal, and all supporting stakeholders should make the appropriate allocation of resources to accommodate household pets, to the extent possible.

Pet embarkation is the responsibility of the owner of the animal in conjunction with the local jurisdiction in which the owner resides. Integration of transportation-assisted pets into the Consolidated Service Sites is essential to allow accurate

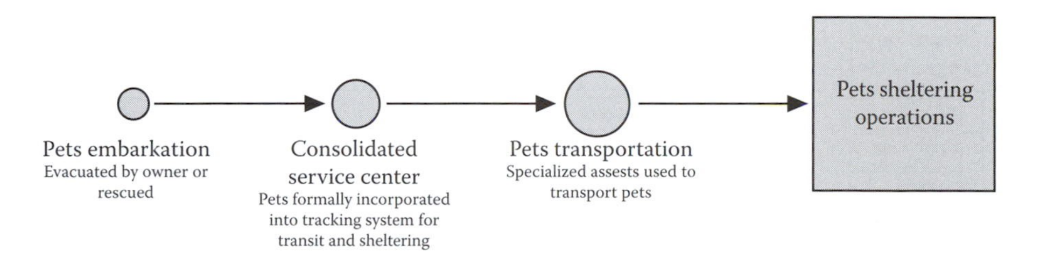

Figure 19.9 Pet sheltering operations.

tracking and evacuation from the impacted area. Local and state agencies should work with local humane societies, other pet service organizations, and veterinarians to accommodate pet evacuation and sheltering.

Conclusion

As stated previously, the functions and responsibilities of ESF-6 are complex and constitute a "big tent" of activities that can challenge even the most prepared jurisdiction. Significant coordination is required with other ESFs, particularly ESF-1, Transportation, and ESF-9, Search and Rescue in the early phases of an event and ESF-7, Logistics and Resource Management, and ESF-8, Public Health and Medical Services, as response operations progress.

Federal resources and responsible agencies are well defined and broad in their capabilities with respect to ESF-6. Local and state organization regarding mass care, sheltering, and human services, in some cases, may be less defined. Figure 19.10 indicates a coordination structure to serve as a starting point for organizing ESF-6 activities in relation to recent guidance and legal findings. Although it does not represent all functions required to support ESF-6 operations, it provides a starting point for addressing challenges such as integrating functional needs support services into shelter and mass care operations, providing evacuation and sheltering of pets, and coordinating use of volunteers and donated goods in a manner that provides the most benefit to victims.

From a victim's perspective, sheltering outside of home is a traumatic experience. Local, tribal, and state agencies in concert with volunteer organizations, NGOs, faith-based organizations, and the private sector can provide relief to victims through effective planning, training and resource allocation to support mass

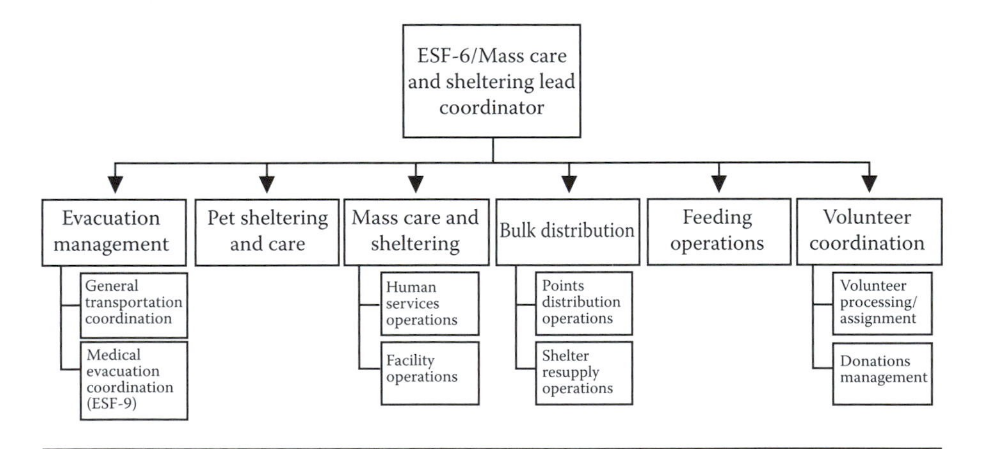

Figure 19.10 ESF-6/mass care coordination structure.

care, sheltering, and human services. Well-planned and executed ESF-6 operations will reassure victims that their lives will return to normal and provide a ray of hope in an otherwise hopeless situation.

References

1. National Response Framework, Emergency Support Function 6 Annex, January 2008, http://www.fema.gov/emergency/nrf/.
2. National Response Framework, Emergency Support Function #6—Mass Care, Emergency Assistance, Housing, and Human Services, Federal Emergency Management Agency, January 2008, page 1. http://www.fema.gov/pdf/emergency/nrf/nrf-esf-06.pdf.
3. Enforcing the ADA: A Status Report from the Department of Justice, U.S. Department of Justice, Civil Rights Division, 2007, page 3. http://www.ada.gov/julsep07.pdf.
4. Guidance on Planning for Integration of Functional Needs Support Services in General Population Shelters, Federal Emergency Management Agency, 2010, pp. 10–11.
5. Sheltering People with Disabilities, Space and Layout Considerations, Connecticut Universal Access Workgroup, February 2007. http://www.ct.gov/demhs/lib/demhs/space__layout_considerations.pdf.
6. Functional Needs of People with Disabilities, Emergency Preparedness Initiative, The National Organization on Disabilities, 2009. http://www.nod.org/assets/downloads/Guide-Emergency-Planners.html.
7. Adapted from Sheltering People with Disabilities, Space and Layout Considerations, Connecticut Universal Access Workgroup, February 2007, p. 8. http://www.ct.gov/demhs/lib/demhs/space__layout_considerations.pdf.

Chapter 20

Children and Disasters

Michael Steinle

Contents

Introduction

In October 2010, the National Commission on Children and Disaster (NCCD) released its *2010 Report to the President and Congress*.[1] The NCCD conducted a comprehensive study to examine and assess the needs of children* relative to

* Children are defined as 0 to 18 years of age.

disaster planning, response, and recovery. The report presents recommendations relative to 11 individual categories:

1. Disaster Management and Recovery
2. Mental Health
3. Child Physical Health and Trauma
4. Emergency Medical Services (EMS) and Pediatric Transport
5. Disaster Case Management
6. Child Care and Early Education
7. Elementary and Secondary Education
8. Child Welfare and Juvenile Justice
9. Sheltering Standards, Services, and Supplies
10. Housing
11. Evacuation

The recommendations presented in the NCCD report have broad implications for emergency planning, management, and response. Emergency Support Functions (ESFs) particularly impacted by the recommendations include ESF-5, Emergency Management; ESF-6, Mass Care, Sheltering, and Human Services; ESF-7, Logistics and Resource Management; and ESF-8, Public Health and Medical Services. This chapter explores some of the key recommendations set forth in the report and offers considerations for practical implementation of key provisions.

Disaster Management and Recovery

Relative to disaster management and recovery, the NCCD report provides four recommendations discussed below.

- Distinguish and comprehensively integrate the needs of children across all inter- and intragovernmental disaster management activities and operations.
- Accelerate development and implementation of the National Disaster Recovery Framework with an explicit emphasis on addressing the immediate and long-term physical and mental health, educational, housing, and human services recovery needs of children.
- The Federal Emergency Management Agency (FEMA) should ensure that information required for timely and effective delivery of recovery services to children and families is collected and shared with appropriate entities.
- FEMA should establish interagency agreements to provide disaster preparedness funding, technical assistance, training, and other resources to state and local child serving systems and child congregate care facilities.

The practical implications of these recommendations are broad, sweeping revisions to federal guidance and doctrine to include children's issues in planning, funding, and response efforts. As an example, revisions to the Target Capabilities List are cited specifically in the report as well as integration of specialists in children's issues within the federal response infrastructure. At the local, tribal, and state levels, specific guidance and funding are recommended within the report to support full integration of preparedness relative to children at all levels of government. These recommendations, and the report overall, may require specific identification of positions to manage issues relative to children and families within the Incident Command System structure during an activation.

Mental Health

The NCCD report contains five recommendations regarding mental health issues for disaster planning and response relative to children. Recommendations direct the Department of Health and Human Services (HHS) to

- Lead efforts to integrate mental and behavioral health for children into public health, medical, and other relevant disaster management activities.
- Enhance the research agenda for children's disaster mental and behavioral health, including psychological first aid, cognitive–behavioral interventions, social support interventions, bereavement counseling and support, and programs intended to enhance children's resilience in the aftermath of a disaster.
- Convene a working group of children's disaster mental health and pediatric experts to review the research portfolios of relevant agencies, identify gaps in knowledge, and recommend a national research agenda across the full spectrum of disaster mental health for children and families.

The ultimate goal of these recommendations is to develop a disaster mental and behavioral health Concept of Operations to formalize disaster mental and behavioral health as a core component of preparedness, response, and recovery activities.

The report also recommends that all federal agencies and nonfederal partners enhance preparedness activities and just-in-time training in pediatric disaster mental and behavioral health, including psychological first aid, bereavement support, and brief supportive interventions, for mental health professionals and individuals such as teachers who work with children. It is likely that future national level exercises and other joint local, tribal, state, and federal preparedness activities will focus on children's issues. Recommendations throughout the report also indicate the possibility that grants may be available in the future to support preparedness efforts relative to children.

Another recommendation relative to mental health issues directly impacts states. The NCCD recommends that FEMA and the Substance Abuse and Mental Health Services Administration strengthen the Crisis Counseling Assistance and Training Program (CCP) to better meet the mental health needs of children and families. Among possible implications, the Immediate Services Program grant application may be simplified to minimize the burden on communities affected by a disaster and facilitate the rapid allocation of funding and initiation of services. In addition, the recommendation includes provisions to establish the position of Children's Disaster Mental Health Coordinator within state-level CCPs.

The report also recommends that Congress establish a single, flexible grant funding mechanism to specifically support delivery of mental health treatment services that address the full spectrum of behavioral health needs of children including treatment of disaster-related adjustment difficulties, psychiatric disorders, and substance abuse.

These recommendations positively impact the focus on mental health as a method to foster full recovery. One of the likely challenges in addressing mental health issues, particularly after a large-scale event, is availability and credentialing of mental health professionals.

Child Physical Health and Trauma

The NCCD report provides six distinct recommendations relative to child physical health and trauma as summarized below:

- Congress, HHS, and FEMA should ensure availability of and access to pediatric medical countermeasures at the federal, state, and local levels for chemical, biological, radiological, nuclear, and explosive threats.
- HHS and the Department of Defense should enhance pediatric capabilities of disaster medical response teams through integration of pediatric-specific training, guidance, exercises, supplies, and personnel.
- HHS should ensure that health professionals who may treat children during a disaster have adequate pediatric disaster clinical training.
- The Executive Branch and Congress should provide resources for a formal regionalized pediatric system of care to support pediatric surge capacity during and after disasters.
- Prioritize the recovery of pediatric health and mental health care delivery systems in disaster-affected areas.
- The Environmental Protection Agency should engage state and local health officials and nongovernmental experts to develop and promote national guidance and best practices on reoccupancy of homes, schools, child care, and other child congregate care facilities in disaster-impacted areas.

The recommendation regarding pediatric medical countermeasures could be a great boon to local, tribal, and state agencies during a disease or bioterrorism scenario. Currently, provision of certain medications for children requires compounding (preparing proportional doses) at the local level, which may add considerable time to the process of mass prophylaxis. Moreover, greater emphasis on pediatric capabilities with the National Disaster Medical System (NDMS) can provide great benefit to impacted regions lacking a sufficient number of pediatric specialists.

These recommendations also have more specific local implications. HHS recommendations include classifying pediatric surge capacity as a required funding capability in the Hospital Preparedness Program. In addition, the report indicates the desire that state and hospital accrediting bodies should ensure all hospital emergency departments stand ready to care for ill or injured children through the adoption of emergency preparedness guidelines jointly developed by the American Academy of Pediatrics, the American College of Emergency Physicians, and the Emergency Nurses Association. Furthermore, implementation of NCCD recommendations may include congressional funding mechanisms to support restoration and continuity of for-profit and nonprofit health and mental health services to children.

The final recommendation regarding guidance and best practices on reoccupancy of homes, schools, child care, and other child congregate care facilities in disaster-impacted areas addresses the need to identify measures to protect children from environmental risk factors such as lead-based paint and asbestos. Properly maintained in homes, certain potential environment hazards pose no immediate threat to children. However, the impacts associated with physical damage, flood waters, high wind, and other impacts can release environmental hazards, which have particularly devastating effects on children. Best practices for reoccupancy can help to prevent exposure to children.

Emergency Medical Services and Pediatric Transport

EMS and pediatric transport recommendations include

- Clear designation and appropriate resourcing of a lead federal agency for EMS to coordinate grant programs, research, policy, standards development, and implementation.
- Improved capabilities of EMS to transport pediatric patients and provide comprehensive prehospital pediatric care during daily operations and disasters.
- Development of a national strategy to improve federal pediatric emergency transport and patient care capabilities for disasters.

These recommendations may lead to a federal grant program to assist local, tribal, and state agencies in improving prehospital EMS disaster preparedness including pediatric equipment and training. They may also lead to congressional funding for the Emergency Medical Services for Children (EMSC) program to ensure that states and territories meet targets and achieve progress in the EMSC performance measures and to support research and development.

As a result of this report, eligibility guidelines for Centers for Medicare and Medicaid Services reimbursement may require first response and emergency medical response vehicles to acquire and maintain pediatric equipment and supplies in accordance with the national guidelines for equipment for Basic Life Support and Advanced Life Support vehicles. In addition, HHS and DHS may establish more stringent pediatric EMS performance measures within relevant federal emergency preparedness grant programs.

Disaster Case Management

The NCCD report recommends appropriate resourcing of disaster case management programs and standards for consistent holistic services that achieve tangible, positive outcomes for children and families affected by the disaster. Aside from the federal funding implications and desire to provide broad based case management, it is possible that clearer definitions for transition from federal to state-led disaster case management programs will be developed. Development of voluntary consensus standards regarding essential elements and method of case management are also likely. Standards may include precredentialing of case managers and training that includes focused attention to the needs of children and families. These enhancements to the national strategy for disaster case management should minimize reunification issues and lead to faster, more complete recovery for families.

Child Care and Early Education

Regarding child care and early education, the report addresses three primary recommendations:

- Improve disaster preparedness capabilities for child care.
- Improve capacity to provide child care services in the immediate aftermath of and recovery from a disaster.
- Require disaster preparedness capabilities for Head Start Centers and basic disaster mental health training for staff.

Although these recommendations are directed primarily at HHS, they have substantial local and state impact. It is likely that states will be required, at some point in the future, to develop statewide child care disaster plans in coordination with state and local emergency managers, public health, state child care administrators and regulatory agencies, and child care resource and referral agencies. FEMA has also been directed to revise Public Assistance regulations to codify child care as an essential service with similar changes possible in the Stafford Act. Thus, certain costs relative to child care and early education may be reimbursable during the recovery period. Grant funding mechanisms may also be available to repair or rebuild private, for-profit child care facilities, support the establishment of temporary child care, and reimburse states for subsidizing child care services to disaster-affected families.

State and local emergency planning agencies, along with public health and mass care representatives, should begin to coordinate with early childhood education centers and schools to begin discussions regarding disaster planning, training, and exercises.

Elementary and Secondary Education

Elementary and secondary education recommendations are as follows:

- Improve preparedness of schools and school districts by providing additional support to states.
- Enhance the ability of school personnel to support children who are traumatized, grieving, or otherwise recovering from a disaster.
- Ensure that school systems recovering from disasters are provided immediate resources to reopen and restore the learning environment in a timely manner and provide support for displaced students and their host schools.

These recommendations include providing disaster preparedness grants to state education agencies to oversee, coordinate, and improve disaster planning, training, and exercises statewide, and to ensure that all districts within the state meet certain baseline criteria. Provisions are also suggested that would provide funds to states to implement training and professional development programs in basic skills to provide support to grieving students and students in crisis including requirements for teacher certification.

Other potential outcomes include congressional establishment of an emergency contingency fund within the Education for Homeless Children and Youth program to expedite grants to school districts serving an influx of displaced children. In addition, the report suggests the need for expert technical assistance and consultation regarding services and interventions to address disaster mental health needs of students and school personnel.

As stated above relative to early childhood centers, state and local emergency planning agencies, along with public health and mass care representatives, should begin to coordinate with elementary and secondary education officials to begin discussions regarding disaster planning, training, and exercises.

Child Welfare and Juvenile Justice

Child welfare and juvenile justice recommendations are as follows:

- Ensure that state and local child welfare agencies adequately prepare for disasters.
- Ensure that state and local juvenile justice agencies and all residential treatment, correctional, and detention facilities that house children adequately prepare for disasters.
- Ensure that juvenile, dependency, and other courts hearing matters involving children adequately prepare for disasters.

These recommendations require a certain amount of assessment before implementation. First, it is important to assess child welfare disaster planning to determine if significant advances have been made since passage of the Child and Family Services Improvement Act of 2006 (CFSIA). The report also suggests that HHS should develop planning guidance to supplement the basic procedures mandated in CFSIA. In addition, assessment is needed regarding appropriate preparedness relative to the juvenile justice system (courts).

At the local level, recommendations in the report will require emergency management and child welfare agencies to work with courts and residential treatment, correctional, and detention facilities that house children to develop effective emergency response plans. Federal agencies, including FEMA, the Department of Justice, and HHS, are directed within the report to provide funding and technical guidance regarding emergency planning for the child welfare and juvenile justice systems.

Sheltering Standards, Services, and Supplies

The report recommends that government agencies and nongovernmental organizations supporting mass care and sheltering (ESF-6) should provide a safe and secure shelter environment for children including access to essential services and supplies. Specifically, the report addresses the need for national standards for mass care shelters specific to children.

Implications at the local, tribal, and state levels include the need to obtain caches of age-appropriate shelter supplies for infants and children for immediate deployment to support shelter operations. In addition, the report suggests implementing criminal background checks to mitigate risks unique to children in shelters such as child abduction and sex offenders. ESF-6 planning coordinators should consider implementing child-centered standards in conjunction with revisions to address functional need support services.

Housing

Relative to housing, the report recommends prioritizing the needs of families with children, especially families with children who have disabilities or chronic health, mental health, or educational needs, in disaster housing assistance programs. The NCCD report and FEMA's "Guidance on Planning for Integration of Functional Needs Support Services in General Population Shelters" (hereinafter referred to as the FNSS Guidance) highlight the importance of ensuring that housing considerations include availability of age appropriate and disability/functional needs-specific services.

Additional local, tribal, and state implications include the concept that FEMA should be authorized to reimburse state and local governments for providing wrap-around services to children and families in community sites. Such reimbursement could be very helpful to state and local governments in providing important and expedited services during the recovery phase. The report also recommends development of innovative programs to expedite the transition into permanent housing for families with children.

Evacuation

Regarding evacuation, the report recommends that federal agencies should provide sufficient funding to develop and deploy a national information sharing capability to quickly and effectively reunite displaced children with their families, guardians, and caregivers when separated by a disaster. A direct response to challenges faced in the aftermath of Hurricane Katrina, this recommendation may lead to the development of a nationwide information technology capability to collect, share, and search data from any patient and evacuee tracking or family reunification system. At this point, it is unclear if the National Mass Evacuation Tracking System will achieve the goals set forth in the report or will require modification.

Building on the FNSS guidance, the NCCD report contains provisions that disaster plans at all levels of government must specifically address the evacuation and transportation needs of children with disabilities and chronic health needs, in coordination with child congregate care facilities such as schools, child care, and health care facilities. Although necessary, certain jurisdictions may find these transportation needs challenging.

Support of Children in ESF-6/Mass Care Operations

From a local, tribal, and state planning standpoint, Appendix F of the NCCD Report, "Supplies for Infants and Toddlers in Mass Care Shelters and Emergency

Congregate Care Facilities," provides perhaps the most directly beneficial information. This Appendix identifies basic supplies necessary to sustain and support 10 infants and children up to three years of age for a 24-hour period. These rates allow an estimation of perishable and nonperishable supplies necessary to support a given number of children. Additional guidelines include:

- Shelters should have supplies to support the care of children for a minimum of 72 hours.
- Supplies should accommodate the potential number of children up to three years of age as determined by assessment of current jurisdictional demographic data.
- If space or other challenges prevent on-site storage, supplies should be available for immediate delivery to the shelter within 3 hours via local vendor agreements, supply caches, interagency mutual aid, etc.

Tables 20.1 through 20.5[2] indicate NCCD supply recommendations. Sample calculations for a hypothetical jurisdiction are also provided.

Using the rates specific in the NCCD resource tables, it is possible to calculate the amount of food necessary to support pediatric needs in a given jurisdiction. Table 20.6 indicates the food, snack, formula, and electrolyte requirements for a five-county area with a total population of children aged 0 to 3 years of 11,147. Once quantities are known, it is possible, given pack size, to calculate pallet loads and truckloads necessary to support pediatric shelter needs in a 24-hour cycle. Similar calculations can be performed for other perishable and nonperishable goods to develop a resupply cycle to support logistics and shelter management.

Role of State and Local Governments

According to the NCCD report, children under the age of 18 years comprise about 25% of our population. Thus, planning to support their critical needs during a disaster is vital to achieve quick and full recovery. Essential planning elements identified by the NCCD for state and local governments include

- Evaluate the demographics of jurisdictional child populations (age 0–18 years) including children with disabilities and special health care needs.
- Identify places children will most likely be when under supervised care such as school, preschool, child care, summer camps, group homes, and juvenile justice facilities.
- Include accommodations for children in disaster training, exercises, and equipment purchases.
- Evaluate performance in meeting needs of children during emergency exercises and in after action reports.

Table 20.1 Recommended Perishable Supplies for Immediate Delivery within 3 Hours

Quantity	*Description*	*Comment*
40 jars	Baby food—Stage 2 (jar size is 3.5–4 oz)	Combination of vegetables, fruits, cereals, meats
1 box (16 oz)	Cereal—single grain cereal preferred (e.g., rice, barley, oatmeal)	Rice, barley, oatmeal, or a combination of these grains
See Note	Diaper wipes—fragrance free (hypoallergenic)	Minimum of 200 wipes
40	Diapers—Size 1 (up to 14 lb)	Initial supply should include one package of each size, with no less than 40 count of each size diaper
40	Diapers—Size 2 (12–18 lb)	
40	Diapers—Size 3 (16–28 lb)	
40	Diapers—Size 4 (22–37 lb)	
40	Diapers—Size 5 (27 lb+)	
40	Pull Ups 4T—5T (38 lb+)	

(*continued*)

Table 20.1 (*Continued*) Recommended Perishable Supplies for Immediate Delivery within 3 Hours

Quantity	*Description*	*Comment*
320 oz	Formula, milk-based, ready to feed (already mixed with water)++	Breastfeeding is the best nutritional option for children and should be strongly encouraged.
64 oz	Formula, hypoallergenic-hydrolyzed protein, ready to feed (already mixed with water)++	
64 oz	Formula, soy-based, ready to feed (already mixed with water)++	
1 quart	Oral electrolyte solution for children, ready-to-use, unflavored (e.g., Pedialyte)—dispensed by medical/health authority in shelter++	Do not use sports drinks. The exact amount to be given, and for how long, should be determined by an appropriate medical authority (doctor or nurse). To be used in the event an infant/child experiences vomiting or diarrhea, and the degree of dehydration.
See Note	Nutritional Supplement Drinks for Kids/Children, ready-to-drink (e.g., Pediasure, Kids Essential/Kids Boost)—dispensed by medical/health authority in shelter	Requirement is a total of 40–120 fl oz per day; in no larger than 8-oz bottles.[a]

Note: See "Supplemental Information" for additional information regarding the items followed by "++."

[a] Not for infants younger than 12 months of age.

Table 20.2 Nonperishable Supplies and Equipment

Quantity	*Description*	*Comment*
25	Infant feeding bottles (plastic only)++	4–6 oz size preferred (to address lack of refrigeration)
30	Infant feeding spoons++	Specifically designed for feeding infants with a soft tip and small width. Can be used for younger children as well.
50	Nipples for baby bottles (nonlatex standard)++	2 per bottle
25	Diaper rash ointment (petroleum jelly, or zinc oxide-based)	Small bottles or tubes
100 pads	Disposable changing pads	At least 13–18 in size. Quantity is based on 8–10 diaper changes per infant per day
10	Infant bathing basin	Thick plastic nonfoldable basin. Basin should be at least 12 × 10 × 4 in
See Note	Infant wash, hypoallergenic	Either bottle(s) of baby wash (minimum 100 oz), which can be "dosed out" in a disposable cup (1/8 cup/day per child) or 1 travel size (2 oz) bottle to last ~48 hours per child.
10	Wash cloths	Terry cloth/cotton—at least one per child to last the 72-hour period

(*continued*)

Table 20.2 (*Continued*) Nonperishable Supplies and Equipment

Quantity	*Description*	*Comment*
10	Towels (for drying after bathing)	Terry cloth/cotton—at least one per child to last the 72-hour period
2 sets	Infant hat and booties++	Issued by medical/health authority in shelter
10	Lightweight blankets (to avoid suffocation risk)	Should be hypoallergenic (e.g., cotton, cotton flannel, or polyester fleece)
5	Folding, portable cribs, or playpens	To provide safe sleeping environments for infants up to 12 months of age
2	Toddler potty seat	That can be placed on the seat of an adult toilet, with handles for support. One each should be located in both a Men's and Women's restroom
1 pack	Electrical receptacle covers	Minimum 30[a] (Note: Prioritize covering outlets in areas where children and families congregate (family sleeping area, children's areas, etc.))

Note: See "Supplemental Information" for additional information regarding the items followed by "++."

[a] Not for infants younger than 12 months of age.

Table 20.3 Other Recommended Perishable Supplies

Quantity	*Description*	*Comment*
40	Baby food—Stage 1 (jar size ~2.5 oz)	Combination of vegetables, fruits, cereals, meats
40	Baby food—Stage 3 (jar size ~6 oz)	Combination of vegetables, fruits, cereals, meats
40	Diapers—preemie Size (up to 6 lbs.)	As needed for shelter population
	Healthy snacks that are safe to eat and do not pose a choking hazard (intended for children 2 years and older)	Should be low sugar, low sodium: yogurt, applesauce, fruit dices (soft) (e.g., peaches, pears, bananas), veggie dices (soft) (e.g., carrots), 100% real fruit bite-sized snacks, real fruit bars (soft), low sugar/whole grain breakfast cereals and/or cereal bars, crackers (e.g., whole grain, "oyster"/mini)

Table 20.4 Other Recommended Nonperishable Supplies and Equipment

Quantity	*Description*	*Comment*
10	Sip cups (support for toddlers)++	

Note: See "Supplemental Information" for additional information regarding the items followed by "++."

- Designate a focal point of responsibility for coordinating children's needs.
- Design an evacuation plan that provides transportation for children with their families and caregivers, especially children with disabilities.
- Include child tracking and family reunification procedures in disaster plans.
- Provide safe, accessible shelter environments for children and families including essential age-appropriate supplies and care for medically dependent children.
- Develop capability of emergency personnel to provide effective prehospital pediatric transport and medical care (training and supplies).
- Work with hospital emergency departments to develop capabilities to provide effective care for children.
- Provide basic psychological first aid training for emergency personnel to assist children.

Table 20.5 Supplemental Information

Description	*Supplemental Notes*
Formula	Use of a powdered formula is at the discretion of the jurisdiction or shelter operator. If using powdered, preparation of the formula should be conducted by appropriately trained food preparation workers. Water used should be from an identified potable water source (bottled water should be used if there is any concern about the quality of tap or well water).
	Hypoallergenic hydrolyzed formula can be provided in powdered form—(1) 400 g can—but only if potable water is accessible.
Infant feeding bottles and nipples	Each time nutritional fluids, formula and/or other infant feeding measures (including breast milk in a bottle) are distributed by trained, designated shelter staff and/or medical professionals, clean, sterilized bottles and nipples must be used. Note: After use, bottles are to be returned to the designated location for appropriate sterilization (and/or disposal). Bottle feeding for infants and children is a 24/7 operation and considerations must be in place to provide bottle feeding as needed. (On average, infants eat at minimum 5–8 times daily).
	Note to staff: Sterilizing and cleaning
	Sterilize bottles and nipples before you use them for the first time by putting them in boiling water for 5 minutes. Nipples and bottles should be cleaned and sterilized before each feeding. If disposable bottles and nipples are not available and more durable bottles and nipples will be reused they must be fully sterilized before each feeding. To the greatest extent possible bottles and nipples should be used by only one child.
	In the event parents want to use their own bottles and nipples, shelter staff should provide support for cleaning these items between feedings. Support such as access to appropriate facilities for cleaning (not public restrooms).

Note regarding all feeding implements for infant/children	There is a specific concern with cleaning and sanitizing of all feeding implements associated with infants and children (infant feeding bottles/nipples, spoons, sip cups, etc). These items will require additional attention by food preparation staff to ensure they are sanitary as a means of reducing food borne illness. Staff medical/health staff should be consulted on best means of raising awareness among shelter residents and enlisting their support for these extra sanitary measures.
	Feeding implements such as spoons and sip cups should be cleaned using hot soapy water provided potable water is available. When the item is being cleaned to give to another child, the item must be sterilized.
For the following items: infant bathing basin, lightweight blankets, diaper rash ointment, wash cloths, and towels	Consider prepackaging the listed items together and providing one package to each family with children. Note: Additional blankets and towels will be necessary for families with more than one child.

Table 20.6 Hypothetical Resource Calculations for Pediatric Food Supplies, 24 Hours

County	Total[a]	Baby Food Stage 1 (jars)	Baby Food Stage 2 (jars)	Baby Food Stage 3 (jars)	Healthy Snacks (cans/cartons)	Cereal Single Grain (16 oz. box)	Formula, Milk-Based, Premixed with Water (oz)	Formula, Hypoallergenic-Hydrolyzed Protein, Premixed (oz)	Formula, Soy-Based, Premixed (oz)	Oral Electrolyte Solution (quart)	Nutritional Supplement Drinks for Kids (8 oz bottles)	Total Pallets	Total Truckloads
County 1	750	3,000	3,000	3,000	1,200	75	24,000	4,800	4,800	75	1,125	23.83	1.08
County 2	2,931	11,724	11,724	11,724	4,690	293	93,792	18,758	18,758	293	4,397	93.15	4.23
County 3	4,760	19,040	19,040	19,040	7,616	476	152,320	30,464	30,464	476	7,140	151.27	6.88
County 4	397	1,588	1,588	1,588	635	40	12,704	2,541	2,541	40	596	12.62	0.57
County 5	2,309	9,236	9,236	9,236	3,694	231	73,888	14,778	14,778	231	3,464	73.38	3.34
Total	11,147	44,588	44,588	44,588	17,835	1,115	356,704	71,341	71,341	1,115	16,721	354.25	16.10

[a] Population by county, 0–3 years of age.

- Support disaster plans, training, and drills for child congregate care providers that include evacuation, reunification, and addressing children with disabilities or chronic health needs.
- Work with jurisdictional stakeholders to develop plans to establish emergency child care.
- Identify resources in county and surrounding counties to address surge capacity relative to children's needs, especially medical and mental health needs.
- Develop a long-term disaster recovery plan that addresses the needs of children and families including housing, continuity of schools and child care, and medical and mental health needs.

Conclusion

Children's needs are substantially different from those of adults and should be treated as such. Studies have shown that the traumatic effects of a disaster, separation from friends and family, and general upheaval in their normal routines can have long-lasting effects on health, mental health, and academic success. Effects are even more dramatic when a child loses a loved one or a close friend. Quick, effective intervention may not relieve the short-term sorrow and bereavement, but can place a child on the pathway to recovery.

The recommendations set forth in NCCD report provide valid and much needed support to the specialized planning and resources necessary to support children in the aftermath of a disaster. Early interaction with the broad spectrum of child stakeholders, serious consideration of child issues at every step of the planning process, and attention to health, medical, and mental health needs will facilitate an effective concept of operations to support children in their time of need.

References

1. National Commission on Children and Disasters. *2010 Report to the President and Congress*. AHRQ Publication No. 10-M037. Rockville, MD: Agency for Healthcare Research and Quality. October 2010.
2. National Commission on Children and Disasters. *2010 Report to the President and Congress*. AHRQ Publication No. 10-M037. Rockville, MD: Agency for Healthcare Research and Quality. October 2010, pp. 166–171.

Chapter 21

Emergency Management and the Media

Randall C. Duncan

Contents

Understanding and working with the media is an important part of an overall emergency management system. This relationship—between the emergency manager and the media—is one that has more opportunity to excel, or fail, than almost any other.

Let us begin our examination of the relationship between emergency management and the media by defining what the media are.

Traditionally, we think of the media as consisting of newspapers, radio, and television. Newspapers have been the media staple since the modern printing press was invented in 1450 by Johannes Gutenberg. Radio and television entered the world of media much more recently, but changed the way media operated and functioned within our society by bringing news on a timelier basis (live reports) and adding the elements of voices (radio) and moving pictures (television).

More recently, the development of the World Wide Web, social network sites, and blogs have led yet another revolution in the way media impacts our lives.

In order to more fully understand the elements of the relationship between emergency management and the media, it is necessary to understand the characteristics of the various types of media. Let us begin our examination with the traditional media forms of newspapers, radio, and television.

Newspapers

Arguably, the first newspaper in the United States was called *Publick Occurrences Both Forreign and Domestick* (National Humanities Center 2006). It was published on September 25, 1690, and edited by Benjamin Harris. It only printed one issue, and was banned four days after publication by the Governor and Council of Massachusetts (Massachusetts Historical Society 2010). The only surviving copy of the newspaper is in the Public Record Office in London (Library of Congress 2009).

Traditionally, modern newspapers have been published on a daily or weekly basis, depending on the size of the reading audience. Circulation of newspapers varies greatly. The newspapers with the three largest weekday circulations in 2009 were the *Wall Street Journal* (circulation 2,024,269), *USA Today* (circulation 1,900,116), and the *New York Times* (circulation 927,851). The Newspaper showing the smallest circulation was the *Medina (NY) Journal Register* (circulation 2117) (Audit Bureau of Circulation 2009).

Newspapers have traditionally been viewed as providing more in-depth coverage than either radio or television because of the amount of space available in which to write the story. Newspapers also provided some of the first coverage of events far removed from the place where they were published by the mechanism of the telegraph (see section "Radio," for more details). This allowed remote correspondents to send a story from far away back to the newspaper home office, and created a style of journalistic writing known as the "inverted pyramid." The inverted pyramid style of writing called for the correspondent to relay the most important facts first, followed by those of lesser importance in the body of the story (Scanlon 2008).

Based on this information, then, we can anticipate what print organizations want in the way of news, as shown in Table 21.1.

Radio

It is not possible to talk about the history of radio without mentioning the wired telegraph system. The telegraph was made practical within the United States by Samuel Morse, who did his first public demonstration of the device in 1838

Table 21.1 What Traditional Print Organizations Want by Way of News

Item	*Explanation*
Details	Print media wants to paint a picture in the reader's mind with words.
Questions	Will be more oriented toward details.
Background Information	How many times did the truck roll over?
	How far away from the edge of the road did it come to rest?
	Were there flames? If so, how high?
	History related to an event.
	Has this ever happened before?
	History of individuals involved in the event.
Deadline	Traditional: usually daily. Current policy may be impacted by Newspaper website.

(Smithsonian Institution 2010). In 1843, Congress provided funding to install a telegraph between Baltimore, Maryland, and Washington, DC. The Whig Party held its nominating convention in Baltimore on May 1, 1844, and selected Henry Clay as their nominee. This was the first news item relayed by telegraph (Smithsonian Institution 2010). In 1901, Guglielmo Marconi began developing what would become broadcast radio—he sent the Morse code signal for the letter S from a wireless transmitter in Poldhu, Cornwall, England, to a wireless receiver in Newfoundland, Canada (Public Broadcasting System 1998b). A few years later—on Christmas Eve, 1906—some wireless telegraph operators onboard ships heard the Christmas carol "Silent Night" and a voice reading bible verses interrupted the Morse code they normally heard (Public Broadcasting System 1998a). This marked the first radio broadcast.

From these humble beginnings, radio had an impact on the way we listened to news and found out about other events. We could sit in our living rooms and hear the voices of presidents, dictators from overseas, and Hollywood stars endorsing commercial products.

Unlike newspapers, radios could bring us the sounds and words of a news event as they happened.

Modern radio stations are separated into various interest groups called "formats." Some of the formats in today's radio broadcasting include news/talk stations, music stations, public radio, and non-English radio. Table 21.2 explains some of the features and news items as pertain to radio media.

Table 21.2 Features and News Items as Pertain to Radio Media

Item	*Explanation*
Details	Radio news utilizes short, concise information in the voice of the newsmaker.
	Typically, they are 10- to 15-second "actualities" or "sound bites."
News/Talk	More stories; a little more depth than other radio formats
Public Radio	Uses "natural sound." Records background of event happening with open mike.
Non-English stations	Help to reach those who speak a language other than English within the community.
Music stations	May or may not carry news. If they do, it typically consists of only short news items.
Deadline	Hourly—depending on schedule of newscasts

Television

The first authorized broadcast of a television in the United States started on July 2, 1928, in Wheaton, Maryland, a suburb of Washington, DC, by C.F. Jenkins (Popular Mechanics 1928). The heyday of television may have come on the evening of March 7, 1955, when one in two Americans watched Mary Martin's portrayal of "Peter Pan" on live television (Bogart 1958, p. 1). Other significant events that marked the impact of television on the way Americans received news included the coverage of such live events as the Kennedy–Nixon debates and mankind's first step on the moon. The addition of live images to go with sound literally brought the world into our homes every night. There are various types of news broadcasts on local television stations (Table 21.3). They may range from spot news/breaking news of particular activities currently in progress to the regularly scheduled news programs. In addition, some local television stations may air special investigative reports or programs. Typically, local television stations will have an affiliation with a network, and will present a network-originated program of national and international news.

Social Network Sites and the World Wide Web

No examination of media would be complete without exploring the impact of the World Wide Web and social media on individuals, as well as the traditional media of newspaper, radio, and television.

Table 21.3 Features and News Items as Pertain to Television Media

Item	*Explanation*
Details	Video of the event may determine whether there is a story.
	A story that otherwise would not make the news may become a story if there is video.
	Similarly, a story of real importance may not make the news if there is no video.
Types of news	Local news includes spot news; regular local news; investigative reports. Also, feature news programs (either local or network); national news; and international news.
Deadline	Various depending on the type of news that will be broadcast. Major deadlines are typically for the evening news broadcast and the late night news broadcast.

The World Wide Web, as we know it today, first became reality in 1990 with the release of a point-and-click hypertext editor called "World Wide Web" (Berners-Lee 1998). In the ensuing years, the number of websites, their functionality, and the pure amount of information has literally exploded. Naturally, with such rapid expansion, there is a need for a "buyer-beware" approach by users. It is as easy to find an academically reputable and accurate source on the World Wide Web as it is the lunatic ravings of fringe elements. It is up to the user to find and place the appropriate value on sources available through this medium.

Aside from the ease of accessing information on the World Wide Web, people soon found that it was becoming a tool for personal communication and networking between friends, leading to the development of social network sites. Social network sites are defined as:

> ...web-based services that allow individuals to (1) construct a public or semipublic profile within a bounded system, (2) articulate a list of other users with whom they share a connection, and (3) view and traverse their list of connections and those made by others within the system. The nature and nomenclature of these connections may vary from site to site (Boyd and Ellison 2007).

We also think of social network sites as social media—a way to convey information to friends and learn about information from friends and others with opinions. Sites such as Twitter, Facebook, MySpace, and blogs have become a way to share information, opinions, and even reflect on news events.

These same sites have had an impact on the way traditional media—newspapers, radio, and television—interact with their viewers and listeners. As a result, newspapers now also shoot video of news stories and provide it to their readers through the mechanism of their website. Radio stations do the same thing. Television stations now write news stories and publish them, similar to newspapers, on their websites. This has had a major impact on the traditional deadlines for the various forms of media.

Yet another World Wide Web–based phenomena has emerged recently—the Weblog, or Blog. This phenomenon has blurred the distinction between traditional press and bloggers.

> In the media world, it used to be clear who was in the news business and who was not. News businesses provided news, nonnews businesses did not. Reporters worked for companies who were in the business to provide news. News businesses got paid, usually by advertisers, to collect, package and distribute information of interest to news audiences. Nonnews businesses or organizations exist for other purposes—perhaps to deliver public service such as environmental protection, or to produce commercial goods such as fertilizer. Providing news for these businesses is simply not their reason for existence. In an instant news world, that distinction is becoming increasingly fuzzy.
>
> One of the most significant trends to come out of the collection of technologies we call the Internet, is the emergence of citizen journalists. "Blogging," from the term "Weblog," which used to describe people who would record and publish what they discovered on the Internet, reflects the ease with which almost anyone who writes today can also publish. As mentioned earlier, some of the bloggers have accumulated audiences in the millions and have influence as great as any of the celebrity journalists that used to be staples of our early evening hours at home (Baron 2006, p. 47).

Because of these factors, the Emergency Manager or spokesperson for the jurisdiction has to keep in mind that there are now more audiences for the information they prepare than the traditional media. There are the families of those directly impacted by the event or emergency, those in the immediate area of the emergency or disaster, and the "traditional media" along with the citizen journalist.

Dealing with the Media in a Crisis

To begin our discussion about dealing with the media in a crisis, we need to understand some of the basics about communication. We first learn to communicate as babies—before we can even begin to say words. We communicate with gestures and sounds, through a thing called nonverbal behavioral clusters. We will discuss this in more detail shortly.

Communications is extremely complicated for such a seemingly simple thing. The act of communications starts as an idea in our brain. That idea wishes to be expressed or communicated to someone. It must then make its way through the filters of our belief system and perceptions. Then it must be encoded (either in speech or writing) and then broadcast to a receiver (a reader or listener). The receiver has to get the message, decode it, and run the decoded material through their own filters of belief systems and perceptions in order to understand the idea we originally wished to communicate to them. An understanding of the complications associated with the process of how we communicate allows for a new appreciation of a statement as seemingly simple as, "Pass the salt, please."

Nonverbal behavioral clusters associated with how we say and express things are more important in conveying meaning than the words we actually say (Blatner 2009). Because these nonverbal behavioral clusters typically convey a larger percentage of our communication than the specific choice of words do, we tend to place more faith in the way the message is expressed. When the message being conveyed to an audience by a speaker's words is in conflict with their nonverbal behavioral clusters, the audience will not believe the speaker. As a simple thought experiment, recall the last time you observed a person on television and your reaction to that person was the thought that you did not believe a word they said. The odds are, you felt that way because there was a conflict between the words of the speaker and their nonverbal behavioral clusters.

The normal process of communications takes a slight detour under a crisis situation. When a crisis is in progress, we need to provide assistance to the elected official or spokesperson to make sure we do not allow circumstances to take away from the messages we need to communicate to the public typically through the media. In other words, we need to avoid media pitfalls. Table 21.4 is adapted from unpublished material from Dr. Vincent Covello, founder and director of the Center for Risk Communiction.

In a crisis situation, the media follows certain patterns. Those patterns include:

- Searching for background information on the incident
- Dispatching reporters to the scene
- Obtaining access to the scene or the official spokesperson
- Dramatizing the situation
- Expecting a briefing complete with written information
- Expects *you* to panic
- Becomes confused by technical information
- Exhausting resources
- Sharing information among themselves
- Acting professional and expecting the same
- Providing filler for stories if credible information is not available

Table 21.4 Media Do's and Don't's

Do	*Don't*
Define all technical terms and acronyms (jargon).	Use language that may not be understood by even a portion of your audience.
If you use humor, direct it at yourself.	Use humor in relation to safety, health, or environmental issues.
Refute negative allegations without repeating them.	Refer to national problems—"This isn't Love Canal."
Use visuals to emphasize key points.	Rely entirely on words.
Remain calm. Use the question or allegation as a springboard to say something positive.	Let your feelings interfere with your ability to communicate positively.
Ask whether you made yourself clear.	Assume you have been understood.
Use examples, stories, and analogies to establish a common understanding.	Talk only in abstractions.
Be sensitive to nonverbal messages you are communicating.	Allow your body language or your position in the room to be inconsistent with your message.
Make them consistent with what you are saying.	Dress inconsistently with your message.
Attack the *issue*.	Attack the person or the organization.
Promise only what you can deliver.	Make promises you cannot keep.

Set, then follow strict deadlines.	Fail to follow up on those items that you promise to follow.
Emphasize achievements made and ongoing efforts.	Say there are no guarantees.
Refer to the importance you attach to health, safety, and environmental issues—your moral obligation to public health outweigh financial considerations.	Don't refer to the amount of money spent as a representation of your concern.
Use personal pronouns (e.g., I, we).	Take on the identity of a large organization.
Take responsibility for your share of the problem.	Try to shift blame or responsibilities to others.
Assume everything you say and do is part of the public record.	Make side comments or "confidential" remarks.
Discuss risks and benefits in separate communications.	Discuss your costs along with risk levels.
Use risk comparisons to help put risks in perspective.	Compare unrelated risks.
Stress that the true risk is between zero and the worst-case estimate.	State absolutes or expect laypersons to understand.
Emphasize performance, trends, and achievements.	Mention or repeat large, negative numbers.
Focus your remarks on empathy, competence, honesty, and dedication.	Provide too much detail or get drawn into protracted technical debates.
Keep presentation to 15 minutes total.	Ramble or fail to plan the time well.
Keep answers to 2 minutes maximum.	Tell people more than they want.

The Public Information Officer

The Public Information Officer (PIO) typically has the responsibility for coordinating the collection, verification, and dissemination of information to the public. These duties may occur as a part of day-to-day organizational operations, or on an emergency basis. Since the focus of deliberations in this discussion is "Emergency Management and the Media," let us concentrate on the roles and responsibilities of a PIO in an emergency.

The PIO has responsibilities to a number of different constituencies. These include

- The Public—This segement is the largest user group for emergency messages. This implies that the PIO should be aware of any special demographic characteristics of the community being served, and have familiarity with the best media channels to distribute information to those who need it.
- The Media—This is one of the most important relationships to establish to make sure information is distributed to those who need it. The PIO will need to understand the traditional and social media outlets within the jurisdiction.
- The Agency—This relationship is the basis of trust within the jurisdiction. The PIO has a duty to positively portray the efforts and successes of the agency they represent. This relationship will be especially important when navigating an agency through the dangerous shoals and reefs of a negative news story.
- The Other Responding Agencies—The PIO needs to have a good working relationship with other agencies responding to the emergency or disaster. These relationships are especially important in helping to avoid conflicting stories or statements. If the incident becomes large enough to engage the Joint Information System (JIS) or the Joint Information Center (JIC), the PIO needs to be able to function in that environment. In addition, the PIO needs to be aware of the possibility that there will be differing priorities among the agencies responding to the emergency or disaster and how to deal with those differing priorities so that "mixed messages" are not given to the public.

We generally make the statement that the PIO provides public information. It would be helpful to provide a definition of what public information is, with respect to emergencies and disasters. Generally, we can conclude that public information is used by people to save lives, reduce injury and harm, and protect property (FEMA 2009). Given that public information covers such a wide territory, it is understood that almost every piece of information coming from your agency or emergency operations center (EOC) could result in the public taking some type of action to protect themselves or others from the effects of a disaster or emergency. This also emphasizes the criticality of the accuracy and timeliness of your information (FEMA 2009).

We generally expect that information communicated to the public through our PIO will result in action by people, provide information, change behaviors or attitudes, or create a positive view of our agency or EOC within the community. Some examples of public information could include

- The current status of the emergency or disaster
- Agency response actions to the disaster
- Information or warnings as conditions change
- Important locations (i.e., where food, water, and shelters are located)
- Specific evacuation information or directions
- Other pertinent information:
  - What is open or closed
    - Government facilities
    - Stores
    - Roads
    - Schools
  - Status of lifeline systems—electricity, gas, water, and sewer
  - Volunteer recruitment
  - Where people can find aid or assistance
  - Public inquiry telephone numbers

In order to perform the job of PIO well, the person in that position needs to have a number of qualities. Some of those qualities include

- Knowledge of the organization they represent. This allows them to speak with credibility about the operations of the organization. It demonstrates to the media and the public that the PIO has access to agency leadership. It also provides the media with opportunities for interviews and briefings about the agency.
- A good working relationship with the organization or EOC. This is a necessary quality for the PIO to have access to the information and resources everywhere within the agency or EOC they are representing.
- A certain amount of aggressiveness. The PIO may need to be able to go directly into the organization and get to decision makers and leaders with minimal delays. In addition, the PIO will undoubtedly be called upon to provide advice to leadership—making it necessary for the PIO to be in the inner circle of the organization or EOC.
- A high level of trust and ability to strategize. The PIO needs to be able to establish trust within the organization or EOC. Essentially, the PIO becomes an advisor that understands how things will be viewed outside of the agency or EOC, and understand the implications of information to which the public will have access. It will be important to understand what the potential negative consequences of these issues are and how to present them, truthfully, in as positive a light as possible.
- Community relations skills are necessary for the PIO. The PIO needs to understand the demographics of the jurisdiction—who lives and works there

and what the prevailing local values, concerns, and interests are. The PIO must also know about organizations within the community and how they work and interact.

- The PIO needs to have good media relations skills. This includes a level of credibility that is usually only developed over time and through hard work.

It is also important for the PIO to have several sets of skills, including

- Writing abilities—organizes clear thoughts in a written format (whether electronic or printed). This includes the ability to develop talking points, guidance, strategy papers, speeches, and general information for management. It is especially important that proper grammar and spelling be utilized in these pieces. This should probably include familiarity with Associated Press Stylebook (http://www.apstylebook.com). Generally, the PIO needs to be able to produce quality documents, whether electronic or paper.
- Other abilities—the PIO should be able to understand the basics of using video as a means of communication. This would include knowledge of the basic elements of photography. In addition, the PIO should be able to clearly communicate and outline ideas in a manner allowing the PIO or other spokesperson to communicate effectively with an audience; the ability to speak effectively and persuasively in front of an audience; and, an awareness of the impact of nonverbal behavioral clusters on the delivery of a message.

The Joint Information System/Joint Information Center

As you have read previously, the National Incident Management System (NIMS) was developed at the direction of Homeland Security Presidential Decision Directive (HSPD)-5. The original NIMS document was developed in 2004, updated in 2006, and updated again in 2008 (FEMA 2008). Public information, and the process of establishing the system to collect, integrate, and coordinate it, is defined as a part of NIMS Component IV—Command and Management (FEMA 2008, pp. 70–74). The overall system is called the JIS, and the specific place where this process happens is called the JIC.

Typically, this process involved a PIO, who supports the incident command structure, and who is also a member of the Command staff. The responsibilities of this position typically include

- Responding to inquiries from the media, public, and elected officials
- Supervising the process of collecting, integrating, and coordinating information for emergency public information
- Supervising the process of collecting, integrating, and coordinating information for warning information

- Monitoring for rumors and responding to them
- Relations with the media
- Creating coordinated and consistent messages through
  - Identifying key information to be communicated to the public
  - Creating the message that provides the key information in a clear and easily understood method
  - Prioritizing messages so the most important gets out first and that the public is not overwhelmed with the amount of information
  - Verifying the accuracy of information
  - Making sure the message gets out through the most effective means available

The overall JIS provides the means to coordinate the messages being released to the public by all elements of government involved in the disaster response—whether multiple local jurisdictions, or local, state, and federal governments; multiple disciplines involved in the response; nongovernment organizations involved with the response; and the private sector. This coordination is particularly important because all the voices involved in the disaster should be providing substantially the same message.

One of the important elements to keep in mind when dealing with a large emergency response situation is that different disciplines may have their own spokesperson or PIO present and different jurisdictions may have their own spokesperson or PIO present as well. It is possible that these PIOs may serve as the basis for the JIC

Table 21.5 Noteworthy Details on JIS and JIC

JIS Provides a Structure and System for	*JIC Provides a Place to*
Developing and delivering coordinated interagency messages	Centrally facilitate operation of the JIS during and after an incident
Creating, recommending and executing public information plans and strategies	Increase information coordination
Advising Incident Commander about incident relevant public affairs issues	Reduce misinformation
Monitoring and correcting erroneous information circulating among the media or the public	Maximize resources for dealing with the public and the media
Be adaptable to the size and scale of the incident—from three PIOs at the scene to 150 PIOs at a major disaster from multiple locations	Provides "one stop shopping" for the media

staff. Remember that each of these officials will have the primary responsibility for making sure the story—as it relates to their agency, discipline, or jurisdiction—gets out to the media. But there is no reason they cannot work together and collaborate in order to establish the JIS and provide the personnel for the JIC.

The JIC is basically an instrument to help facilitate the processes that take place within the JIS—much like the EOC is an instrument to help facilitate the processes that take place within the Local Emergency Operations Plan (LEOP). As a result, there are other parallels between these two elements of Public Information. The elements of the JIS must be worked out well in advance of the occurrence of a disaster.

Table 21.6 Different Types of JIC

JIC Type	*Description*
Incident	Typically, an incident-specific JIC is established at a single, on-scene location in coordination with federal, state, tribal, and local agencies or at the national level, if the situation warrants. It provides easy media access, which is paramount to success. This is a typical JIC.
Virtual	A virtual JIC is established when a physical colocation is not feasible. It connects PIOs through e-mail, cell/landline phones, faxes, video teleconferencing, Web-based information systems, etc. For a pandemic incident where PIOs at different locations communicate and coordinate public information electronically, it may be appropriate to establish a virtual JIC.
Satellite	A satellite JIC is smaller in scale than other JICs. It is established primarily to support the incident JIC and to operate under its direction. These are subordinate JICs, which are typically located closer to the scene.
Area	An area JIC supports multiple-incident ICS structures that are spread over a wide geographic area. It is typically located near the largest media market and can be established on a local, state, or multistate basis. Multiple states experiencing storm damage may participate in an area JIC.
Support	A support JIC is established to supplement the efforts of several Incident JICs in multiple states. It offers additional staff and resources outside of the disaster area.
National	A national JIC is established when an incident requires federal coordination and is expected to be of long duration (weeks or months) or when the incident affects a large area of the country. A national JIC is staffed by numerous federal department and/or agencies, as well as state agencies and nongovernment organizations.

Plans			
Do you have systems and procedures for:		Yes	No
✓	Developing an emergency response or crisis communications plan for public information and media relations?	☐	☐
Does your emergency response or crisis communications plan have systems and procedures for:		Yes	No
✓	Designating and assigning line and staff responsibilities for the public information team?	☐	☐
✓	Identifying and updating current contact numbers for PIO staff and other public information partners in your plan?	☐	☐
✓	Identifying and updating current contact numbers for regional and local news media (including after-hours news desks)?	☐	☐
✓	Establishing the JIC at the Emergency Operations Center (if activated)?	☐	☐
✓	Securing needed resources (space, equipment, people) to conduct the public information operation during an incident 24 hours a day, using such mechanisms as Memorandums of Understanding, contracts, etc.?	☐	☐
✓	Creating messages for the news media and the public under severe time constraints, including methods to clear these messages within the emergency response operations of your organization (including multijurisdiction and/or agency cross-clearance)?	☐	☐
✓	Disseminating information to news media, the public, and partners (e.g., website capability 24/7, listservs, broadcast fax, printed news releases, door-to-door leaflets)	☐	☐
✓	Verifying and clearing/approving information prior to its release to the news media and the public?	☐	☐
✓	Operating a public inquiry hotline with trained staff available to answer questions from the public and control rumors?	☐	☐
✓	Activating the Emergency Alert System, including the use of prescripted messages?	☐	☐
✓	Coordinating your public information systems planning activities with other response organizations?	☐	☐
✓	Testing the plan through drills and exercises with other response team partners?	☐	☐
✓	Updating the plan as a result of lessons learned through drills, exercises, and incidents?	☐	☐

Figure 21.1 JIC readiness assessment checklist.

People			
Do you have systems and procedures for:		Yes	No
✓	Identifying staffing capabilities needed to maintain public information operations for 24 hours per day for at least several days? (Note: Staff may include regular full- and part-time staff as well as PIOs from other agencies or departments, disaster employees, volunteers, etc.)	☐	☐
✓	Establishing and maintaining agreements for acquiring or borrowing temporary staff? (Note: Such agreements may be mutual aid arrangements or Memorandums of Understanding.)	☐	☐
✓	Granting emergency authority to hire or call up temporary staff or those on loan from other organizations?	☐	☐
✓	Establishing and maintaining job descriptions and qualifications for individuals serving as your organization's PIO and other roles during an incident?	☐	☐
✓	Assigning a staff member and at least one to alternate the role and responsibilities of PIO?	☐	☐
✓	Determining if the assigned PIO(s) is qualified? Sample qualifications include: ○ Experience and skills in providing general and emergency public information. ○ Ability to represent your organization professionally (can articulate public information messages well when dealing with the media and the public, and can handle on-camera interviews). ○ Written and technical communication skills (writing/editing, photography, graphics, and Internet/Web design proficiency). ○ Management and supervision experience and skills needed to run a JIC.	☐	☐
✓	Establishing and maintaining a list of language translators available to assist with public information? (Note: Such a network should include sign language interpreters and individuals capable of writing and speaking the non-English language(s) used by individuals in your jurisdiction.	☐	☐
✓	Establishing and maintaining working relationships with PIO partners from other organizations that you might need to work with during an incident (e.g., PIOs from other jurisdictions, other government agencies or departments, nongovernmental organizations, and private entities)?	☐	☐
✓	Developing and maintaining working relationships with your local and regional media, and established procedures for providing information to those media entities effectively and efficiently during incidents?	☐	☐

Figure 21.1 **(*Continued*)**

Logistics			
Do you have a go-kit for PIO use during an incident, including:		Yes	No
✓	Laptop computer capable of linking to the Internet/e-mail?	☐	☐
✓	Cell or satellite phone, pager, and/or PDA/palm computer with wireless e-mail capability?	☐	☐
✓	Digital camera, photo storage media, and charger/backup batteries?	☐	☐
✓	Flash drives, CDs and/or disks containing the elements of the crisis communication plan (including news media contact lists, PIO contact lists, and information materials such as topic-specific fact sheets, backgrounders, talking points, and news release templates)? REMEMBER: Redundancy is important in case the computer you are using doesn't have a USB port, CD, or floppy drive?	☐	☐
✓	Office supplies such as paper, pens, self-stick notes, etc.?	☐	☐
✓	Manuals and background information necessary to provide information to the media and the public (e.g., your Smart Book)? (Note: A Smart Book is a compilation of factual information assembled about your jurisdiction, such as population, number of schools and hospitals, size and description of geographic or infrastructure features, etc.)?	☐	☐
✓	Hard copies of all critical information?	☐	☐
Do you have systems for:		Yes	No
✓	Acquiring and maintaining go-kits with a funding mechanism (e.g., credit card) that can be used to purchase operational resources? (Note: A go-kit is a mobile response kit that allows PIOs to maintain communications in the event that they are working outside of their normal place of operation.)	☐	☐
✓	Ensuring PIOs can access the go-kit when serving at an incident?	☐	☐
✓	Acquiring and maintaining portable communications equipment, critical up-to-date information, and supplies?	☐	☐
✓	Acquiring and maintaining essential media production equipment (cameras, digital storage, laptops, etc.)?	☐	☐
✓	Acquiring and maintaining a Smart Book (or equivalent technologies) to assist PIOs in accurately informing the media and the public during an incident?	☐	☐
✓	Identifying a dedicated location to house the JIC? (Note: The location selected must be wired for telephone, internet access, cable, etc.)	☐	☐
✓	Securing and maintaining the necessary JIC equipment and supplies to allow information to be disseminated to the media and the public?	☐	☐
✓	Inventorying and restocking the PIO go-kit after an incident?	☐	☐

Figure 21.1 **(*Continued*)**

Logistics (Continued)			
✓	Inventorying and restocking JIC equipment and supplies after an incident?	☐	☐
✓	Periodically updating your Smart Book with current information?	☐	☐
Do you have equipment and supplies needed for a JIC, including:		Yes	No
✓	Computers on a LAN with Internet access and e-mail listserves designated for news media and partner entities?	☐	☐
✓	Laptop computers?	☐	☐
✓	Electric and manual typewriter(s) in case of power outage or other problems that interfere with computer/printer usage?	☐	☐
✓	Fax machine preprogrammed for broadcasting fax releases to news media and partner entities?	☐	☐
✓	Printers and copy machines, with supplies such as toner and paper?	☐	☐
✓	Paper shredder and trash bags?	☐	☐
✓	Televisions with access to cable hookups and VHS VCRs or other recording media?	☐	☐
✓	Cell or satellite phones, pagers, and/or PDAs/palm computers with wireless e-mail capability?	☐	☐
✓	Digital camera, photo storage media, and charger/backup batteries?	☐	☐
✓	Audio recorder and batteries?	☐	☐
✓	Flash drives, CDs, and/or disks containing the elements of the crisis communication plan (including media contact lists, PIO contact lists, and information materials such as topic-specific fact sheets, backgrounders, talking points, and news release templates)?	☐	☐
✓	Office furniture/accessories such as desks, chairs, file cabinets, bulletin boards, white boards, trash cans, lights, in/out baskets, landline phones, clocks, large calendars, etc.?	☐	☐
✓	Audio equipment and furniture necessary for conducting news conferences (e.g., wireless microphones, lectern, multibox, etc.)?	☐	☐
✓	Office supplies (e.g., white and colored paper, pens, self-stick notes, folders, blank tapes, binders, overnight mail supplies, tape, poster board, erasable and permanent markers, chart paper, easels, staplers and staples, press kit folders, binders, computer disks/CDs, hole punch, organization logo on stickers, letterhead, postage stamps, etc.)?	☐	☐
✓	Manuals, directories, and background information necessary to provide information to the media and the public (e.g., your Smart Book)?	☐	☐
✓	Hard copies of all critical information?	☐	☐

Figure 21.1 (*Continued*)

The system plans and processes, much like the roles and responsibilities within the LEOP, must be worked out and understood in advance of their application.

There should also be consideration given as to what kind of triggers might initiate the activation of the JIC. Some suggestions might include:

- The creation of a standard operating procedure or guideline that defines the opening of the facility. This SOP/SOG could be modeled after the existing document for the activation of the Emergency Operations Center.
- An analysis of the potential impact of the incident.
- An analysis of the potential media interest in the incident.
- The potential duration or the response and recovery phases of the emergency or disaster.

Other noteworthy items about the JIS and JIC are outlined in Table 21.5.

Table 21.6 presents a number of different types of JICs. It is adapted from information found in FEMA G290 Basic Public Information Officer Course (FEMA 2009).

We close out the chapter with a useful JIC Readiness Assessment form in Figure 21.1.

References

Audit Bureau of Circulation. 2009. U.S. Newspapers—Search Results. Retrieved April 15, 2010, from Audit Bureau of Circulation: http://abcas3.accessabc.com/ecirc/newstitle searchus.asp.

Baron, G. R. 2006. *Now is Too Late 2: Survival in an Era of Instant News*. Bellingham, WA: Edens Veil Media.

Berners-Lee, T. 1998. The World Wide Web: A Very Short Personal History. Retrieved April 23, 2010, from World Wide Web Consortium (W3C): http://www.w3.org/People/Berners-Lee/ShortHistory.

Blatner, A. 2009. *About nonverbal Communications*. Retrieved April 21, 2010, from Adam Blatner's website: http://www.blatner.com/adam/level2/nverb1.htm.

Bogart L. 1958. *The Age of Television*. New York: F. Unger Publishing Company.

Boyd, D. M., and N. D. Ellison. (2007). Social Network Sites: Definition, History, and Scholarship. Retrieved April 21, 2010, from Journal of Computer-Mediated Communication: http://jcmc.indiana.edu/vol13/issue1/boyd.ellison.html.

FEMA. 2009. *G290 Basic Public Information Officer Training*. Washington DC: Federal Emergency Management Agency.

FEMA. 2008. National Incident Management System. Retrieved April 23, 2010, from Federal Emergency Management Agency: http://www.fema.gov/pdf/emergency/nims/NIMS_core.pdf.

Library of Congress. 2009. *Eighteenth-Century American Newspapers in the Library of Congress*. Retrieved April 15, 2010, from Library of Congress: http://www.loc.gov/rr/news/18th/200.html.

Massachusetts Historical Society. 2004. Premier issue of the Boston News-Letter. Retrieved April 15, 2010, from Massachusetts Historical Society, http://masshist.org/objects/2004.april.cfm.

National Humanities Center. 2006. *Publick Occurrences Both Forreign and Domestick.* Retrieved April 15, 2010, from National Humanities Center: http://nationalhumanitiescenter.org/pds/amerbegin/power/text5/Publickoccurrences.pdf.

Popular Mechanics. 1928. What television offers you. *Popular Mechanics* 50(5): 820–824.

Public Broadcasting System. 1998a. KDKA begins to broadcast 1920. Retrieved April 16, 2010, from A Science Odyssey: People and Discoveries: http://www.pbs.org/wgbh/aso/databank/entries/dt20ra.html.

Public Broadcasting System. 1998b. Marconi receives radio signal over Atlantic 1901. Retrieved April 16, 2010, from A Science Odyssey: People and Discoveries: http://www.pbs.org/wgbh/aso/databank/entries/dt01ma.html.

Scanlon, C. 2008. The Inverted Pyramid Structure. Retrieved April 16, 2010, from Purdue Online Writing Lab: http://owl.english.purdue.edu/owl/resource/735/04/.

Smithsonian Institution. 2010. History Wired: A few of our Favorite Things. Retrieved April 16, 2010, from National Museum of American History, Smithsonian Institution, http://historywired.si.edu/detail.cfm?ID=324.

Chapter 22

Impact of Social Media on Emergency Management

Adam S. Crowe

Contents

Power and Purpose

As the second decade of the twenty-first century dawns, the biggest challenge for emergency managers is the need to modify long-standing philosophies on how citizens are communicated with regarding emergency preparedness and management issues that might affect them. This communication includes pre-event preparedness and planning as well as responsive crisis communications necessary during the emergency or disaster.

Traditional outreach has included providing educational pamphlets and flyers during local presentations or community events where citizens may receive the information along with a plethora of other materials. Unfortunately, those citizens may or may not be interested in receiving that information and consequently will not be impacted enough to consider necessary behavioral changes such as personal preparedness and/or prompt response when directed by emergency officials. Moreover, establishing trust between governmental representatives and the general public is challenging and can again lead to lack of behavior change. These traditional approaches and common challenges have been the cornerstone of some of the most significant communication issues faced by emergency managers.

These challenges can be overcome (in many ways) through the use of social media. Regardless of the system, it creates a shared connection of people and/or organizations with common values and interests that choose to engage in the exchange of information for the common good. It also creates an inherently higher trust factor for information because of the shared network of friends, contacts, and organizations. It is noteworthy, however, that this engagement of information can be good or bad, and this is why it is critical for emergency managers to understand and engage in the collaboration that is social media.

For instance, Facebook currently has more than 500 million active accounts, which means that Facebook has more users than the entire populations of Russia, Japan, and Mexico combined and has nearly twice as many as that of the United States.[1] Or put another way, if Facebook were its own country, it would be the third most populous in the world behind China and India.[2] In contrast to the traditional model of education and outreach that is offered where citizens may or may not be spending their time, Facebook is where communication and relationships are happening. Specifically, an average Facebook user has 130 friends, connection to more than 80 community groups, and shares 90 pieces of personal content each month.[3] This type of pervasive establishment of community is also common in other social media outlets such as Twitter, YouTube, and several different blog sources.

Twitter is arguably the second most important social media site related to emergency management practitioners. While currently maintaining 105 million users (21% of whom are active), media outlets at all levels and all types actively utilize this system to seek out and distribute newsworthy information at a tremendous pace.[4] This pace of information dissemination is exponentially increased because of the significant levels of mobile use of Twitter and its redistribution (or retweet) functionality of liked or trustworthy information. Specifically, numerous third-party applications (e.g., HootSuite and TweetDeck) allow for various forms of utilization of Twitter including monitoring and direct messaging.

Although impressive, many governmental leaders and local emergency managers worry about the credibility of systems such as Twitter, which lack verifiable accounts (on most governmental accounts) and are often admittedly filled with insignificant and/or irrelevant information. Alternative communication sites such as Nixle have been developed to address these issues related to emergency management's use of Twitter as a public education and information dissemination tool.

Unfortunately, the number of users (particularly media) of these alternative systems pales in comparison to Twitter, and therefore is highly unlikely to be an impactful tool to emergency management even though it allows for verifiable accounts.

Nixle is just one of many social media systems that have been specifically built to draw in people for specific purposes—often related to emergency management and preparedness. For instance, in early 2009, Microsoft announced the creation of Vine, a social media system intended to serve as an emergency notification and monitoring system by friends and family. However, by September 2010, Microsoft announced the discontinuation of this system.[5] Citizens were not finding Vine beneficial presumably because people were not actively engaged on that network—especially as compared to Facebook and Twitter. Based on the statistics already mentioned, Facebook and Twitter have established their supremacy and should be treated as such by emergency management. Emergency management must be careful to distribute their messages where local citizens are spending time, not where they want them to be, to avoid repeating the mistakes already established through traditional outreach approaches.

This premise is also true in how people seek out information through the Internet. According to one major technology publication, the Web (as local governments have utilized for the last decade) is dead. Citizen activity has moved away from static browsing for information toward applications and mobile browsing.[6] Additional studies have indicated that mobile browsing will overtake traditional browsing by 2015.[7] In the United States, the Federal Emergency Management Agency (FEMA) and the National Weather Service have both implemented new mobile and/or application-based outreach.[8,9] The impact of this mobile browsing and notification has had exponential growth due to the ability of many jurisdictions to automate emergency alert messages to established social media outlets such as Facebook and Twitter.[10]

Interestingly, the availability of practical emergency preparedness mobile applications is not limited to formal offerings from governmental and quasigovernmental sources. There are a variety of apps offered for free or minimal fees including step-by-step guides for first aid, CPR, pet preparedness, and personal allergies[11] as well as function-based software including flashlights[12] and emergency dispatch feeds.[13] This type of information continues to carry common messages expressed by emergency managers and/or creates a transparency toward activities and direction.

Leveling the Playing Field

As outreach philosophies are beginning to change in the emergency management field, the benefits of such a change have to be understood. These benefits are broad and multifaceted, affecting components ranging from public education to public communications to response tools. Moreover, traditional approaches to project management, technological development, training, and public involvement are on the verge of revolutionary change due to the inclusion of social media and other Web 2.0 concepts.

Specifically, there are three fundamental rules of social media application in emergency management: (1) conversations are key, (2) no more middleman, and (3) it has got to be free. These rules allow a leveling of the proverbial playing field between emergency management programs of all sizes and at all levels of government.

Perhaps the most significant change relates to the cost management of activities related to emergency management projects. Social media and Web 2.0 concepts are often eliminating the need for costly development of systems to manage emergency management concepts such as planning, exercise management, and response mechanisms. This cost saving is possible because of the establishment of robust networks, servers, and infrastructure by nearly all social media outlets (e.g., Facebook, Twitter, YouTube, and Flickr) that allows for a high level of confidence when used by emergency management or other secondary sources.

A second reason these systems are beginning to replace traditional mechanisms is the implementation of crowdsourcing. Crowdsourcing allows tasks typically performed by employee to now be performed by a collection of individuals within a crowd who have no particular connection outside of the ability to perform the desired function. Within emergency management, crowdsourcing has been used numerous times, but most recently was utilized by BP during the oil spill in the Gulf of Mexico to collect suggestions about possible ways to stop the spill. BP received more than 20,000 suggestions that were categorized into not possible, already planned, or feasible. As a result, they identified nearly 100 options that were feasible to stop one of the largest oil spills ever.[14] These response concepts and ideas were made accessible to decision makers and emergency responders in a more timely manner and ultimately may have contributed to the resolution of the incident in a more quick and efficient manner.

Additionally, a free crowdsourcing website called Ushahidi has been utilized during several international emergencies including the Haiti and Chile earthquakes in early 2010. Ushahidi provided Web-based or mobile connectivity to collect (from "the crowd") information about the incident. This included Web-based maps that provided real-time crowd-generated information about health conditions, infrastructure damages, and localized emergencies.[15] The speed and accuracy of this type of information aggregation is impossible by governmental or first responder agencies utilizing current systems. The application of Ushahidi in these situations is strong support that the public's growing expectation of speed and breath of information is much faster than official government communication channels are currently able to provide.

Not only can emergency managers utilize public gathering and collection of information, they can also self-define preparedness and response messages as well as certain operational processes. Specifically, traditional media outlets (television, radio, and print) are vital partners in public dissemination of emergency management messages; however, these groups inherently filter the message. This type of message adjustment can be positive or negative, but inherently happens for a variety of reasons ranging from media bias due to time (or space) limitations based on the format utilized for distribution. Social media helps eliminate and/or control this process and allows emergency managers to have an outlet for an unfiltered and fully developed preparedness or response message, which is critical to ensure that public citizens receive clear and consistent information.

Likewise, the operational processes such as donations and volunteer management have been significantly improved because of the involvement and application of social media. For instance, in 2009, the City of Fargo, North Dakota, was responding to significant flooding from the Red River and was having difficulty arranging for enough volunteers to support efforts during the middle of winter. The community implemented a Facebook group and generated interest in volunteerism that was roughly equal to 5% of their local population, which significantly improved their response capabilities.[16]

Similarly, donations management has successfully moved into the social media and Web 2.0 realms after the American Red Cross utilized donations through text messaging in support of the 2010 earthquake in Haiti. Specifically, the Red Cross was able to generate a grand total of $5 million in donations within the first 48 hours[17] and $30 million within 10 days of the disaster.[18] This figure ultimately accounted for approximately 10% of the total funds donated to the relief funds. This figure represented a significant reduction in the time commitment and resources often necessary to collect, manage, and process donations generated in response to an emergency or disaster.

Another example of a cost-effective, direct access training and public education venue is the U.S. Centers for Disease Control and Prevention's (CDC) utilization of Second Life. Second Life is an online virtual world where users create avatars (or digital likenesses) and within the virtual world can engage in the physical environment and communicate openly on various topics of interest. The CDC created an area within Second Life in the spring of 2008 with the hope of providing information exchange and health education on a variety of issues supported by the CDC.[19]

The CDC's support of Second Life has also ventured into emergency preparedness and response. For instance, in the spring of 2010, the CDC held a virtual talk in Second Life that was later captured on video and shared via blogs and YouTube.[20] Likewise, Second Life has also been utilized by the University of Illinois–Chicago School of Public Health to simulate POD sites and distribution of prophylaxis materials after an anthrax attack.[21] This type of systematic utilization of a virtual environment has the opportunity to ultimately decrease the cost of trainings and exercises by minimizing costs related to physical setups and elimination of perishable items necessary for public health emergency preparedness training and exercise activities.

The Future Is Right Here, Right Now

There are many examples of how both disasters and emergency management have been impacted by social media and Web 2.0 concepts. These include the utilization of Facebook, Twitter, and YouTube, to name a few systems, as well as a strong push to redefine the relationships between local governments, the media, and their citizens. Although these issues may continue to be developed and/or redeveloped in the near future, the emergency management community will also have the immediate opportunity to begin to utilize various online, Web 2.0 tools that are available for free.

These tools are free because of a shared network of servers, computers, networks, and interrelated systems often referred to as the "cloud." This cloud is utilized by all online social media and Web 2.0 service providers to ensure robust networks that are both redundant and sufficient to meet the needs of the end user. When this robustness fails, the social media community has often abandoned the system or come up with colloquial monikers such as Twitter's "Fail Whale."[22] This sector represents numerous operational and response tools that can (and will) be utilized by emergency management as a cost-effective alternative to many current systems commonly used.

For instance, real-time collaborative editing tools would be of great value to emergency managers and first responders who are creating planning and public information documents during an event. This type of tool would allow for multiple users to be simultaneously creating a document rather than the document being written, reviewed, edited, and then reviewed before distribution or implementation. The time necessary for review and approval for press releases and other operational documentation could also be minimized and/or eliminated because of the simultaneous reading, writing, and editing of a document. Again, this type of functionality is critical in ensuring clarity and consistency of public messages, which is necessary to ensure the public provides safe and expected response behavior.

The largest and most ambitious version of this type of tool was Google Wave, which was initially released in May 2009. This system promised to have collaborative editing with time-stamped tracking of information management, which was projected as a possible new technology for implementation in Joint Information Centers and other information management sources. Unfortunately, Google was unable to address issues identified during its beta testing and ultimately shut down Wave in August 2010. Since that time, other software and browser-based collaborative editing systems have been released with TypeWith.Me showing the strongest possibility for implementation similar to what was initially projected for Google Wave.

Additional "cloud" technologies that may impact emergency management include those systems that support information management, organization, and distribution. Specifically, there is a group of social media systems referred to as social bookmarking that allow for a Web-based listing and categorization of

Internet links that can be privately accessed through login/password combinations and/or shared publically. Not only is this type of social media an excellent opportunity for planning and operational response, it also allows for a free and robust redundancy for many emergency managers and emergency operations centers (EOC). Likewise, there are similar online systems that allow for free online storage of files and online materials with almost no limit to size or type of file. Sources of these services are sometimes fee-based, but there are several robust free services such as Drop.io, Evernote and MyOtherDrive that could be utilized by emergency management in this fashion.

Another powerful type of social media tool available for emergency management is referred to as social geolocation systems. These tools include FourSquare, GoWalla, Google Latitude, and most recently, the implementation of Facebook Places. Although all built on slightly different formats, these social media systems are all based on the concept of utilizing mobile telephone devices to determine the geographic location of individuals. This geographic location, which is based on WiFi and GPS signaling, allows for the individual user to be virtually engaged in the actual environment that surrounds him. For instance, if friends and/or favorite restaurants were geographically close they would appear in these systems and allow for the establishment and/or increased level of social interaction.

Emergency management utilization of social geolocation systems is in its infancy because of the relatively recent establishment of this technology. However, there are numerous operational applications that could be considered for usage including weather spotting, search and rescue, damage assessment, and debris management. These emergency management functions are dependent on field operations at diverse geographic locations that are managed from one central command location. This makes communications, documentation, and technological implementation a necessity.

For instance, debris management operations, because of the necessity of contracted labor, are extremely vulnerable to abuse and misreporting. Significant levels of process accountability are required to eliminate duplicate trips, weighted trucks, and other abuses. The utilization of social geolocation systems would allow impacted jurisdictions to require contracted workers to identify themselves geographically over certain intervals, which could then be recorded and reviewed by emergency management staff to ensure proper actions were maintained.

Likewise, weather spotters, damage assessment teams, and search and rescue teams would be able to be deployed to certain geographic areas and report back real-time observed information. Although this reported information would typically be done through radio communications and/or traditional paper documentation, social geolocation systems allow for instantaneous reporting and capturing of the data for faster processing of the information being provided by the field teams. Having this type of information faster and with greater reliability would be an extremely valuable tool for efficient and effective emergency management and resource coordination.

Although emergency managers have not yet fully implemented social geolocation systems for operational usage, many emergency response agencies (at all levels of government) have more thoroughly implemented Web 2.0 mapping for geographic information analysis and information sharing. This was particularly utilized by media outlets and educators to show the progression of events such as the BP Oil spill in the summer of 2010.[23] Perhaps the most powerful utilization of socially interactive maps was Google's Crisis Response Center that integrated publicly generated YouTube videos, visual mapping, oil spill forecasts from the National Weather Service, spill berm locations provided by the state of Louisiana, and satellite images of the spill provided by NASA all into the Google Map technology that is free to all. The need for crowd-sourced, real-time mapping for emergencies and disasters is so well accepted that Google has established MapMaker to help facilitate just this concept.[24]

The power of text messaging for the improvement of donations management has already been discussed in relation to the American Red Cross's fundraising efforts for the Haiti Earthquake in 2010. However, text messaging has also become an extremely impactful tool for mobile and portable communications. According to the latest Pew Internet research, 72% of American adults and 87% of American teenagers who use cellular phones text message on a regular basis, which is 9% more than one year ago.[25] This high level of utilization is extraordinarily beneficial to local emergency managers because of the relatively easy, cost-effective, and robust mechanism to communicate emergency public information notifications to their citizens.

As was established for Facebook and Twitter, emergency managers would be doing their local citizens an injustice by ignoring the presence and growth of text messaging within a community. Many jurisdictions (especially schools and higher education institutions) have utilized private companies to perform an automated text messaging service; however, there are a few that have tied Twitter's capability for text notification to provide these services at a fraction of the cost and with similar efficacy.[26]

Slaying the Giant

Even with the numerous examples of the impact of social media on emergencies and disasters, most local emergency management communities have yet to adopt comprehensive use of either social media or the communication concepts inherent in its use. The application of social media and Web 2.0 options falls into three categories: (1) proactive, (2) reactive, and (3) inactive. Proactive utilization includes the active usage of social media systems such as Facebook and Twitter to disseminate information and monitor public comments regarding their agency and/or community event. Progressive is the most complicated use of social media and requires the most time and resources to master. Reactive utilization of social media only disseminates and/or monitors public comments, but not both. This is the most common application within emergency scenarios because of its more reasonable utilization of resources. The last and final category covers organizations that are completely inactive in social media. This inactive status is probably the most dangerous to emergency managers because it ignores the significant impact of social media on emergencies and disasters.

According to a recent study by the American Red Cross, citizens are now seeking out and utilizing social media and Web 2.0 systems to send and receive information. Specifically, the online survey found that 20% of adults who could not reach 911 would try to contact responders through a digital means such as e-mail or social media. Moreover, 44% stated that they would ask other people in their social networks to contact local authorities on their behalf, 35% would post a direct request on a response agency's Facebook page, and 28% would send a direct Twitter message to responders.[27] More alarmingly, to inactive first response agencies, this study also found that 69% of respondents felt that first responders should be monitoring social media sites to send help quickly and nearly 74% expected emergency help to come in less than one hour after a post to Twitter or Facebook.[27] This survey clearly states that the public expects proactive social media usage by emergency management and first response agencies. Based on these findings, it is operationally, ethically, and politically irresponsible for local emergency management organizations to simply try and ignore social media's impact on their response.

These survey findings are fully supported by BP's experience during the 2010 oil spill with a lack of presence in social media. BP's presence on Twitter immediately after the oil spill in the spring of 2010 was not fully developed and lacked the content and added value that the general public was seeking out regarding this disaster. Unfortunately for BP, this social media void was field with a satirical twitter account presenting itself as an official source of BP public relations that posted inflammatory comments. Interestingly, the number of followers of the fake BP account outpaced the real BP account by nearly tenfold.[28] As already established, social media users who ultimately are local citizens are seeking out information via social media systems and will seek that information until they find a source that appears or implies legitimacy.

In addition to the reactive versus proactive challenge, another problem that emergency managers and emergency public information officers face is the balance between style and substance. Social media system users expect informal, conversational tones and language that often includes colloquialism, slang, abbreviations, and misspelled terms. Unfortunately, this level of informality is extremely uncommon in official government information releases. Consequently, emergency managers must decide what level of modification they are willing to accept. For instance, the U.S. government released its social media guidelines, which included maintaining active voice, present tense, speaking directly to constituents, utilization of key words, as well as the avoidance of colloquialism, slang, and governmental jargon.[29]

Additionally, some emergency management agencies are overwhelmed with the process implementation required for utilization of social media. This implementation includes the identification of personnel to oversee social media as well as vigorous and realistic policies. Many emergency management offices are often small with only part-time or volunteer support staff, which makes new concepts such as the application of social media challenging if leadership is not passionate about its use. Rather than seeking out creating and innovative ways to be proactive or at least reactive, some emergency managers have taken the stance that social media is simply a fad and will pass along if it is ignored long enough.

Unfortunately, this attitude is shortsighted. The utilization of analytics and monitoring measurement tools such as Tweetdeck, Google Analytics, and Monitter will show nearly a constant social media discussion on various issues impacting a local jurisdiction. This social media conversation will simply grow exponentially during an emergency or disaster and be occurring all around emergency managers whether they acknowledge it or not.[30] These monitoring tools are free and dynamic enough to search for certain terms, concepts, and associations to determine how the public is discussing certain issues, which will ultimately lead to more effective communications with the public regarding the incident in question.

Lastly, social media is also significantly impacting operational response systems such as the National Incident Management System (NIMS) in the United States that help define a uniform and coordinated response to emergencies and disasters. Specifically, methods such as NIMS define processes to include the collection, analysis, and distribution of emergency public information through a command and control system in which all messages are ultimately approved by a single person with ultimate authority for the overall operations (e.g., Incident Commander or EOC Manager).[31] However, this review-and-approval process is antagonistic to the speed and formality (or lack thereof) of social media systems such as Facebook and Twitter.[32] No system exists that effectively and efficiently blends operational models with social media systems, which will continue to be a challenge to emergency managers until adjustments are made to the operational responses systems that maintain levels of accountability and control without eliminating the benefit of utilizing social media systems.

Conclusion

Although the utilization of social media systems by emergency management professionals is in its infancy, the future benefit and application is nearly boundless. Emergency managers cannot deny the fact that social media is already being utilized by numerous citizens and media outlets for the monitoring and distribution of emergency public information. Because of the ever-changing nature of information related to disasters, social media thrives in emergencies and must be considered in all phases of emergency management including preparedness, response, recovery, and mitigation.

Social media must begin to be used by emergency managers in conjunction with traditional outreach to provide a comprehensive and thorough mechanism for the distribution of education and emergency public information. Only this type of approach will effectively ensure that emergency management professionals are providing information in a timely and effective manner via mechanisms where citizens are not where they have been or are hoped to be.

Additionally, emergency managers have a tremendous opportunity to implement social media and Web 2.0 systems as operational response tools. Many of these systems potentially provide greater accountability and safety as well as redundant systems to store documentation, resources, and other vital response components. This application is nearly always free and easily integrated into or through mobile browsing and/or applications, which allows for significant mobility and portability of these new operational tools.

The future of emergency management is right here, or rather, right together. Social media's ability to improve real-time collaboration via cloud networking or social geolocation systems is already becoming a valuable tool and will continue to be so as more emergency management professionals learn more about these systems and creatively apply their uses to current response systems in joint information centers and mapping centers.

Implementation of social media is occurring at many different levels and in many different ways. These organizations should be profiled and examined for best practices and ideal application for certain emergencies and/or crisis situations.[33] Moreover, significant work must be completed related to the recommended implementation of social media into emergency public information systems to ensure they can be utilized as a tool and ultimately benefit the clear and consistent review of public messages. These best practices, along with a little creativity and ingenuity, will help drive the future of social media's continued application in emergency management. Social media is not going away, nor are disasters. Therefore, it is paramount for emergency management—from the individual to the industry—to find ways to understand and embrace how social media is impacting their lives.

References

1. "Battle of the sizes: Social network users vs. country populations." Pingdom. Accessed on September 8, 2010. http://royal.pingdom.com/2009/03/13/battle-of-the-sizes-social-network-users-vs-country-populations/.
2. "Facebook's Half a Billion Users: Fun Facts." *PC World*. Accessed on September 8, 2010. http://www.pcworld.com/article/201650/facebooks_half_billion_users_fun_facts.html.
3. "Facebook Statistics." Accessed on September 9, 2010. http://www.facebook.com/press/info.php?statistics.

4. "Twitter User Statistics Revealed." Huffington Post. Accessed on September 9, 2010. http://www.huffingtonpost.com/2010/04/14/twitter-user-statistics-r_n_537992.html.
5. "Microsoft Vine." Microsoft. Accessed on September 10, 2010. http://www.vine.net/.
6. "The Web is Dead. Long Live the Internet." *Wired*. Accessed on September 10, 2010. http://www.wired.com/magazine/2010/08/ff_webrip/all/1.
7. "Mobile Web to Overtake Desktop by 2015." Switched. Accessed on September 10, 2010. http://www.switched.com/2010/04/14/mobile-web-to-overtake-desktop-by-2015-facebook-fans-worth-3-6/.
8. "FEMA: Mobile." Federal Emergency Management Agency. Accessed on September 12, 2010. http://m.fema.gov.
9. "Interactive NWS." National Weather Service. Accessed on September 12, 2010. http://inws.wrh.noaa.gov/.
10. "JOCOAlert." Johnson County Emergency Management and Homeland Security. Accessed on September 12, 2010. www.jocoem.org.
11. "7 iPhone Apps That Could Save Lives." Mashable. Accessed on September 29, 2010. http://mashable.com/2009/07/11/iphone-save-lives/.
12. "Flashlight for iPhone." Apple iTunes. Accessed on September 29, 2010. http://itunes.apple.com/us/app/flashlight/id285281827?mt=8.
13. "911 Dispatch App Puts Emergency Data in Hands of Citizens." *Emergency Management Magazine*. Accessed on September 29, 2010. http://www.emergencymgmt.com/safety/911-Dispatch-App-Emergency-Data.html.
14. "IT Trends: Gulf Oil Spill Crowdsourcing." Government Technology. Accessed on September 14, 2010. http://www.govtech.com/e-government/IT-Trends-Gulf-Oil-Spill-Crowdsourcing.html?topic=117673.
15. "Haiti." Ushahidi. Accessed on September 16, 2010. http://haiti.ushahidi.com.
16. "Fargo Uses Social Networks to Fight Floodwaters." MSNBC. Accessed on September 18, 2010. http://www.msnbc.msn.com/id/29901184/.
17. "Red Cross Raises $5,000,000+ for Haiti through Text Messaging Campaign." Mashable. Accessed on September 18, 2010. http://mashable.com/2010/01/13/haiti-red-cross-donations/.
18. "Mobile Giving to Help Haiti Exceeds $30 Million." MSNBC. Accessed on September 18, 2010. http://today.msnbc.msn.com/id/34850532/ns/technology_and_science-wireless/.
19. "Virtual Worlds—eHealth Marketing." CDC. Accessed on September 20, 2010. http://www.cdc.gov/healthmarketing/ehm/virtual.html.
20. "CDC Discusses Swine Flu in Second Life." Ill Clan Animation Studios. Accessed on September 20, 2010. http://www.illclan.com/ill-blog/35-ill-blog/124-cdc-discusses-swine-flu-in-second-life.
21. "First Responders Meet in Second Life: Public Health Enters the Virtual World." Medill–Northwestern University. Accessed on September 20, 2010. http://news.medill.northwestern.edu/chicago/news.aspx?id=114473.
22. "The Story of the Fail Whale." ReadWriteWeb. Accessed on September 21, 2010. http://www.readwriteweb.com/archives/the_story_of_the_fail_whale.php.
23. "Tracking the Oil Spill in the Gulf." *New York Times*. Accessed on September 24, 2010. http://www.nytimes.com/interactive/2010/05/01/us/20100501-oil-spill-tracker.html.
24. "Lalitesh Katragadda: Making Maps to Fight Disaster, Build Economies." Posted by TED: Ideas Worth Spreading. Accessed on September 26, 2010. http://www.ted.com/talks/lalitesh_katragadda_making_maps_to_fight_disaster_build_economies.html.

25. "Cell Phones and American Adults." Pew Internet. Accessed on September 26, 2010. http://pewinternet.org/Reports/2010/Cell-Phones-and-American-Adults.aspx.
26. "JOCOAlert." Johnson County Emergency Management and Homeland Security. Accessed on September 27, 2010. http://www.jocoem.org/CIT/jocoalert.shtml.
27. "Web Users Increasingly Rely on Social Media to Seek Help in a Disaster." American Red Cross. Accessed on September 28, 2010. http://www.redcross.org/portal/site/en/menuitem.94aae335470e233f6cf911df43181aa0/?vgnextoid=6bb5a96d0a94a210VgnVCM10000089f0870aRCRD.
28. "Fake BP Twitter Account Followers with Oil-Spill Satire." *The Wall Street Journal*. Accessed on September 29, 2010. http://blogs.wsj.com/digits/2010/05/24/fake-bp-twitter-account-draws-followers-with-oil-spill-satire/.
29. "Social Media Style and Editorial Guide for USA.gov." U.S. Government. Accessed on September 28, 2010. http://www.usa.gov/webcontent/documents/socmed_editorial_guidelines_041210.pdf.
30. "Top 10 Twitter Trends This Week." Mashable. Accessed on September 30, 2010. http://mashable.com/2010/05/29/top-10-twitter-trends-this-week-chart-3/.
31. "IS-702.a—National Incident Management Systems (NIMS) Public Information Systems." FEMA's Emergency Management Institute. Accessed on October 3, 2010. http://training.fema.gov/EMIWeb/IS/is702a.asp.
32. "The Elephant in the JIC: The Fundamental Flow in Emergency Public Information within the NIMS Framework." *Journal of Homeland Security and Emergency Management*. http://www.bepress.com/jhsem/vol7/iss1/10/.
33. White, Connie, and Plottnick, Linda. "A Framework to Identify Best Practices: Social Media and Web 2.0 Technologies in the Emergency Domain." Slideshare.net. Accessed on October 4, 2010. http://www.slideshare.net/conniewhite/a-framework-to-identify-best-practices-social-media-and-web-20-technologies-in-the-emergency-domain.

Afterword

Emergency managers, today, face a growing challenge to continuously exhibit the required leadership to bring together various individuals and organizations in order to develop an effective all-risk emergency management plan for their community. The perception of "it won't happen here," is a formidable obstacle to the development of teamwork and collaboration required to write, exercise, and quickly and efficiently implement the community's plan. Today, communities must understand that it is not a question of "if" they will implement their emergency plan to protect citizens and property from any of a myriad of likely risks and hazards, but "when." The successful emergency manager will have honed the skills and ability necessary to create effective teams of public, private, and nongovernmental organizations who all understand the need, value, and critical importance of having a practical plan in place to deal with expected and unexpected situations.

As such, today's emergency manager is tasked with the responsibility to

- Develop an effective planning process
- Organize the various resources and funding elements required in the process
- Identify all the necessary agencies and individuals to be represented
- Provide leadership and direction during the process
- Communicate openly and effectively to all participants
- Delegate responsibility where appropriate to facilitate ownership of the plan
- Supervise, motivate, and direct the planning team through the process
- Justify adequate financial and support resources to develop an effective final plan
- Recognize the efforts and participation of the planning team members
- Implement the completed and approved plan in the community

Following the prescribed process should develop and command the desired reaction from all the various agencies and individuals, whether police officer, firefighter, elected official, business owners, or citizens. Completion of the process will result in not only a reasonable and practical community plan, but also attention and respect of the community to all the participants tasked with keeping the community protected.

The successful emergency manager is tasked with the tremendous responsibility for the safety and continuity of operation of local government. This responsibility includes the safety and security of a community's residents and property from terrorism, natural and man-made disasters, and other daily all-risk emergency situations. Although there will be a diverse group of planning partners from different government and private sector organizations involved in the process, the emergency manager is the key human resource required for success.

Being a leader is not a part-time assignment. To be effective in today's changing environment, the emergency manager must ascribe to a philosophy of "continuous learning." Through this ongoing educational commitment, emergency managers will lead their individual organizations ethically, professionally, technically, and personally. Emergency managers should, on a daily basis, read the words of Albert Einstein who said: "Example is not another way to teach, it is the only way to teach."

The foremost goal of the emergency manager must be competence.

This is the ability to stand among your peers and colleagues with a certain aura, an attitude that personifies every aspect of a stalwart and concerned leader. Those attributes that most commonly contribute to competence and a sense of professionalism are a leader's appearance, confidence, knowledge, mannerisms, charisma, firmness, situational awareness, expertise, personal integrity, and character. A professional emergency manager will draw others to him or her naturally. Everyone wants to follow these types of leaders. All of these attributes are necessary in order to effectively train and lead our subordinates, peers, and others in the organization and the community.

Knowing one's own emergency management occupational competence is paramount to achieving the operational commitments and goals of elected community officials and city leaders. Occupational competence is also critical in gaining the confidence of our local government peers and subordinates. As effective leaders, we must know our jobs to the fullest extent possible. If we are not competent and proficient in our emergency and nonemergency capabilities, how can we manage, supervise, and lead our communities technically. When a question is asked, we must know the answer, or, more important, know where to find the answer.

This book is aimed at a wide audience in the field of emergency management broadly tasked with playing a key role in community preparedness. The information in this book, and others, will help achieve this vision of the effective emergency management leader. It has been my distinct honor and pleasure to personally know both Mike Fagel and Shane Stovall for a number of years. Both of these emergency management leaders bring a wealth of knowledge and experience, gained over a number of years at both public and private organizations, to the task of sharing their knowledge with others. This book will help guide both novice and experienced emergency managers through the technical issues they are likely to face both currently and in the years to come. The information contained in these pages will help those who have a sense of what is needed in their organization or community, but may not be exactly sure how to get there. The *Principles of Emergency Management: Hazard Specific Issues and Mitigation Strategies* is an emergency manager's professional resource for those who are faced with today's, and tomorrow's challenges. This book is for those who need and desire to implement practical solutions to common problems and issues. The information is critical to being able to develop the emergency management results that our communities want at a price they are willing to accept.

William (Bill) Peterson CEM, CFOD, FIFireE
Former Local Emergency Manager and
Regional Administrator DHS/FEMA Region 6

Index

Page numbers followed by b indicate boxes, f indicate figures, n indicate notes, t indicate tables.

G

H

O

P

Q

R

U

V

W

Y

Z